Planetary Landscapes

RONALD GREELEY

Department of Geology and Center for Meteorite Studies,
Arizona State University

London
ALLEN & UNWIN
Boston Sydney

George Allen & Unwin (Publishers) Ltd,
40 Museum Street, London WC1A 1LU, UK

George Allen & Unwin (Publishers) Ltd,
Park Lane, Hemel Hempstead, Herts HP2 4TE, UK

Allen & Unwin Inc.,
8 Winchester Place, Winchester, Mass. 01890, USA

George Allen & Unwin Australia Pty Ltd,
8 Napier Street, North Sydney, NSW 2060, Australia

First published in 1985

British Library Cataloguing in Publication Data

Greeley, Ronald
Planetary landscapes.
1. Planets—Surfaces
I. Title
523.4 QB601
ISBN 0-04-551080-6

Library of Congress Cataloging in Publication Data

Greeley, Ronald
Planetary landscapes.
Bibliography: p.
Includes index.
1. Planets—Surfaces. 2. Solar system—Exploration.
I. Title.
QB603.S95G74 1985 559.9′2 85-1415
ISBN 0-04-551080-6 (alk. paper)

Set in 10 on 12 point Times by Mathematical Composition Setters Ltd, Salisbury, Wiltshire
and printed in Great Britain by Anchor Brendan, Tiptree, Essex

Preface

The objective of this book is to introduce the surface features of the planets and satellites in the context of geomorphic processes. Introductory chapters include the "hows" and "whys" of Solar System exploration and a review of the primary processes that shape our planet, Earth, and which appear to be important to planetary sciences. The remaining chapters describe the geomorphology of the planets and satellites for which data are available. For most of these objects, the general physiography and terrain units for each are introduced, then the geomorphic processes that are inferred for the development of their surfaces are described. Each chapter then ends with a synopsis of the geologic evolution of the surface. The principal sources of information on the geomorphology of the planets and satellites are spacecraft photographs. These are usually shown in the book oriented so that the illumination is from the upper left or left side (so that craters will appear correctly as holes to most readers!), although this means that north orientation may not be toward the top.

Because the level of knowledge is not uniform for all of the objects in the Solar System, the individual treatment varies among the chapters. For example, it was difficult to decide what to leave out of the chapter on Mars because so much is known about the surface, whereas data are rather limited for Venus and Mercury.

In addition to introducing the geomorphology of planetary objects, this book is intended to be a "source" for obtaining supplemental information. References are cited throughout the text. However, these citations are not intended to be exhaustive but rather are given to provide a "springboard" for additional literature surveys.

Finally, it must be pointed out that planetary sciences are in their infancy and the techniques for analyzing the geomorphology of extraterrestrial objects are still evolving. I hope that the reader will find these extraterrestrial worlds as fascinating and as exciting as I do and, together with my colleagues, we are pleased to share the results of Solar System exploration with our readers.

R. Greeley
March 1984

Acknowledgements

Preparing a book is a formidable project. Without the help of many people it is doubtful that this project would have been completed. I wish to acknowledge the following individuals and thank them for their contributions: Jeffrey Moore and Curtis Manley for research assistance for all phases of preparation; Loretta Moore for helpful comments on early drafts of the manuscript; Daniel Ball, Joo-Keong Lim and Joseph Riggio for photographic support; and Maureen Schmelzer for her accurate and speedy typing of countless draft versions.

Numerous colleagues very kindly provided photographs and illustrations as noted in the figure captions. The Regional Planetary Image Facilities aided greatly in providing images for this book and I would especially like to thank Linda Jaramillo (Arizona State University), Leslie Pieri (Jet Propulsion Laboratory), Gail Georgenson (University of Arizona) and Jody Swann (US Geological Survey).

Reviews of individual chapters and helpful discussions were provided by Paul Spudis (US Geological Survey), Alan Peterfreund and James Garvin (Brown University), George McGill (University of Massachusetts), Peter Schultz (Lunar Planetary Science Institute) Michael Malin (Arizona State University), Victor Baker (University of Arizona) and Peter Cattermole (University of Sheffield).

Finally, I acknowledge with gratitude the editorial assistance of Cynthia Greeley.

To Don Gault and Steve Dwornik,
who introduced me to the Moon and planets

Contents

List of tables

1 Introduction

The mid-1960s witnessed two fundamental revelations that resulted in the beginning of an era which continues to have profound effects on the geologic sciences. Although the basic ideas of continental drift had been proposed nearly a half century earlier by Alfred Wegener, it was not until the late 1950s and early 1960s and the development of modern instruments to measure sea-floor spreading, to date rocks radiometrically, and to conduct accurate geophysical surveys, that the concept of crustal plate tectonics was accepted. At the same time as the gradual acceptance of plate tectonics, a similar, equally profound, view of the Solar System was emerging. Just as the ideas of continental drift were hampered by the lack of data, Solar System studies were also limited until the space age. Prior to the return of results from space probes sent to the Moon and planets, views of most planetary objects (except the Moon) were limited to little more than fuzzy blurs or tiny pin-points of light, even when viewed with the most powerful Earth-bound telescopes (Fig. 1.1).

Primitive by today's standards, the first probes sent to the Moon by the Soviets ushered in a new scientific era which has brought the surfaces of the planets within the grasp of study. For the most part, the study of planetary surfaces has passed from the astronomer to the geologist and has resulted in the establishment of the new discipline, *planetary geology*. Planetary geology is defined as the study of the origin, evolution and distribution of matter which forms the planets, natural satellites, comets and asteroids. The term "geology" is used in the broadest sense and is considered to mean the study of the solid parts of the planets. Aspects of geophysics, geochemistry, geodesy and cartography are all included in the general term.

This book is concerned with one aspect of planetary geology – the geomorphology of the planets. For simplicity, the term "planet" is also applied to natural satellites, such as the Moon. Geomorphology, a discipline long established in the earth sciences, seeks to determine the form and evolution of Earth's surface.

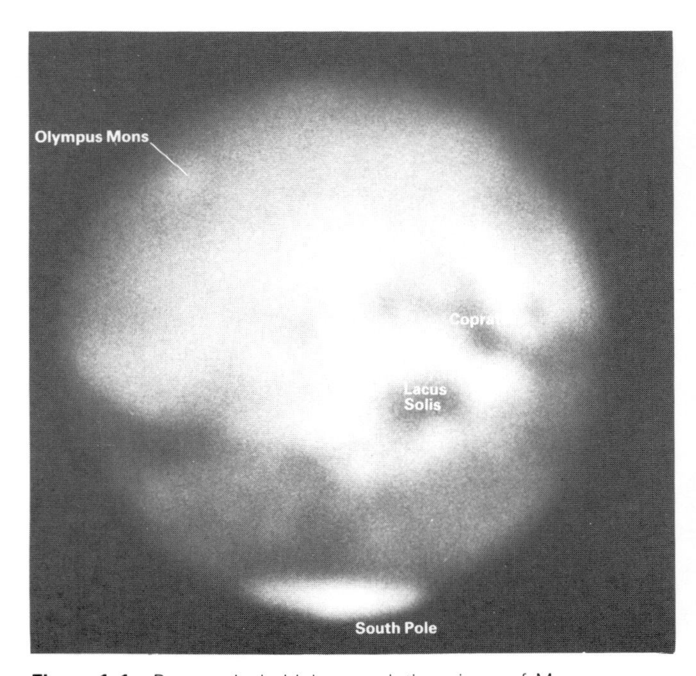

Figure 1.1 Progressively higher-resolution views of Mars.

(a) Earth-based telescopic photograph of the western hemisphere of Mars taken 16 August 1971 by the University of Arizona, Lunar and Planetary Laboratory using the Catalina Observatory Telescope (courtesy of the Lunar and Planetary Laboratory).

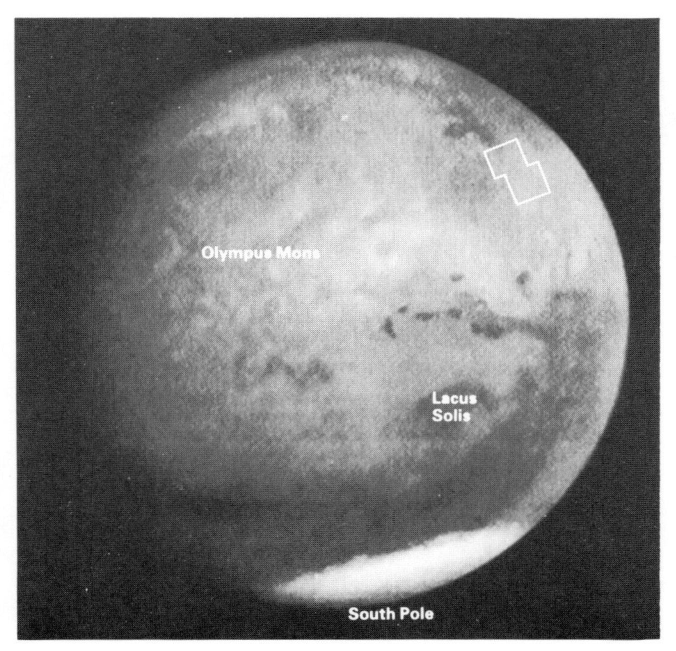

(b) A global view of Mars similar to Figure 1.1a, imaged by Mariner 7 on 4 August 1969 from a distance of 495 086 km; outline shows location of Figures 1.1c and d (Mariner 7 7F71).

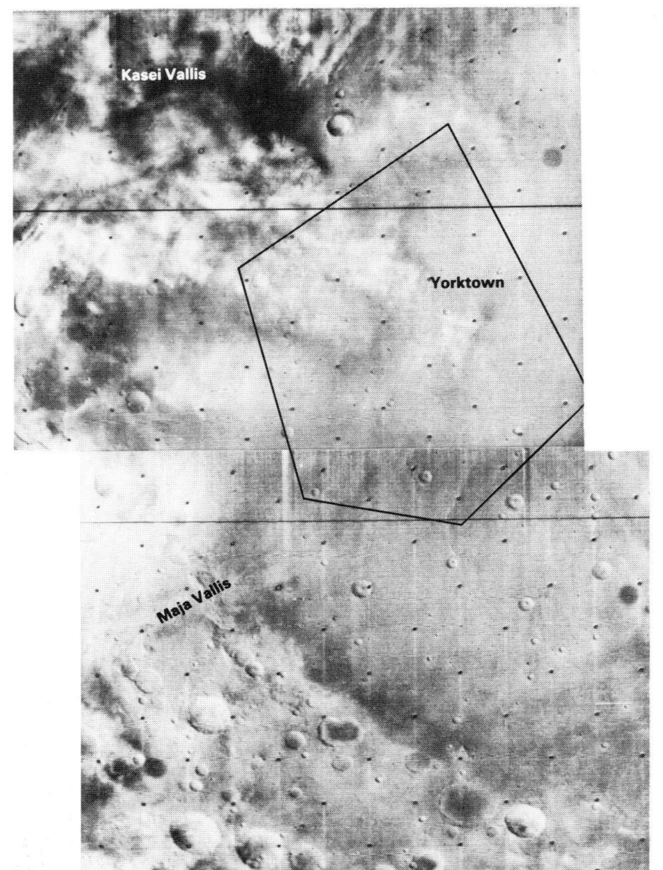

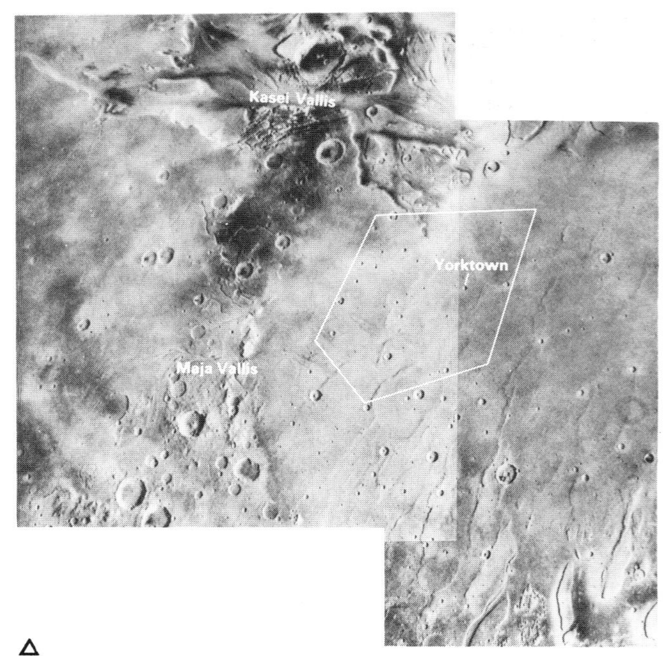

△

(d) Viking orbiter images of about same region as shown in Figure 1.1c; the significant improvement in overall picture quality results primarily from an improved imaging system. Area shown is about 1200 km across. The outline locates Figure 1.1e (Viking Orbiter 666A04, 666A06).

(c) Mariner 9 images of the Kasei region of Mars taken from 2835 km; resolution ~1.5 km/pixel; area shown is 650 km by 1000 km; outline shows location of Figure 1.1e (Mariner 9 DAS 07543588 and DAS 07543518).

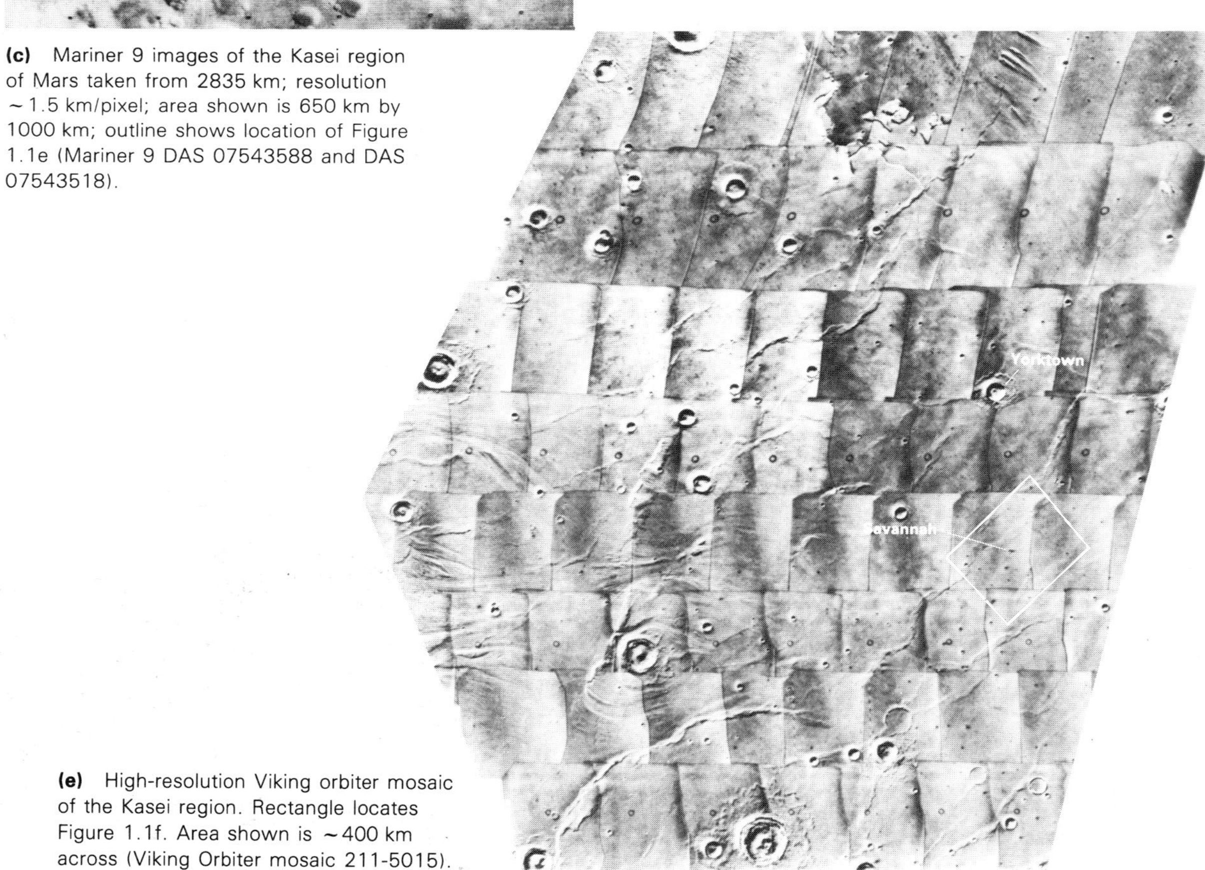

(e) High-resolution Viking orbiter mosaic of the Kasei region. Rectangle locates Figure 1.1f. Area shown is ~400 km across (Viking Orbiter mosaic 211-5015).

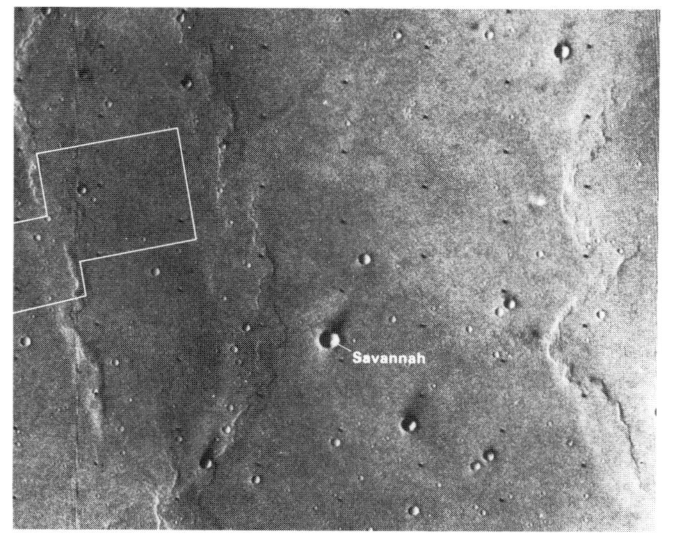

(f) High-resolution Viking orbiter image of the Viking 1 landing site taken from 1551 km; resolution ~350 m/pixel. The polygon outline locates Figure 1.1g. Area shown is 46 km across (Viking Orbiter 27A33).

(g) Highest-resolution (25 m/pixel) orbiter images of Viking 1 landing site. Craters near the site can be seen in the lander panorama (Fig. 1.1h). Area shown is about 12 km across (Viking Orbiters 452B11 and 452B10).

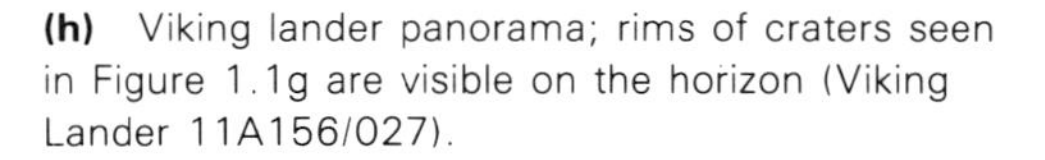

(h) Viking lander panorama; rims of craters seen in Figure 1.1g are visible on the horizon (Viking Lander 11A156/027).

This same goal applies to all the planets. Because much of our knowledge of the planets has been gained by remote sensing – primarily spacecraft pictures – we are, in fact, using landforms to interpret much of the geology of the planets. The approach commonly employed is first to determine the physiography of each planet and then to interpret the processes that have shaped the landforms that are observed on the pictures.

1.1 Objectives of Solar System exploration

The question is often asked, "Why study the Solar System?" During the mid-1960s, the United States National Academy of Sciences addressed this question and defined three principal goals for the exploration of space: (a) to determine the origin and evolution of the Solar System; (b) to determine the origin and evolution of life; and (c) to clarify the nature of the processes that shape humankind's terrestrial environment. The US National Aeronautics and Space Administration (NASA) has refined these objectives and derived a plan for Solar System exploration through the year 2000 (SSEC 1983).

Planetary geology and geomorphology figure prominently in all three of these goals, as discussed in detail by Greeley and Carr (1976). Let us consider the first goal. There are two ways to approach the study of the origin and evolution of the Solar System. The first is to model the possible conditions of Solar System formation and then follow the evolution through a series of stages leading up to the present time. This approach is typically employed by astronomers and uses observations of stars that are thought to represent various stages of evolution. The second approach is to begin with the present state of the Solar System and try to work backward in time. This is the approach typically taken by the planetary geologists – it is basically a geologic approach. Of course, the final goal will probably be reached through a combination of these two approaches. Astronomical modeling is probably the better approach for the early stages of formation and for that part of the early record missing in geologically derived histories; the geologic approach is better for the later stages of Solar System evolution represented by the rock record preserved on the surfaces of planetary objects.

The geology of planetary surfaces also relates to the second goal of Solar System studies, the origin and evolution of life. The planetary environment, including rock and mineral compositions and active geologic processes, has a direct bearing on the starting conditions for life and influences the process of natural selection.

And what of the third goal? Many fundamental geologic problems on Earth might be solved by detailed comparisons with other planets where the relative effects of different sizes, compositions and atmospheres on the evolution of the planets could be assessed. For example, very little is known of the early history of the Earth. Only the last 0.5 aeons of the estimated 4.6 aeon history is readily available for study because so much of our planet is constantly attacked and altered by tectonic, weathering and erosional processes, and the remainder is covered by water. On the other hand, because the Moon has no erosive atmosphere and the crust has been stable for several aeons, it displays a surface that is commonly five to eight times older than most of the Earth's surface. In this older surface is stored and available for study the early history of the Moon and probably the Earth as well. Thus, through the study of the Moon, we are better able to understand the processes that contributed to the early history of the Earth (Fig. 1.2).

To achieve the goals of Solar System exploration, data are obtained through a series of steps that begins with Earth-based observations, progresses through reconnaissance missions and ultimately ends with manned exploration. Obviously, we are a long way from completing this sequence, even for a small fraction of the Solar System, as indicated in Table 1.1. Nevertheless, sufficient data are available to begin analyses and to formulate ideas on the origin and evolution of the planets and their geologic histories.

Table 1.1 Sequence of planetary exploration.

Type of mission	Mercury	Venus	Moon	Mars	Jupiter	Saturn	Uranus	Neptune	Pluto	Comets	Asteroids
earth-based	×	×	×	×	×	×	×	×	×	×	×
flyby	×	×	×	×	×	×	(1)	(2)		(4)	(3)
orbiter		×	×	×	(3)						
unmanned landers		×	×	×							
sample return			×								
manned landings			×								

(1) Anticipated Voyager flyby, 1986.
(2) Anticipated Voyager flyby, 1989.
(3) Anticipated Galileo mission, 1989.
(4) Anticipated by European, Soviet and Japanese missions, 1987.

Figure 1.2 Apollo 17 view of the eastern part of the Moon ("farside" is to the right) showing heavily cratered uplands and dark mare areas (circular zone in upper left is Mare Crisium). The samples from the uplands have been dated by radiometric techniques as having been formed in excess of 4 aeons ago and thus represent geologic events from the earliest history of the Solar System. Because of the proximity of the Moon, it is presumed that Earth experienced a similar period of heavy cratering, but because of tectonic and erosional processes, most of the early record on Earth has been lost (Apollo 17 AS 17-152-23308).

1.2 The geologic approach

With each step in the exploration of the Solar System, three primary geologic questions are asked: (a) What is the present state of the planetary object observed? (b) What is its geologic history? (c) How do the present state and geologic history compare with other objects in the Solar System? The keys to addressing these questions involve the interpretation of planetary processes, the derivation of geologic histories through mapping and comparative planetology.

1.2.1 Present state

An understanding of the present state of a planetary object requires knowledge of its composition, interior properties, exterior environment and the geologic processes which may be currently active. Knowledge of the composition – at least of the surface – can be obtained directly by measurements made on returned samples (Fig. 1.3), or *in situ* by various instruments on landed spacecraft (Fig. 1.4), or indirectly by various remote-sensing techniques (Fig. 1.5). Information on the interior, such as lithosphere–mantle–core configurations, can be obtained from instruments, such as seismometers, or from geophysical models based on knowledge of the planetary density, size, moment of inertia and shape. Information on the exterior environment includes the temperature range, mechanical and thermal properties of the surface materials, the influence of various external processes (such as impact cratering), solar wind flux and the presence or absence of an atmosphere and, if present, its composition and density. Finally, knowledge is required of the various active processes which may be operating on the planetary object, such as volcanism.

1.2.2 Past state of the planets

This objective is to trace the processes, events and characteristics of the planets from their origin to the present; in other words, to determine their geologic histories. The approach for meeting this goal is met primarily through geologic mapping, in which surface materials (rock units) are identified and placed in a time sequence. In planetary geology, mapping is accomplished primarily by photogeologic techniques; however, as we shall see later, this method is not without some problems. Despite the difficulties, the level of mapping which is attainable at least provides a broad framework for the derivation of geologic histories.

1.2.3 Comparative planetology

Once knowledge is gained – even if incomplete – of the present and past states of the planets, it is then possible to begin to compare the planets to see how they are similar to and different from each other. Such comparisons help to meet the objectives of Solar System exploration by enabling a better understanding of the evolution of all the objects in the Solar System and of the processes involved in their formation and evolution.

Figure 1.3 Apollo 11 astronaut on Moon, showing various experiments deployed around the lander. Samples returned from the Moon by the Apollo 11 crew enabled for the first time detailed compositional determination and dating of the ages of emplacement of mare lavas (Apollo 11 69-HC-898).

1.3 Relevance of geomorphology

Ideally, the various geologic objectives of Solar System exploration would be met by carrying out the entire sequence of missions and measurements, as indicated in Table 1.1. Limited resources prevent the completion of this strategy for Solar System exploration, at least in the foreseeable future. Meanwhile, nearly all previous and near-term anticipated missions have been equipped with **imaging systems** to acquire pictures of planetary surfaces. Aside from the obvious appeal of pictures, it is generally recognized that imaging science can yield answers to a broader range of questions than any other single instrument which might be carried on board spacecraft.

Planetary pictures also play a key role from the standpoint of engineering. During mission operations, flight engineers use pictures taken by the spacecraft for celestial navigation in guiding the probe in its journey through space. For missions involving landers, pictures of potential landing sites play a key role in the selection of safe sites.

The first phase of most planetary geologic studies involves the classification of various landforms observed through images and the production of **physiographic**

Figure 1.5 Apollo 15 photograph of the Service Module taken from the Landing Module, showing the Scientific Instrument Module (SIM) containing various remote-sensing experiments, including cameras, alpha, gamma-ray and X-ray spectrometers (Apollo 15 AS 15-88-11972).

Figure 1.4 View of the martian surface from Viking Lander 1, showing meteorology boom (left side, extending out of view toward top) used to measure wind speed and direction, temperature and atmospheric pressure, and sample arm (lower middle) used to obtain samples for analysis. Arm could dig into surface (trench for sample acquisition is marked with arrow) to retrieve the sample and dump it into a hopper on the spacecraft for analysis.

maps (Fig. 1.6). The next step is to derive geologic maps using various photogeologic techniques. Color, **albedo** (reflective properties of the surface), texture and other remotely sensed characteristics are used to distinguish possible rock formations. Then, using various geometric relationships, such as superposition, embayment and cross-cutting relations, the identified formations are placed in a relative sequence. These techniques have been long-established in photogeologic studies of the Earth. An additional technique useful in planetary geology for relative dating of some surfaces is to establish the size-frequency distribution of impact craters. The idea is that surfaces act as impact "counters": the older the surface, the higher the frequency of superposed craters (Fig. 1.7). This concept will be discussed in more detail in Chapter 3.

A true geologic map should show the distribution of *material* units, that is, three-dimensional rock units, or

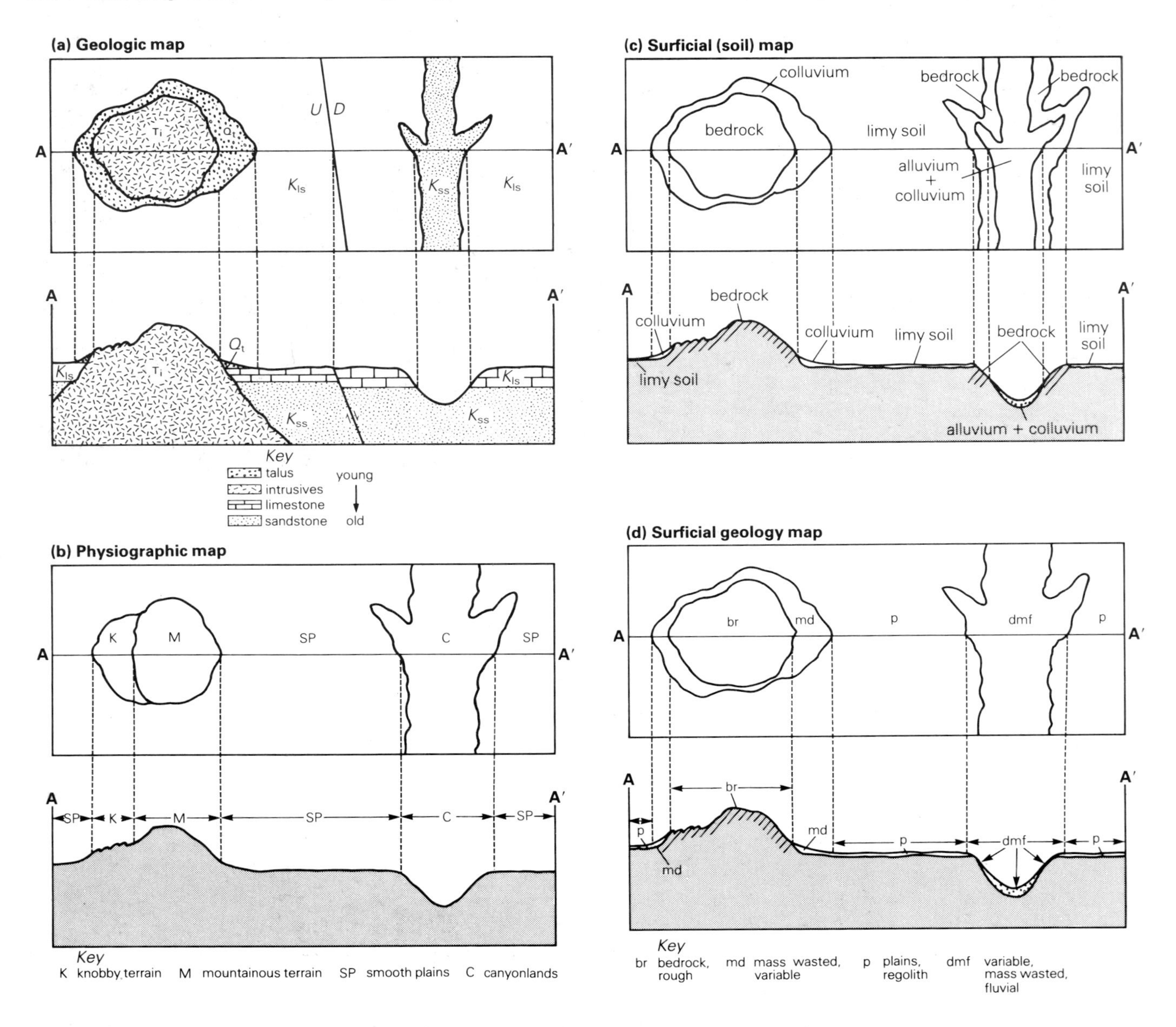

Figure 1.6 Diagrams showing types of maps commonly used to portray planetary surfaces (from Spudis & Greeley 1976). (a) *Geologic maps* show three-dimensional rock units in space and time, plus structural features such as folds and faults. They portray the surface of the planet as though all vegetation, soil and other surficial materials were stripped away; sequences of geologic events can be derived from this type of map. (b) *Physiographic maps* show landforms such as hills, valleys and plains. (c) *Surficial (soil) maps* show the distribution only of surficial covers or the lack thereof and do not show topography or terrain types. (d) *Surficial geology maps* which characterize the local geologic and surficial geology of a given area. Note that this map presents all the data of the physiographic (terrain) and surficial (soil) maps, but with a substantial reduction in map complexity.

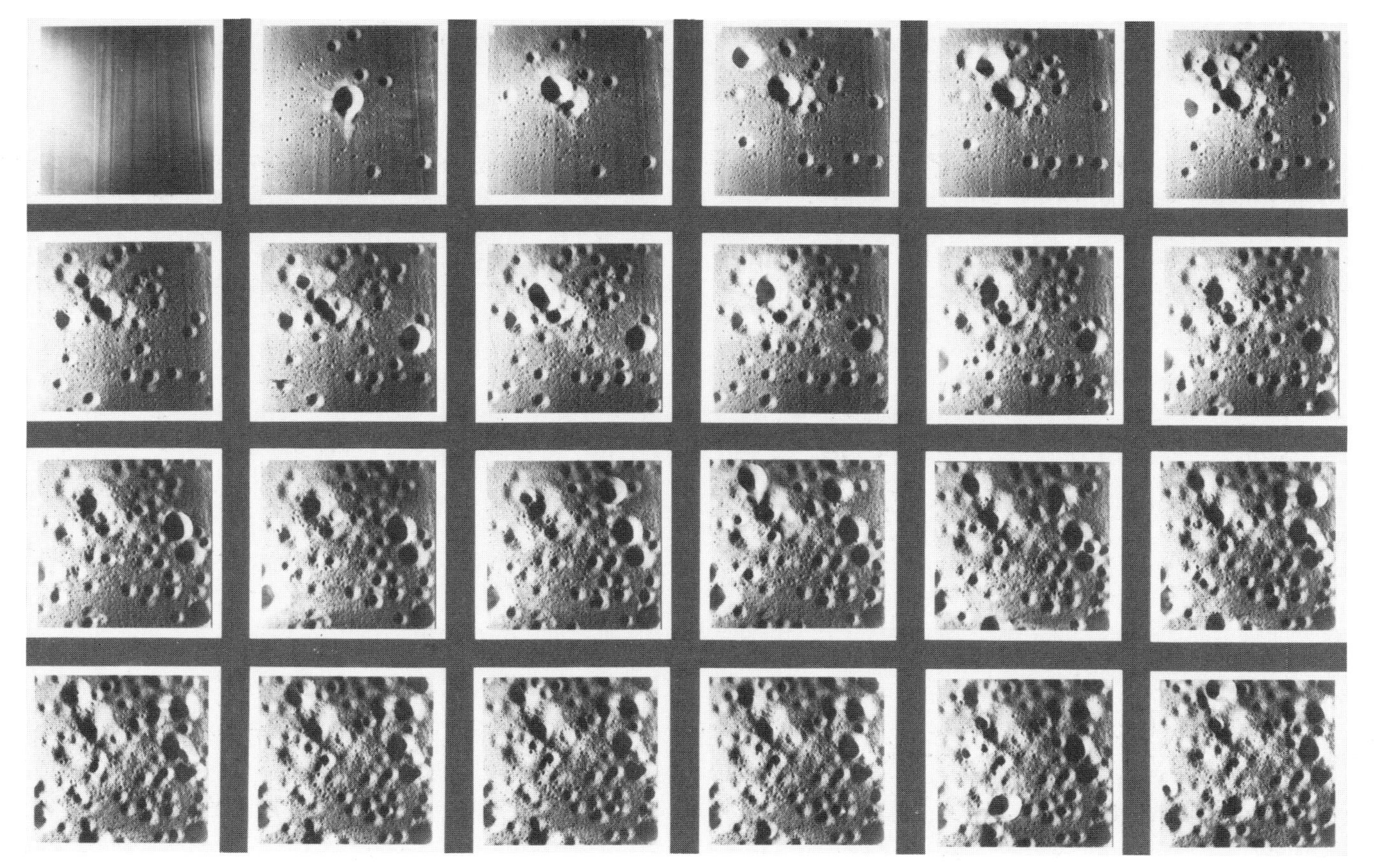

Figure 1.7a Sequence of photographs from laboratory simulations showing evolution of a cratered surface, beginning in upper left with smooth, uncratered surface. As time progressed, the total numbers of craters increased and statistically there was a greater chance for large craters to form. Thus, by comparing the size-frequency distribution of craters among different surfaces, it is possible to obtain relative dates for those surfaces. For example, frame 3 is sparsely cratered in comparison to frame 10, and is therefore younger. Note, however, that the last six frames all appear to be about the same; this surface is said to be *in equilibrium*, in which craters of a given size are being destroyed at the same rate of formation (NASA-Ames photograph AAA-312-23, from Gault 1970).

formations. The test of a geologic map is to draw a cross section through the map; the units should appear to have a finite thickness (Fig. 1.6). In mapping some planetary surfaces, often it is not possible to construct true geologic maps owing to lack of adequate data; rather, the maps show *surficial geology* or are some combination of geologic and terrain maps as shown in Figure 1.6.

Once features and units are classified and tentatively identified, interpretations are attempted to determine the processes involved in their formation. This is often the most difficult and controversial part of planetary geology. Typically, a three-fold approach is used – results from spacecraft data analysis, modeling and studies of terrestrial analogs are combined to derive the best possible answers for the problem at hand. Figure

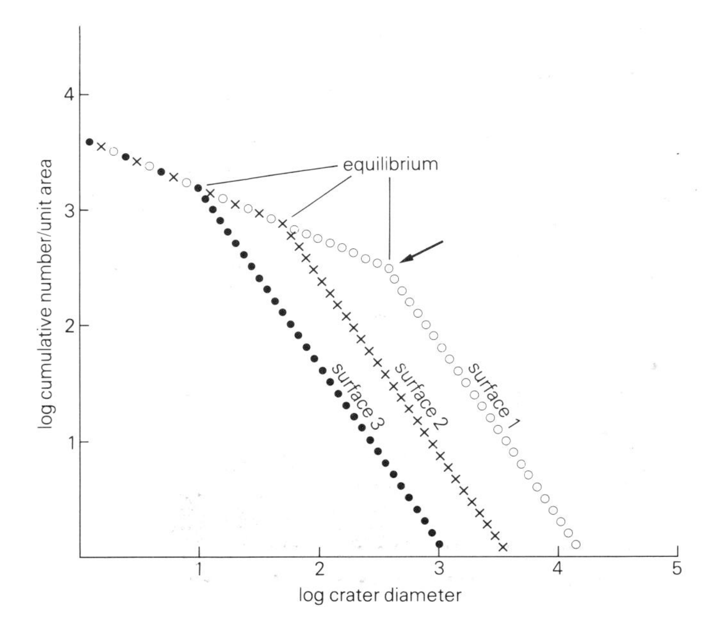

Figure 1.7b Stylized diagram indicating method of displaying crater statistics in which cumulative numbers of craters in given size ranges are plotted per unit surface area. "Break" in slope (arrow) marks upper crater size that has reached "equilibrium" (see Gault 1970). As a surface "ages", the total number of craters shifts toward larger sizes and greater numbers; surface 1 (oldest), surface 3 (youngest), assuming all craters are of impact origin.

1.8 illustrates how this approach has been used in planetology to understand the process of impact cratering.

First, the problem is defined from the study of spacecraft images; in the case illustrated here (Fig. 1.8a), the objective is to understand impact cratering as a geologic process. Next, impact cratering is physically modeled in the laboratory where the various parameters, such as the velocity of the impacting object, can be controlled, isolated and studied to determine their effects in the process (Figs 1.8b, c & d). Numerical modeling is also carried out, based on theory, to fill in detail and to gain insight into parts of the problem that are difficult to simulate in the laboratory. In general, physical and numerical modeling are first carried out for environments appropriate for conditions on Earth. Then, natural impact craters on Earth (Fig. 1.8e) are studied geologically as terrestrial analogs to the features observed on planetary surfaces. These field studies serve as important checks on the modeling results and provide direct information on the full-scale, complex process. Results from the field studies then can be used to modify, correct or adjust the modeling techniques. Once confidence is established in the ability to model the terrestrial case, the simulations can be carried out for the planetary case under study, using values for parameters, such as gravity, that are appropriate for the planet involved. The final results can then be applied to the interpretation of the planetary problem first defined, thus completing the study cycle.

The approach outlined above requires a multidisciplinary team effort, because any one individual seldom has the combined talents and background of geology, chemistry, physics and other fields to handle all aspects of the problem. The ultimate result of the approach is not only the solution of planetary problems but also leads to an understanding of various processes from a "universal" perspective, regardless of planetary object.

Planetologists may not think of this approach in geomorphic terms, but the underlying theme of much of the field is, in fact, geomorphology – be it in terrain mapping or gaining an understanding of surface-forming and surface-modifying processes. Sharp (1980) and Baker (1984) provide summaries of the role of geomorphology in the study of planetary surfaces.

1.4 Sources of data

The results from Solar System exploration are presented in nearly all scientific journals and popularized serials. However, geologic results are more commonly found in a relatively restricted set of publications. Preliminary results from US missions are commonly published in

Figure 1.8a Impact craters occur on nearly all solid-surface objects that have been observed in the Solar System. Understanding the mechanics of impact cratering and the geomorphic characteristics in terms of geologic processes are of key importance in planetary science. This view, obtained by the Apollo 15 astronauts from orbit, shows the craters Autolycus (nearest crater) and Aristillus, a 50 km diameter crater having a central peak; both craters show enlargement by slumping of the walls (Apollo 15 metric frame AS 15-1539).

Figure 1.8b View of the Vertical Ballistic Gun (VBG) facility at NASA-Ames Research Center used for impact cratering experiments. Projectiles can be fired down the gun tube at velocities up to 7.5 $km\,s^{-1}$ into a vacuum tank to impact targets. Gun tube can be elevated in 15° increments to enable crater experiment to be conducted for a range of incidence angles. High-speed motion pictures (exceeding 10^6 frames per second) obtained during the cratering event enables analysis of cratering sequence (NASA-Ames photograph A33996, courtesy of D. E. Gault).

Figure 1.8c Photograph inside chamber of VBG showing typical experiment; crater 40 cm in diameter was formed by a 6 mm glass projectile launched at 6.35 $km\,s^{-1}$ at an incidence angle of 45° (entering from the right); note the circular crater form and distribution of ejecta (throw-out from crater) despite the oblique impact angle.

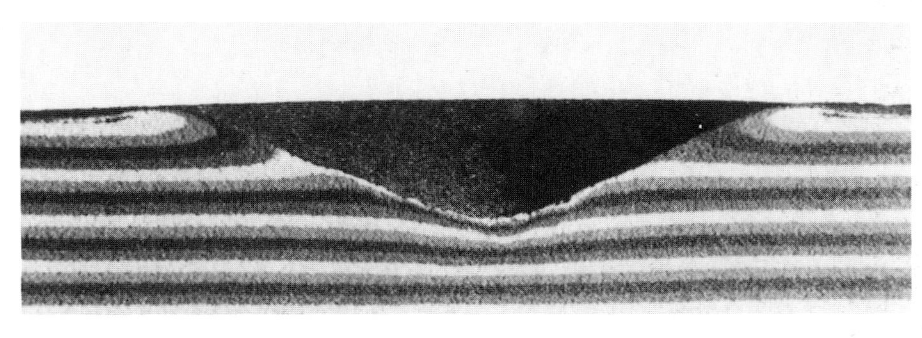

Figure 1.8d Cross section of impact crater in VBG formed in non-cohesive target of quartz sand with colored layers, revealing overturned strata in the rims and compressed strata beneath the crater (from Gault 1974).

Figure 1.8e Oblique aerial photograph of Meteor Crater in northern Arizona, showing eroded ejecta deposits (mottled patterns around crater and blocks on the rim and flanks) and polygonal outline which resulted from pre-crater joints in the target. Field studies of this well preserved 1.2 km diameter crater provide important "checks" on results from cratering simulations and contribute to the overall knowledge of the cratering process (photograph by Malin 1976).

Science. Subsequent results typically are published later in journals such as *Icarus*, *Nature*, the *Proceedings of the Lunar and Planetary Science Conference*, *Earth, Moon and Planets* (formerly *The Moon*), *Earth and Planetary Science Letters* and the *Journal of Geophysical Research*. Table 1.2 lists some of the more important issues devoted to results of planetary exploration. Books introducing planetary geology include King 1976, Guest *et al*. 1979, Murray *et al*. 1981, Beatty *et al*. 1981, Glass 1982, Taylor 1982 and Hartmann 1983.

NASA publishes planetary results through several series and often produces collections of photographs from various missions (Table 1.3). Lunar and planetary maps and charts have been produced by both the US Department of Defense and the US Geological Survey. The Planetary Data Facility, US Geological Survey, Flagstaff, Arizona, provides up-to-date listings of available maps and charts for planets and satellites.

Images from United States' lunar and planetary missions are available from the National Space Sciences Data Center (requests from within the United States) and the World Data Center (requests from outside the United States), at Goddard Spaceflight Center, NASA, Greenbelt, Maryland, USA; both will supply catalogs of images upon request. In addition, various centers (Table 1.4) have been established to provide access to planetary images by serious investigators.

In the chapters that follow, we will first consider the Solar System in the geologic context and review the processes involved in the formation and evolution of the planets. The physiography of each planet and satellite of geologic importance is then presented in separate chapters.

Table 1.2 Special journal issues containing planetary geology references.

Planet	Mission	Journal
Mercury	Mariner 10	*Science*, 1974, **185**, no. 4146
Mercury	Mariner 10	*J. Geophys. Res.*, 1975, **80**, no. 17
Mercury	Mariner 10	*Phys. Earth Planet. Int.*, 1977, **15**, nos 2 & 3
Mercury	Mariner 10	*Icarus*, 1976, **28**, no. 4
Venus	Mariner 10	*Science*, 1974, **183**, no. 4131
Venus	Pioneer	*Science*, 1979, **203**, no. 4382
Venus	Pioneer	*Science*, 1979, **205**, no. 4401
Venus	Pioneer	*J. Geophys. Res.*, 1980, **85**, no. A13
Venus	general	*Icarus*, 1982, **51**, no. 2
Venus	general	*Icarus*, 1982, **52**, no. 2
Moon	Apollo 11	*Science*, 1970, **167**, no. 3918
Moon	Apollo	*The Moon*, 1974, **9**
Moon	general	*Rev. Geophys. Space Phys.*, 1974, **12**, no. 1
Moon	general	*The Moon*, 1975, **13**, nos 1, 2 & 3
Mars	Mariner 6 & 7	*J. Geophys. Res.*, 1971, **76**, no. 2
Mars	Mariner 9	*Icarus*, 1972, **17**, no. 2
Mars	Mariner 9	*Icarus*, 1973, **18**, no. 1
Mars	Mariner 9	*J. Geophys. Res.*, 1973, **78**, no. 20
Mars	Mariner 9	*Icarus*, 1974, **22**, no. 3
Mars	Viking 1	*Science*, 1976, **193**, no. 4255
Mars	Viking 1 & 2	*Science*, 1976, **194**, no. 4260
Mars	Viking	*J. Geophys. Res.*, 1977, **82**, no. 28
Mars	Viking	*Icarus*, 1978, **34**, no. 3
Mars	general	*J. Geophys. Res.*, 1979, **84**, no. B14
Mars	general	*Icarus*, 1981, **45**, nos 1 & 2
Mars	general	*Icarus*, 1982, **50**, nos 2 & 3
Mars	general	*J. Geophys. Res.* 1982, **87**, no. B12
Jupiter	Pioneer 11	*Science*, 1975, **188**, no. 4187
Jupiter	Voyager 1	*Nature*, 1979, **280**, no. 5725
Jupiter	Voyager 1	*Science*, 1979, **204**, no. 4396
Jupiter	Voyager 2	*Science*, 1979, **206**, no. 4421
Jupiter	Voyager	*J. Geophys. Res.*, 1981, **86**, no. A10
Saturn	Pioneer 11	*Science*, 1980, **207**, no. 4429
Saturn	Voyager 1	*Nature*, 1981, **292**, no. 5825
Saturn	Voyager 1	*Science*, 1981, **212**, no. 4491
Saturn	Voyager 2	*Science*, 1982, **215**, no. 4532
Saturn	Voyager	*Icarus*, 1983, **53**, no. 2

Table 1.3 NASA publications of relevance to planetary geology.

Object	Year	Serial	Title	Notes
Moon	1964	SP-61	*Ranger VII photographs of the Moon*, Part I, Camera A series	
Moon	1965	SP-62	*Ranger VII photographs of the Moon*, Camera B series	photographic collection
Moon	1965	SP-63	*Ranger VII photographs of the Moon*, Camera P series	photographic collection
Moon	1966	SP-111	*Ranger VIII photographs of the Moon*, Cameras A,B,P	photographic collection
Moon	1966	SP-112	*Ranger IX photographs of the Moon*	photographic collection
Moon	1966	SP-126	*Surveyor I: a preliminary report*	

Table 1.3 *Continued.*

Object	Year	Serial	Title	Notes
Moon	1966	JPL-TR 32-800	*Ranger VIII and IX experimenters' analysis and interpretations*	mission description and science results
Moon	1969	SP-184	*Surveyor: program results*	
Moon	1969	SP-214	*Apollo 11 preliminary science report*	mission description with photographs; includes descriptions and results from various experiments on board the command module and/or lander spacecraft
Moon	*1969*	*SP-201*	*Apollo 8, photography and visual observations*	
Moon	1970	SP-242	*Guide to Lunar orbiter photographs*	explains camera system and gives footprints
Moon	1970	SP-200	*The Moon as viewed by Lunar orbiter*	mission description and photographs
Moon	1970	SP-235	*Apollo 12 preliminary science report*	mission description with photographs; includes descriptions and results from various experiments on board the command module and/or lander spacecraft
Moon	1971	SP-241	*Atlas and gazetteer of the near side of the Moon*	lunar orbiter photograph with place names
Moon	1971	SP-206	*Lunar orbiter photographic atlas of the Moon*	photographic collection
Moon	1971	SP-232	*Apollo 10, photography and visual observations*	mission description with photographs; includes descriptions and results from various experiments on board the command module and/or lander spacecraft
Moon	1971	SP-238	*Apollo 11 mission report*	
Moon	1971	SP-272	*Apollo 14 preliminary science report*	
Moon	1971	SP-246	*Lunar photographs from Apollos 8, 10, 11*	
Moon	1972	SP-315	*Apollo 16 preliminary science report*	
Moon	1972	SP-289	*Apollo 15 preliminary science report*	
Moon	1972	SP-284	*Analysis of Surveyor 3 material and photographs returned by Apollo 12*	
Moon	1972	SP-306	*Compositions of major and minor minerals in five Apollo 12 crystalline rocks*	description of samples
Moon	1973	SP-330	*Apollo 17 preliminary science report*	mission description with photographs; includes descriptions and results from various experiments on board the command module and/or lander spacecraft
Moon	1974	EP-100	*Apollo*	public information booklet
Moon	1974	SP-341	*Atlas of Surveyor 5 television data*	photographs with short captions
Moon	1975	SP-350	*Apollo expedition to the Moon*	mission description with photographs, general public
Moon	1977	SP-418	*Lunar sample studies*	

Table 1.3 *Continued.*

Object	Year	Serial	Title	Notes
Moon	1978	SP-362	*Apollo over the Moon*	summary of Apollo missions to the Moon; color photographs, science discussions
Mars	1968	SP-179	*The book of Mars*	
Mars	1971	SP-263	*The Mariner 6 and 7 mission to Mars*	mission description and photographic collection
Mars	1974	SP-334	*The Viking mission to Mars*	
Mars	1974	SP-329	*Mars as viewed by Mariner 9*	photographs with science
Mars	1974	SP-337	*The new Mars, the discovery of Mariner 9*	photographs with science
Mars	1978	SP-425	*The martian landscape*	mission description and photographs for Viking landers
Mars	1979	SP-438	*Atlas of Mars, the 1:5 000 000 map series*	map and photomosaic collection
Mars	1980	SP-444	*Images of Mars – the Viking extended mission*	photograph collection
Mars	1980	SP-441	*Viking orbiter views of Mars*	photo collection with science
Mars	1980	CR-3326	*The mosaics of Mars as seen by the Viking lander cameras*	lander photomosaics with discussion of assembly
Mars	1981	SP-429	*Viking site selection and certification*	
Mars	1982	CR-3568	*Viking lander atlas of Mars*	lander photographs and maps
Mars	1983	RP-1093	*A catalog of selected Viking orbiter images*	photomosaics based on Mars charts; gives frame locations
Mercury	1978	SP-423	*Atlas of Mercury*	synopsis of Mariner 10 results, collection of photographs and USGS charts
Mercury	1978	SP-424	*The voyage of Mariner 10*	mission description
Venus	1975	SP-382	*The atmosphere of Venus*	
Venus	1983	SP-461	*Pioneer Venus*	popularized account of mission with discussion of science results
Jupiter	1971	SP-268	*The Pioneer mission to Jupiter*	
Jupiter	1974	SP-349	*Pioneer Odyssey – encounter with a giant*	mission description and photographic collection
Jupiter	1980	SP-439	*Voyage to Jupiter*	popularized account of mission with discussion of science results
Jupiter/Saturn	1977	SP-420	*Voyager to Jupiter and Saturn*	
Jupiter/Saturn	1980	SP-446	*Pioneer: first to Jupiter, Saturn, and beyond*	
Saturn	1974	SP-340	*The atmosphere of Titan*	
Saturn	1974	SP-343	*The rings of Saturn*	
Saturn	1980	JPL-400-100	*Voyager 1 encounters Saturn*	public information booklet
Saturn	1982	SP-451	*Voyages to Saturn*	popularized account of mission with discussion of science results

Table 1.3 *Continued.*

Object	Year	Serial	Title	Notes
general	1971	SP-267	*Physical studies of the minor planets*	
general	1976	SP-345	*Evolution of the Solar System*	
general	1981	EP-177	*A meeting with the universe*	popularized account of Solar System exploration and other space science activities

Table 1.4 Regional and branch space image facilities.

Brown Regional Planetary Data Center, Brown University, Providence, RI 02912

Space Imagery Center, Lunar and Planetary Lab, University of Arizona, Tucson, AZ 85721

Spacecraft Planetary Imaging Facility, Cornell University, 317 Space Sciences Bldg, Ithaca, NY 14853

Planetary Image Facility, Jet Propulsion Laboratory, Bldg 264, Rm 115, 4800 Oak Grove Drive, Pasadena, CA 91103

Planetary Image Facility, Lunar and Planetary Institute, 3303 NASA Road One, Houston, TX 77058

Flagstaff Planetary Data Facility, USGS, 2255 N Gemini Drive, Flagstaff, AZ 86001

Regional Planetary Image Facility, Department of Earth and Planetary Science, Washington University, St Louis, MO 63130

Space Photography Laboratory,* Department of Geology, Arizona State University, Tempe, AZ 85287

Space Image Library,* Louisiana State University, Department of Geology, Baton Rouge, LA 70803

Regional Planetary Image Facility, University of Hawaii, Hawaiian Institute of Geophysics, Honolulu, HI 96822

Regional Space Image Library, University of London Observatory, Mill Hill Park, 2QS SW7, UK

Fotote La-archivio Immagini E Dati Planeteri, Reparto Planetogia, Viale Universite 11-00185, Roma, Italy

Space Image Library, Laboratoire de Géologie dynamique interne (Bat 509), University of Paris Sud, 91405 Orsay, France

Space Image Library, Institut für Allgemeine Unde Angewandte Geologie, Luisenstrasse #37, de-8000 Munchen 2, FRG

Regional Planetary Image Facility, Smithsonian Institute, Washington, DC 20560

*Designates branch facilities not generally open to the public.

2 Geologic exploration of the Solar System

Earth-based telescopic observations of the planets and satellites began in the early 17th century. Through the years these observations have enabled some of the general properties of the planets, such as sizes, densities and orbital characteristics, to be determined. Space probes have added substantially to this knowledge, particularly in regard to the outer planet satellites – some of which were unknown prior to their discovery on spacecraft images. In this chapter we will examine the general properties of the planets and satellites important for geologic analyses. We will also consider the types of space probes involved in Solar System exploration and discuss the means by which geologic data – primarily images – are acquired.

Spacecraft images are extremely important to planetary geology because they are the primary source for understanding the geomorphology of the planets and satellites. Unless the researcher is familiar with how the images are acquired and processed, spacecraft pictures can easily be misinterpreted. Thus, in the last section of this chapter, we will discuss camera systems that have been utilized on various missions along with the techniques that are employed for the analysis of spacecraft images.

Figure 2.1 Comparison of terrestrial planet (Earth) with two jovian planets (Saturn and Jupiter) all shown to same scale (NASA photograph 83 H 202).

2.1 General planetary characteristics

Astronomers have long recognized that the planets can be classified into two groups, the *terrestrial planets* and the *jovian planets*. As the names imply, terrestrial planets are "Earthlike"; these are relatively small, dense objects, composed mainly of silicate materials that have solid surfaces (Fig. 2.1). The terrestrial planets, also called the inner planets from their position in the Solar System, are Mercury, Venus, Earth and Mars. Most investigators consider Earth's Moon to be a terrestrial planet because of its large size and high density (Table 2.1).

Jove is another name for the Roman god Jupiter; thus, jovian planets share the characteristics of the planet Jupiter (Fig. 2.1). They are large, low-density objects composed mostly of hydrogen and helium. They all apparently lack solid surfaces and, thus, are not amenable to geologic study. The jovian planets are Jupiter, Saturn, Uranus and Neptune. These are also known as the *outer planets* by their position in the Solar System. Pluto is also an outer planet, but its small size suggests that it may be more like a terrestrial planet and hence it is not included with the jovian planets. Although the outer planets are not of direct geologic interest, their satellites are solid-surface objects and thus can be studied in the geologic context.

Observations of the surfaces of the martian moons Phobos (~22 km in diameter) and Deimos (~14 km in diameter) by Mariner 9 and the Viking orbiters introduced a new class of object to planetary geomorphology. Although not defined by a specific size, this class is generally referred to as *small bodies* and includes asteroids and the smaller satellites of the outer planets.

The satellites of Jupiter and Saturn (and probably the satellites of the other outer planets as well) display a wide range of sizes and characteristics. The Galilean satellites of Jupiter (so named after their discoverer, Galileo Galilei, in the early 1600s) are respectable-size objects in their own right. The two innermost moons, Io and Europa, are comparable in size and density to Earth's Moon and might also be considered as terrestrial planets. The outer two Galilean satellites are Ganymede and Callisto. They are roughly the size of Mercury, but their low density suggests that they are composed of a mixture of water and silicates.

Table 2.1 Data for selected planets and satellites (from Beatty *et al.* 1981, Carr 1981, Fimmel *et al.* 1977, Hanel *et al.* 1981, Kozai 1976, Lunine *et al.* 1982, Murray & Burgess 1977, Newburn 1978, Newburn & Matson 1978, Orton 1978, Poirier 1982, Smith *et al.* 1982, Tyler *et al.* 1981, Veverka & Thomas 1979).

	Diameter (km)	Mass (Earth = 1)	Density (g cm^{-3})	Surface temperature (K)	Atmospheric composition	Atmospheric surface pressure (bar)	Sidereal period of revolution	Equatorial surface gravity (cm s^{-2})
Mercury	4 880	0.055	5.44	~700	—	—	$87^d23^h15^m$	370
Venus	12 100	0.8150	5.269	~740	CO_2	93	$224^d16^h49^m$	890
Earth	12 756	(5.9733×10^{24} kg)	5.517	~290	N_2, O_2	1.00137	1 year	978
Moon	3 476	0.0123	3.343	~350	—	—	$27^d7^h43^m11^s$*	162
Mars	6 794	0.1074	3.945	~220	CO_2	0.008	1.88 years	371
Phobos	27 × 21 × 19	1.6×10^{-9}	1.9 ± 0.5	~225	—	—	$7^h39^m25^s$*	0.3–0.5
Deimos	15 × 12 × 11	3.3×10^{-10}	1.5 ± 0.5	~225	—	—	$1^d6^h18^m02^s$*	0.3
Jupiter								
Amalthea	240 ± 60	?	—	~150	—	—	$11^h57^m22^s$*	?
Io	3 632	0.01492	3.53	~130	Na, SO_2, S	$\sim 5 \times 10^{-5}$	$1^d18^h27^m33^s$*	179
Europa	3 126	0.00815	3.17	~130	—	—	$3^d13^h13^m42^s$*	132
Ganymede	5 276	0.0249	1.99	~150	—	—	$7^d3^h42^m33^s$*	142
Callisto	4 820	0.0178	1.76	~150	—	—	$16^d16^h32^m11^s$*	122
Saturn								
Mimas	390	7.61×10^{-6}	1.2	~100	—	—	$22^h37^m5^s$*	6.6
Enceladus	500	1.41×10^{-5}	1.2	~100	—	—	$1^d8^h53^m7^s$*	7.7
Tethys	1 050	1.26×10^{-4}	1.2	~100	—	—	$1^d21^h18^m26^s$*	14.7
Dione	1 120	1.76×10^{-4}	1.4	~100	—	—	$2^d17^h41^m10^s$*	22.4
Rhea	1 530	4.17×10^{-4}	1.3	~100	—	—	$4^d12^h25^m12^s$*	28.4
Titan	5 120	0.02252	1.88	~ 90	N_2, CH_4	~1.6	$15^d22^h41^m27^s$*	135
Hyperion	410 × 260 × 220	?	—	~100	—	—	$21^d6^h38^m24^s$	—
Iapetus	1 440	3.15×10^{-4}	1.2	~100	—	—	$79^d7^h56^m23^s$*	14.9
Phoebe	~200	?	—	~100	—	—	$550^d10^h48^m$	—

*Synchronously rotating with respect to primary.

Figure 2.2 View northward obtained by the Apollo 16 astronauts showing Imbrium sculpture northwest of the crater Ptolemaeus. Sculpture resulted from gouging of ejecta from the Imbrium basin, visible at top of photograph (Apollo 16 AS 16-1412).

2.2 Pre-space-age planetology studies

Although the first telescopes used to view the Moon and planets were rather crude, technology advanced rapidly and by the mid-to-late 1600s maps of the Moon based on telescopic observations were remarkably detailed. In 1651, Giovanni Riccioli, a Jesuit priest, made a comprehensive map of the nearside of the Moon – the side that always faces Earth. The smooth, flat, dark areas were interpreted to be bodies of water and were named accordingly as *maria* (Fig. 1.2), the Latin word for seas; the large craters are named after people, such as Copernicus and Kepler. Hevelius, a German astronomer, also mapped the Moon and named lunar mountain chains after mountain ranges on the Earth, so that today we find the lunar Alps and Apennines on modern maps of the Moon.

With the development of photography, the lunar nearside was extensively documented at resolutions as good as a few kilometers. The variety of lunar surface features that were discovered through observations and photographs sparked considerable controversy as to their origin and much of this controversy was not laid to rest until well into the space age.

Craters are the most prominent surface featured on the Moon and have long generated curiosity as to their origin. Green (1965) provides an interesting historical account of the controversy surrounding lunar craters. He notes that the American geologist G. K. Gilbert was one of the first investigators to consider lunar craters in the geologic context and supported earlier speculations that the craters were of impact origin. In the late 1800s, he recognized surface textures and radial patterns around Mare Imbrium (Fig. 2.2) and interpreted the features to be the result of gouging of the surface by material ejected from a huge impact event (Gilbert 1893). Gilbert also carried out experiments in attempts to simulate the impact cratering process (Fig. 2.3).

The 1940s to early 1960s saw the development of ideas to explain not only lunar craters but also other aspects of the general geology of the Moon. Ralph Baldwin's books (1949 & 1963) and the works of Harold Urey (1952), Gerard Kuiper and Robert Dietz described a wide variety of surface features; Spurr published a series

Figure 2.3 Photograph by G. K. Gilbert taken *c.* 1891 showing his impact cratering experiments in which balls of clay impacted a slab of clay; he noted that the crater morphology was dependent upon the impact velocity (US Geological Survey photograph, G. K. Gilbert No. 842).

of books from 1944 to 1949 (Spurr 1944, 1945, 1948 & 1949) detailing various aspects of lunar geology. His work was followed by that of Gilbert Fielder, an English astronomer who attributed most lunar features to volcanism. Although Fielder's (1961, 1965) ideas on volcanic origins for lunar craters are not now accepted, many of his hypotheses regarding lunar lava flows and possible volcanic domes appear to be correct.

With the development of the United States' space program in the early 1960s came a commitment to carry out scientific studies of the Moon and planets. Primarily in support of the Apollo program, the Branch of Astrogeology was created as part of the US Geological Survey. Among other responsibilities, this organization was charged with the task of mapping the geology of the Moon from Earth-based observations. Eugene Shoemaker and R. J. Hackman (1962) demonstrated the feasibility for making geologic maps of the Moon. It was recognized that such maps would play a key role in selecting future landing sites and for plotting the scientific results for lunar exploration within a logical framework. This work was considered so important that an observatory was constructed near Flagstaff, Arizona – home of the Branch of Astrogeology – to provide unhindered observation time for lunar geologic mapping.

Aside from the Moon, Mars is the only other planetary object which can be observed from Earth with sufficient resolution to study surface features. Descriptions of canals on Mars have appeared in writings for more than 100 years, but the canals failed to appear on high-resolution pictures of Mars obtained by spacecraft and are now considered to have been figments of the observers' imaginations. However, Mars displays various surface markings and atmospheric features which appear regularly in concert with the martian seasons. In the mid-1950s Dean McLaughlin speculated that these features, long observed telescopically, might be the result of dust storms. He wrote a series of papers (1954) documenting these and other phenomena on Mars and even published a map showing wind circulation patterns based on tracking of certain albedo patterns now known to be clouds of dust. These and similar features have been seen on hundreds of spacecraft images and many of the ideas proposed by McLaughlin regarding wind processes appear to be valid.

2.3 Lunar and planetary missions

Table 2.2 lists the important "firsts" in spacecraft missions of geologic relevance and includes the first lunar flyby of the Soviets in 1959. This was the first of several series of unmanned missions designed to obtain critical engineering and scientific data.

Ranger spacecraft, first of the unmanned US lunar missions to return geologic data, were launched on trajectories that carried them to direct impact on the Moon. After a series of disappointing failures, Ranger 7 successfully obtained the first close-up pictures of the Moon in the summer of 1964; Rangers 8 and 9 followed a short time later and again returned high-resolution images. In flight these spacecraft took a continuous stream of television pictures of progressively higher resolution as the probes approached the surface, right to the moment of impact. Having a spatial resolution of about

1 m, these last pictures showed that there was a continuum of crater sizes down to the limit of resolution (Fig. 2.4). Analysis of these images showed that the number of progressively smaller craters increases exponentially.

The year 1966 marked the beginning of two important series of US unmanned spacecraft, the *Surveyor* series and the *Lunar orbiter* series. As the name implies, Lunar orbiters were placed in orbit around the Moon and had the principal objective of obtaining high-resolution photographs to be used for the selection of landing sites for the forthcoming manned landings. The first three Lunar orbiters, designated LO I, II and III, were placed in equatorial orbits and were so successful in obtaining high-quality images of candidate sites that the two remaining orbiters were devoted toward scientific objectives. Both were placed in polar orbits (enabling any part of the Moon to be observed); Lunar Orbiter IV was placed in a relatively high orbit and obtained images that were used for making global maps; Lunar Orbiter V was placed in a lower orbit and was commanded to obtain high-resolution images of geologically interesting sites.

Lunar orbiter, nick-named the "flying drugstore" because of its ability to develop its film in flight, returned a wealth of photographs (Fig. 2.5) and documented in exquisite detail not only features which had been so tantalizing from Earth-based views, but also an array of totally unexpected features. Nearly 99 percent of the Moon was photographed by the Lunar orbiters and these images are still an important resource for lunar geologic studies.

Surveyor spacecraft were "soft" landers, designed to function after touchdown. Although the primary purpose of the Surveyors was to obtain pictures of the surface (Fig. 2.6) and was aimed toward gathering information on the physical properties of the lunar surface, later spacecraft in the series were equipped with instruments for determining compositions of the lunar surface. An alpha particle scattering device on board Surveyors V and VI showed that the composition of the lunar maria was similar to that of basalt, lending support to interpretation that mare areas were of volcanic origin.

The Apollo missions to the Moon mark a technological and scientific achievement unparalleled in history, and hundreds of accounts of this milestone in exploration have been written. The Apollo series included manned orbital excursions to the Moon as precursors to the landing. During these pre-landing missions, photographs from Apollos 8 and 10 played a key role in documenting not only the engineering aspects of the flight, but also in providing for the first time high-resolution color images of the Moon and the Earth from deep space.

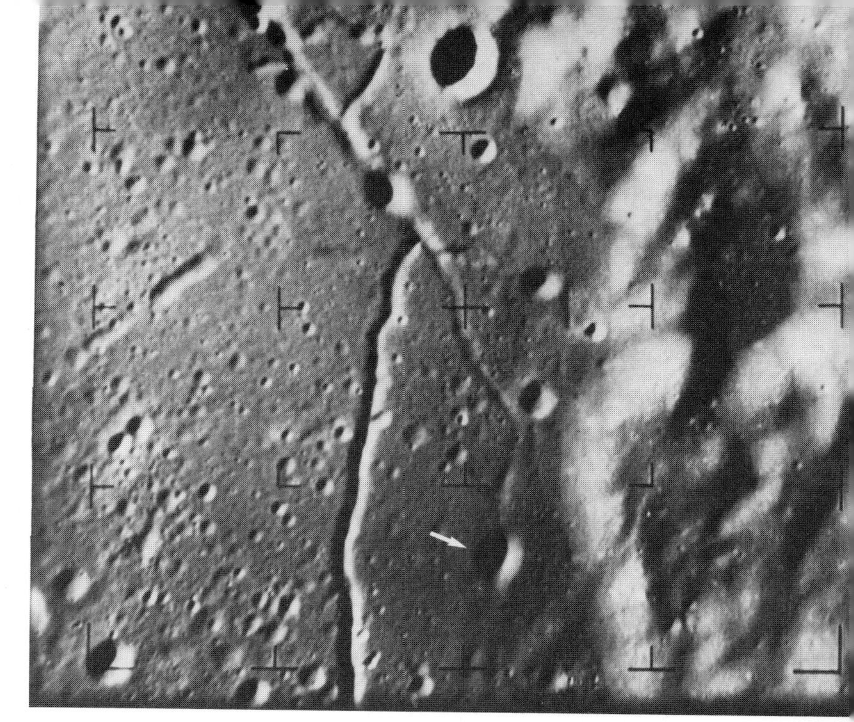

Figure 2.4 High-resolution television picture returned by Ranger 9 at an altitude of 173 km above the lunar surface 1 minute 12 seconds before impact in the crater Alphonsus. The rim of the crater is on the right; prominent fissures cut the crater floor. Elongate, dark-halo craters on the fissures (arrow) are thought to be volcanic vents. Field of view is 26 by 31 km (Ranger 9 frame B75).

Table 2.2 Milestones in geologic exploration of the Solar System.

Encounter Date		Spacecraft	Event
4 Oct.	1957*	Sputnik 1	first spacecraft
12 Sept.	1959*	Luna 2	lunar landing (impact)
4 Oct.	1959*	Luna 3	photograph of lunar farside
26 Aug.	1962	Mariner 2	Venus flyby
1 Nov.	1962*	Mars 1	Mars flyby
28 July	1964	Ranger 7	lunar "hard" landing
26 Nov.	1964	Mariner 4	Mars flyby
16 Nov.	1965*	Venera 3	Venus landing (impact)
31 Jan.	1966*	Luna 9	soft landing and pictures
31 Mar.	1966*	Luna 10	lunar orbiter
30 May	1966	Surveyor 1	controlled soft landing
10 Aug.	1966	Lunar Orbiter 1	lunar photographic orbiter
15 Sept.	1968*	Zond 5	first life forms to Moon and back
21 Dec.	1968	Apollo 8	first men to Moon, no landing
16 July	1969	Apollo 11	first manned lunar landing
17 Aug.	1970*	Venera 7	soft landing on Venus
12 Sept.	1970*	Luna 16	unmanned sample return
19 May	1971*	Mars 2	Mars orbiter
28 May	1971*	Mars 3	Mars lander
30 May	1971	Mariner 9	Mars orbiter
3 Mar.	1972	Pioneer 10	Jupiter flyby
5 Apr.	1973	Pioneer 11	Saturn flyby
3 Nov.	1973	Mariner 10	Mercury flyby
8 June	1975*	Venera 9	surface pictures
20 July	1976	Viking 1	Mars soft landing
20 May	1978	Pioneer Venus	Venus orbiter
5 Mar.	1979	Voyager 1	Jupiter flyby
13 Nov.	1980	Voyager 1	Saturn flyby

*Soviet missions.

Figure 2.5 Lunar orbiter image of the 900 km Orientale basin showing prominent rings – Inner Rook Mountains (A); Outer Rook Mountains (B); and Cordillera Mountains (C) – and the central filling by mare lavas. Only the far right (eastern) side of this basin can be seen from Earth. Note the radial pattern formed by ejecta from the impact basin. The outer two mountains rise more than 3 km above the adjacent terrain and are some of the highest features on the Moon. Spudis (1982) estimates that the transient excavation cavity was about 570 km across. Lunar orbiter images can be recognized from the *framelets* (strips) which have been mosaicked in the image restoration process after transmission from the spacecraft (Lunar Orbiter IV-M187).

Figure 2.6 Mosaic of Surveyor 7 images taken on the surface of the Moon in January 1968. Surveyor images were the first to show the lunar surface in high resolution. Surveyor 7 landed on the rim of the crater Tycho (see Fig. 2.7). Largest rock in view is about 60 cm across; it is probably a block of ejecta from Tycho and was responsible for the small crater to its right (NASA photograph 83 H 40).

Although the first three manned landings, Apollos 11 (Fig. 1.3), 12 and 14, were primarily engineering missions, their returned samples from the Moon were intensively analyzed in laboratories on Earth. Apollos 15, 16 and 17 carried much more extensive scientific payloads, both for the landers (including the lunar "rovers" which gave the astronauts greater mobility on the surface) and for the command module (CM) which remained in lunar orbit during the lander phase. While in orbit, the CM obtained high-quality photographs and utilized an array of other remote-sensing instruments to obtain geochemical and geophysical data (Fig. 1.5).

Concurrent with the US manned lunar program, the Soviets carried out a series of successful unmanned lunar missions which included automated roving spacecraft – the Lunakhod – and returned samples of the Moon. Some of these spacecraft were equipped with drills which enabled cores to be taken from the subsurface. Figure 2.7 shows all lunar landings, manned and automated, from US and USSR missions.

The combined results of the US and USSR missions to the Moon provided a remarkable wealth of data during the early 1970s. Never before had so many scientists and laboratories been involved in such a focused study. Most of the results of the lunar program can be found in the annual Proceedings of the Lunar and Planetary Science Conference (a meeting held annually at the Johnson Space Center in Houston), published as a supplement to *Geochimica et Cosmochimica Acta* by Pergamon Press (1970–1981) and as a supplement to the *Journal of Geophysical Research* from 1982.

Concurrent with the lunar program, both the US and the USSR began exploration of planets beyond the Moon. Partly by chance, partly by design, and partly as a result of various mission failures, the US tended to focus on Mars while the Soviets concentrated on Venus, at least in the early stages of exploration. *Mariner* and *Pioneer* constitute two families of spacecraft that are the main US "buses" to carry instruments throughout the Solar System. Mariner spacecraft are 3-axis stabilized probes which can carry high-resolution imaging systems as well as other instruments; Pioneer probes are spin-stabilized spacecraft which are used primarily for non-imaging experiments.

Between 1964 and 1971, Mariners 4, 6, 7 and 9 successfully flew to Mars and returned more than 8000 images of the surface, some with resolutions better than 100 m. In 1976, the Viking mission marked the first successful landing on Mars. Consisting of two Mariner-class orbiters and two "soft" landers, the Viking mission has returned nearly 60 000 pictures from orbit (Fig. 1.1), 4000 pictures from the two landers and vast quantities of data from other experiments.

The Soviet efforts to Venus began with the first successful landing in1970 by Venera 7. In the ensuing years, landings have been accomplished by Veneras 8, 9, 10, 11, 12, 13 and 14 and the Soviets have been able to obtain images of the surface (Fig. 2.8), determine chemical compositions of surface materials and measure surface winds and temperatures. These are remarkable achievements given the extremely hostile environment on the venusian surface; temperatures average 753 K, atmospheric surface pressure is about 90 bar, and the atmosphere is laced with sulfuric acid. In the fall of 1983, Veneras 15 and 16 were placed in orbit and began returning high-resolution (~ 1–3 km) radar images of the venusian surface.

In the geologic context, the US Pioneer Venus mission was particularly important because it provided our first global "view" of Venus. Since dense clouds completely hide the surface from imaging by conventional camera systems, radar imaging is the only means for "seeing" the surface of Venus from above the atmosphere. Although not specifically designed to obtain high-resolution images, the radar system on board Pioneer Venus was adapted to acquire topographic data which could then be transformed into pseudo pictures of the surface.

Mercury has received relatively little attention in Solar System exploration. In 1973, Mariner 10 made the first reconnaissance flyby and obtained high-resolution images. From serendipitous orbital geometry and clever manipulations by the engineering teams at the Jet Propulsion Laboratory, Mariner 10 was able to loop around the Sun and make not only a second flyby of Mercury, but a third pass as well, acquiring images and other data each time. In all, slightly less than half of the

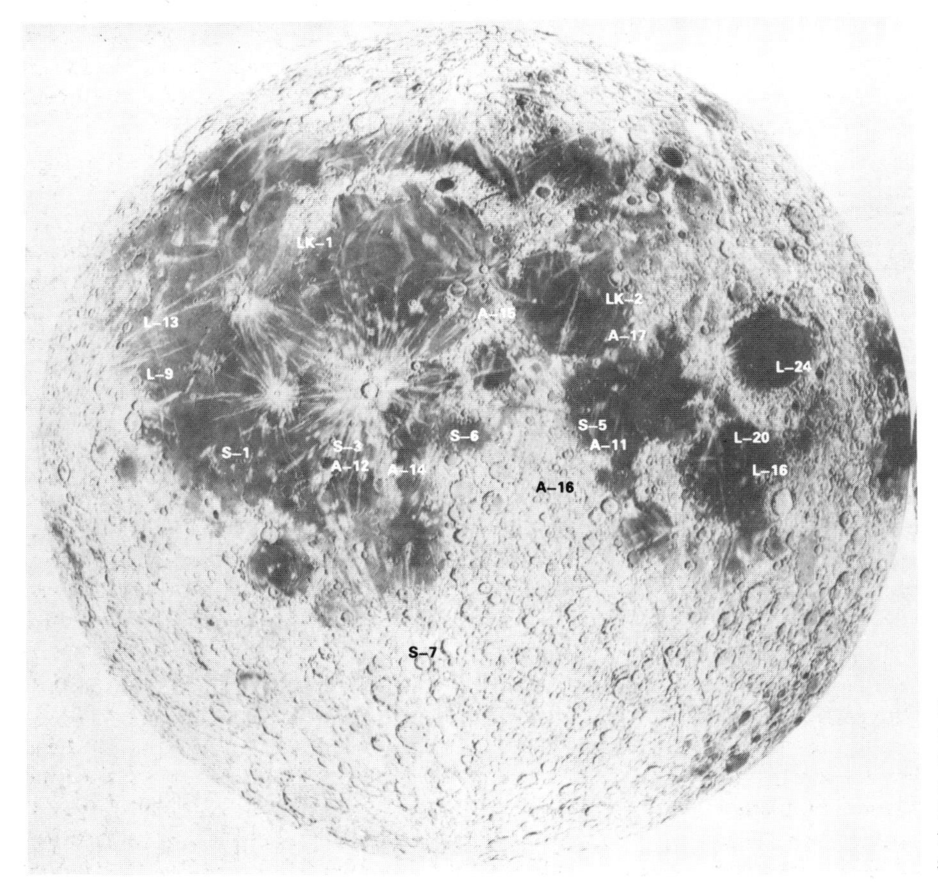

Figure 2.7 Map of the lunar nearside showing various landing sites; includes five Surveyor(s) landings (S), six Apollo sites (A), two Lunokhod roving vehicles (LK), five Luna missions (L), of which three (Lunas 16, 20, 24) returned samples to the Earth.

Figure 2.8 Venera 13 image of the surface of Venus (courtesy NASA-Jet Propulsion Laboratory).

planet was observed by Mariner 10, the only mission to Mercury.

Thus far, only the US has reached beyond Mars to explore the outer Solar System. In 1972, Pioneer 10 made the first flyby of Jupiter and was followed 12 months later by Pioneer 11. In 1979, Pioneer 11 reached Saturn; both Pioneers obtained images during the planetary encounters. Although crude, these images showed variations in surface markings for some of the Galilean satellites and piqued scientific curiosity.

Voyager is by far the most successful mission in the reconnaissance of the Solar System. Consisting of two Mariner-type spacecraft, *Voyager 1* and *Voyager 2*, this mission began with flybys of Jupiter in 1979 and Saturn in 1980 and 1981 and continues with expected flybys of Uranus and Neptune in 1986 and 1989 respectively. More than 20 000 images were obtained of Jupiter and its satellites, providing the first good views of the complex nature of the Galilean satellites. Flybys of Saturn resulted in more than 18 500 images, which revealed the complexities of Saturn's rings and the wide array of satellites, nearly doubling the number of planetary objects available for geologic analysis.

Solar System exploration has made remarkable advances since the first simple probes were sent to the Moon. Although the depth of knowledge is highly variable, sufficient data have been gathered from these missions to describe the major physiographic provinces of about 60 percent of the planets and their satellites.

2.4 Planetary images

The missions described in the last section carried a wide variety of remote sensing systems on their journeys of exploration. It is not the intent to discuss here all the methods and instruments employed in remote sensing – Colwell (1983) provides such a discussion – but rather to review systems important in planetary geology. Basically, most imaging systems consist of three elements: an optical system (i.e. a lens) to focus the image, a shutter to expose the image and a sensor. Sensors are generally of two types, film systems or various electronic devices such as television tubes. Davies and Murray (1971) provide an excellent discussion of the various types of imaging devices used in early Solar System exploration.

Among the many important considerations in designing imaging systems, **spatial resolution** (how small an object can be seen) and **spectral sensitivity** (how much energy from a given part of the electromagnetic spectrum is required to see an object) are the most important. Detail is progressively lost as resolution is degraded. Table 2.3 shows resolutions required to detect various geomorphic features on planets.

2.4.1 Film systems

By far the highest resolution images are produced on photographic film. To utilize the high-resolution capability of film, however, the film must be returned to Earth. Thus far in planetary exploration, this has been achieved only from Soviet Zond and the manned Apollo missions to the Moon. Of the unmanned US lunar and planetary missions, only Lunar orbiter utilized film sensors, but the film was not returned to Earth; it was developed on board and then the image was transferred to Earth as an electronic signal.

Lunar orbiter systems consisted of two cameras which used a common supply of 70 mm film; one camera had a telephoto lens (designated *H* for high resolution) and the other camera had a wide angle lens (designated *M* for medium resolution). Film processing was accomplished with the Kodak Bimat diffusion technique in which a gelatin developer was laminated with the film. After processing, the negative film was passed through a system that converted the images to electrical signals by scanning the film with a microscopic spot of high intensity light, as shown in Figure 2.9. The intensity of the

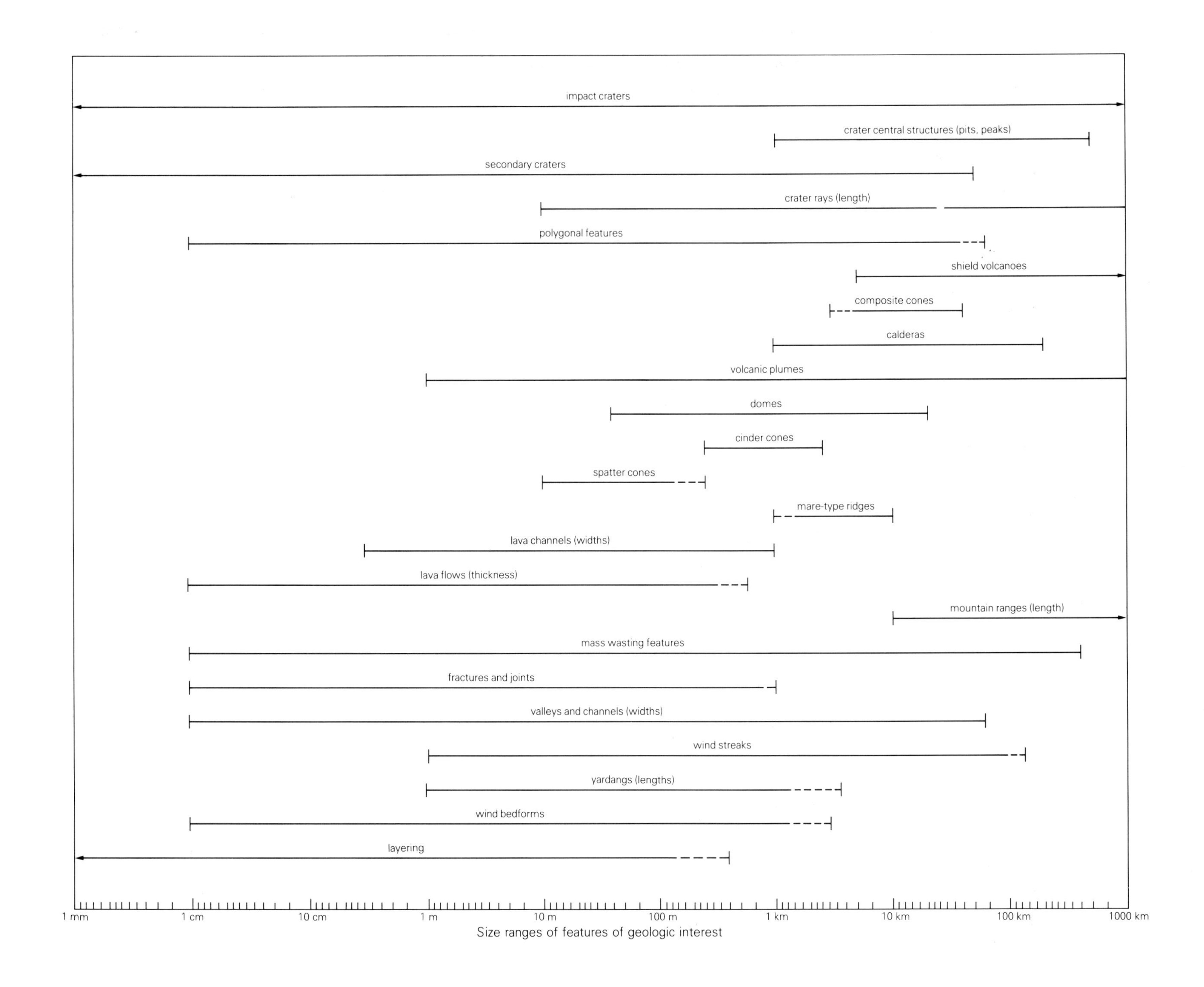
impact craters
crater central structures (pits, peaks)
secondary craters
crater rays (length)
polygonal features
shield volcanoes
composite cones
calderas
volcanic plumes
domes
cinder cones
spatter cones
mare-type ridges
lava channels (widths)
lava flows (thickness)
mountain ranges (length)
mass wasting features
fractures and joints
valleys and channels (widths)
wind streaks
yardangs (lengths)
wind bedforms
layering
1 mm
1 cm
10 cm
1 m
10 m
100 m
1 km
10 km
100 km
1000 km
Size ranges of features of geologic interest

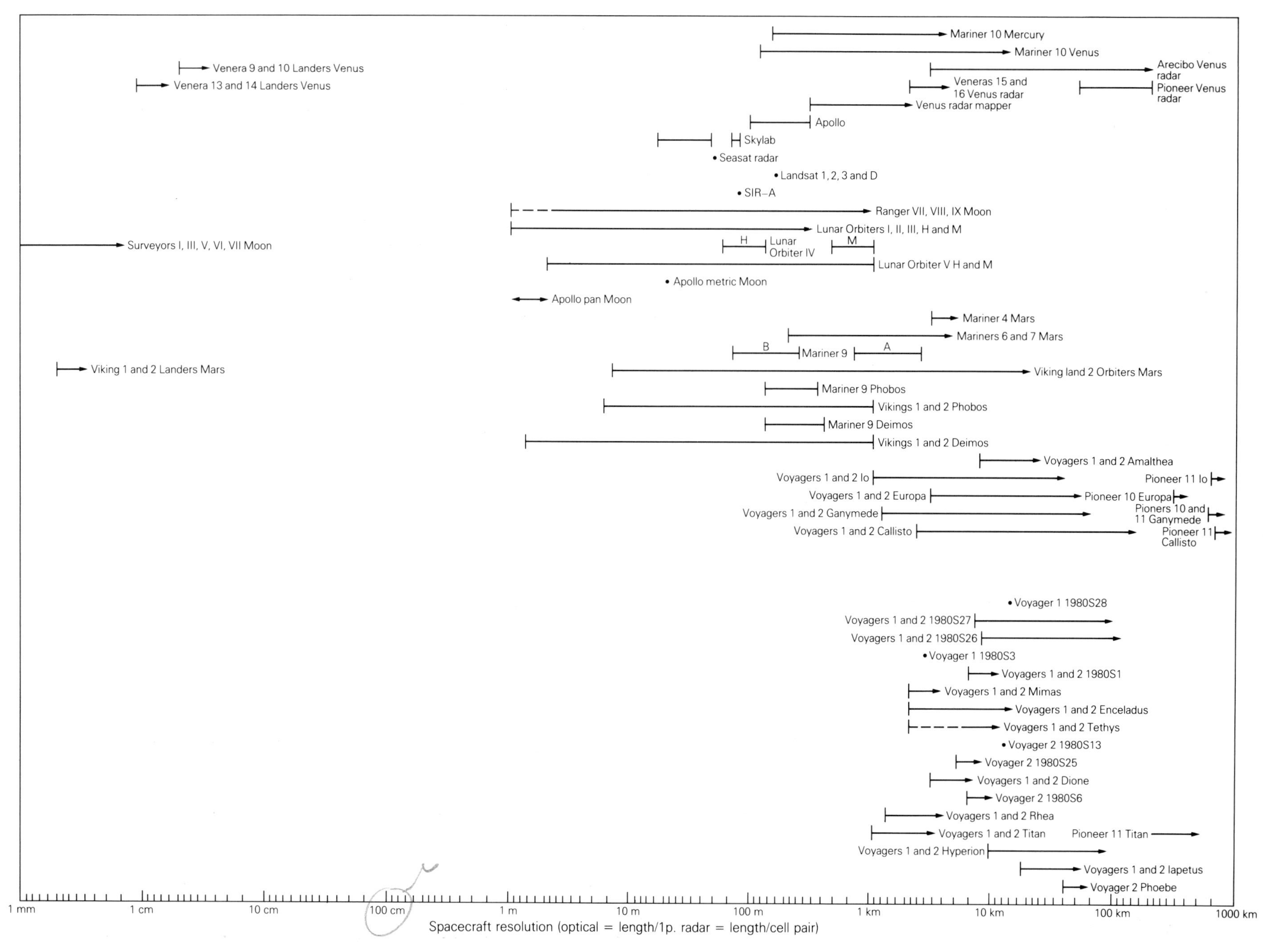

Table 2.3 Resolution required to detect various surface features compared with ranges of resolution from lunar and planetary missions.

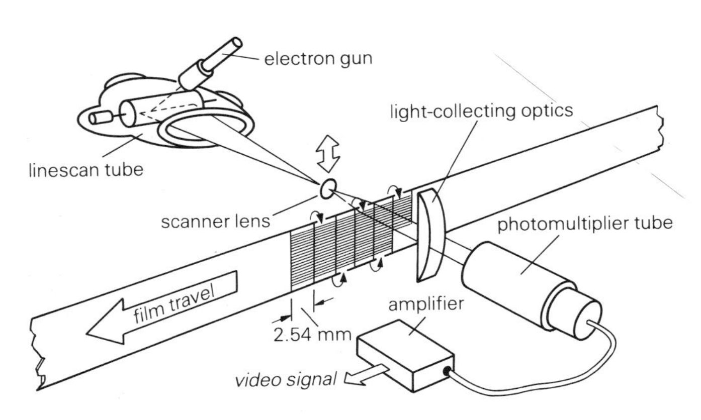

Figure 2.9 Lunar orbiter film scanning system (courtesy NASA).

light reaching a photomultiplier tube was modulated by the density (gray tone) of the image on the film.

The electronic signal from the photomultiplier tube was transmitted to Earth where the signal was used to illuminate a glow tube which exposed fresh film, thus generating a duplicate of the image on the spacecraft. Strips of 35 mm film were used in this process to correspond to the scanning sequence of the original image. Consequently, Lunar orbiter photographs are characterized as being composed of strips of images (Fig. 2.5). Although some stereoscopic views of the lunar surface were made from the Lunar orbiters, the continuity between strips was lost in the electronic transformations, resulting in a stair-stepping effect when viewed stereoscopically.

The Apollo missions to the Moon all involved camera systems using film which was returned to Earth. Returned film enabled analysis without the complexities of electronic transformation and provided reliable stereoscopic models useful for photogrammetry. Most simple were the hand-held cameras using 70 mm film. These were used both from orbit and on the ground (often used to make panoramic views from overlapping frames which were mosaicked) and involved both color and black and white film.

Although a sophisticated mapping camera was carried on the Apollo 14 mission, a malfunction early in the mission resulted in very little usable data. Apollos 15, 16 and 17 all carried two high-quality camera systems, a panoramic camera for detailed geologic studies, and a metric (mapping) camera which enabled topographic data to be derived. The "pan" camera, using a 610 mm focal length lens, was highly sophisticated and obtained stereoscopic views of the surface with resolutions as good as 1 m from an orbital altitude of 100 km. The camera rotated continuously in a direction across the path of the orbiting spacecraft, giving a panoramic view (Fig. 2.10).

The metric camera was a mapping system which consisted of two cameras, one pointing downward and one pointing away from the Moon. Both used 76 mm lenses, with the upward-pointing camera being used to fix precisely the geometric position of the spacecraft in relation to the star field, with the Moon's surface being photographed by the other camera. Although the spatial resolution of metric frames is only about 20 m, the precision of location enabled high-quality maps to be produced.

2.4.2 Television systems

The imaging system most widely used thus far in planetary exploration involves television, or Vidicon, cameras. Vidicon systems employ a small electron gun and a photoconductor. The image is optically focused on the photoconductor so that a beam of electrons from the gun is transformed into a current which varies with the intensity of the light reflected from the scene. This current can be in either analog (flown on early systems, such as Ranger) or digital format and can be stored either on tape on board the spacecraft, or transmitted directly to Earth.

Figure 2.10 Panoramic camera view from Apollo 16 showing horizon to terminator view over typical lunar highlands terrain; the swirling pattern to the south (lower right) of crater Al-Birani remains enigmatic (Apollo 16 AS 16-5533).

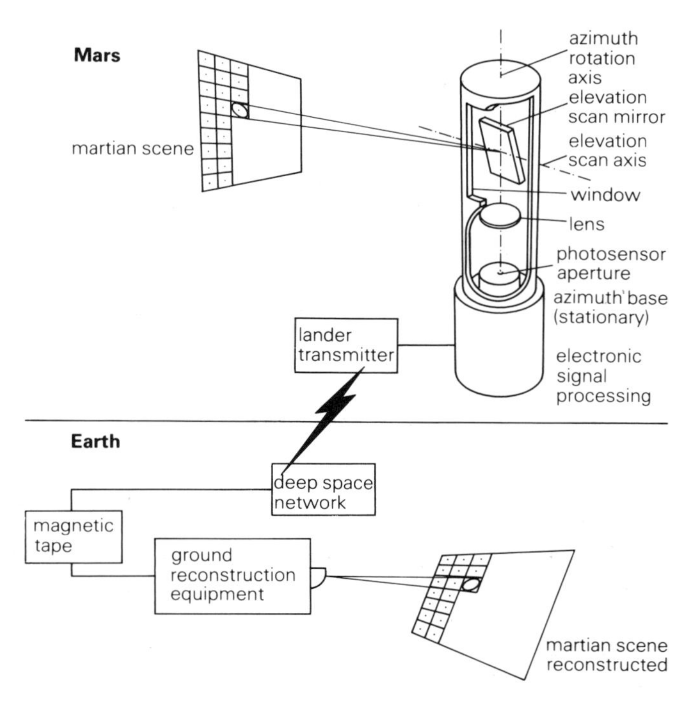

Figure 2.11 Diagram of facsimile camera and related system for the Viking Lander camera flown to Mars (courtesy NASA).

Vidicon images from the Mariner and Voyager missions were compiled as arrays of picture elements (pixels), arranged in lines. The spatial resolution of a Vidicon image is determined by the distance of the spacecraft from the object, the characteristics of the lens system and the size of the individual pixels.

2.4.3 *Facsimile systems*

Facsimile systems are used every day in countless newspaper offices around the world to transmit pictures over telephone lines. Basically, a scene is divided into a grid of pixels and each pixel is assigned a gray level – essentially the same idea as for Vidicon pictures. However, unlike the Vidicon system in which the whole image is focused onto the tube at one time, facsimile cameras scan across the scene by rotating a slit across the scene as a panorama, shown in Figure 2.11. Facsimile cameras were used on the two martian Viking lander spacecraft (Fig. 1.4) and on the Venera landers (Fig. 2.8).

2.4.4 *Radar-imaging systems*

Dense clouds completely obscure two planets of geologic interest – Venus and Titan (one of the moons of Saturn) – and hide their surfaces from imaging by any of the optical systems described above. Using long wavelength energy capable of penetrating clouds, radar-imaging systems have been developed to explore Venus and possibly Titan from spacecraft.

Radar is considered an "active" form of remote sensing because the imaging system generates energy as a radar beam which is then reflected from the scene and received by the sensor. Thus, in effect, radar-imaging systems provide their own "light" and can operate regardless of time of day.

Seasat and SIR-A were two radar-imaging experiments flown in orbit around Earth and provided important insight into the types of images that might be anticipated for Venus. Figure 2.12 shows a typical radar image and the type of geologic information that can be obtained.

2.4.5 *Charge-coupled devices*

Charge-coupled devices (CCD) were invented in 1969 by Bell Laboratories and are used in a variety of communication systems and as a solid-state imaging device. CCD "chips" consist of one layer of metallic electrodes and one layer of silicon crystals, separated by an insulating layer of silicon dioxide. When used as an imaging system, the CCD chip is structured as an array of

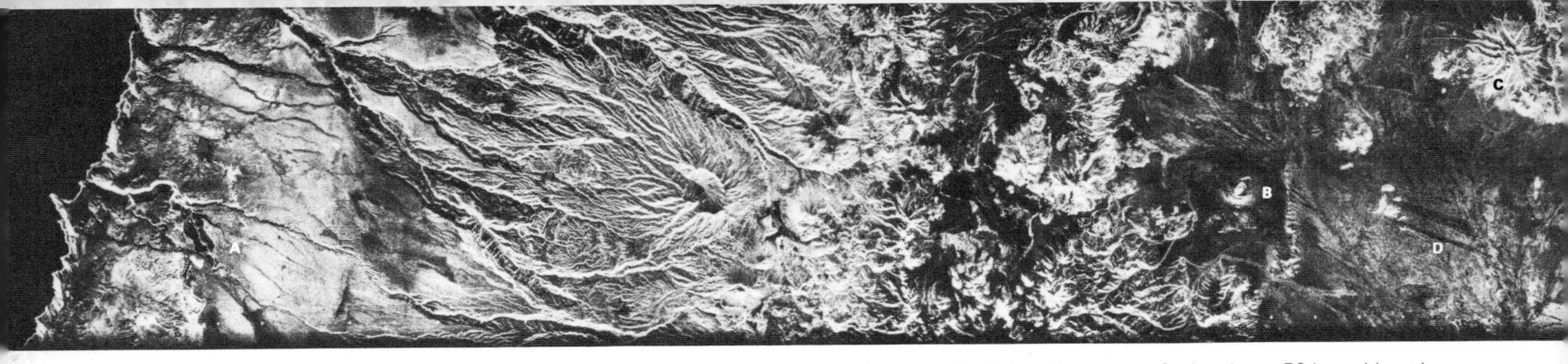

Figure 2.12 Radar image from the SIR-A (Shuttle Imaging Radar) system flown on the Columbia spacecraft showing a 50 km wide strip from the Pacific shoreline of Chile (left side) across the western Andes and the Altiplano of Bolivia. Features visible include stream patterns (A), dome volcano with summit crater (B), stratovolcano (C), and wind streaks (D) formed on smooth sedimentary plains (courtesy NASA, Jet Propulsion Laboratory).

elements; light focused on the chip by a lens causes a pattern of electrical charges to be created. The charges on each element are proportional to the amount of light and provide an accurate representation of the scene. Each charge can be transmitted separately and then reconstructed as pixels using conventional image-processing techniques.

2.5 Digital image processing

Vidicon and facsimile imaging systems used by Mariner, Viking and Voyager missions and the CCD systems to be used by the Space Telescope and flown on Galileo all include digital signal processing for transmission to Earth. Because each pixel is in digital format, the images can be manipulated by computers using various image-processing techniques. Image processing has become an extremely important aspect of planetary exploration, and knowledge of the fundamentals is critically important for geologic utilization of spacecraft digital images. Although many books have been written on the subject, an excellent brief introduction to the fundamentals of digital image processing is given by Condit and Chavez (1979).

A digital picture is a two-dimensional array of pixels, each of which is assigned a value representing the light brightness (called **DN** or **digital number**). The accuracy of the brightness level is a function of the digital encodement; for example, 8-bit (2^8) encodement allows 256 levels of gray to be assigned to each pixel (0 = black, or below the threshold of sensitivity; 255 = white, or saturation). This range of gray levels far exceeds the dynamic range of any photographic film, which means that digital images contain much more information than can be recorded in camera systems using film.

Horizontal rows of pixels are called **lines** and vertical columns are called **samples**. The maximum pixel format used in any planetary mission flown thus far was the Viking orbiter camera system involving 1056 lines by 1204 samples, with each pixel transmitted in 8-bit encodement. Thus, each complete image consists of 10.2 million computer bits of data.

Computer programs for image processing can be classified into three types for geologic analysis: (a) image correction programs, (b) image enhancement programs, and (c) multispectral utilization programs.

2.5.1 Image correction programs

Image correction programs are used to remove electronic noise, replace lost data, correct for variations in sun angle, and make calibration corrections for the camera system. For example, during electronic transmission of the image data, "noise" may be introduced which superimposes a pattern on the image, as shown in Figure 2.13, or data for individual pixels, or lines of pixels, may be lost. In order to generate a "whole" picture, DN values are assigned to the "noisy" or "lost" pixels by determining the average value from the surrounding pixels and the image is then filled in using these values.

2.5.2 Geometric corrections

Geometric distortions in images also can be corrected through image-processing techniques. For example, Figure 2.14 shows an oblique view of part of Mercury. Through knowledge of the camera geometry and location of the spacecraft, individual pixels can be shifted in their position on the image to correspond to a vertical view in which the scale of the image is orthogonal (Fig. 2.15a). Such geometric transformations are important for analyses of sizes and shapes of landforms, and for mosaicking sequences of pictures. Geometric transformations can be made for almost any cartographic format, including Mercator, Lambert conformal or polar stereographic projections.

2.5.3 Enhancement techniques

Enhancement programs are used to emphasize different aspects of the scene on digital images. Although numerous programs have been developed for image enhancement, the most common programs involve **stretching** and **spatial filtering**.

Figures 2.16 and 2.17 show a series of image-processing routines involving both image-correction and image-enhancement techniques. The surfaces of some planets are relatively low in contrast and "stretching" of the DN values can enhance the contrast on the image to bring out the surface detail, as shown in Figures 2.16a and b for a typical scene on Mars. In this view, most of the levels of gray are within the same narrow range, shown by the DN histogram. These gray levels can be "stretched" over a wider range to produce the view shown in Figure 2.16c. The "stretch" can be accomplished using any one of a number of algorithms, including linear stretches (an even spread of the histogram over a given range of DN values), or in various gaussian distributions.

Spatial filtering changes pixel DN values according to the values of neighboring pixels, either as averages, or as differences. **High-pass filters** amplify rapid changes from one pixel to another and are used to sharpen edges

Figure 2.13 Mariner 9 image showing "zipper" pattern (A; camera "noise") and white streaks (B) generated from the reseau marks; both of these artifacts can be removed by image processing. The dark spot (C) results from a dust speck in the camera system and is too large to be "removed" by processing (Mariner 9 4233-31).

Figure 2.14 Specially enhanced Mariner 10 image of Mercury (courtesy M. Davies; Mariner 10 FDS 27321).

Figure 2.15a Same frame as shown in Figure 2.14 which has been corrected for camera shading and variation in scene luminance due to sun angle and transformed to a Mercator projection; processing was also done to enhance albedo markings (courtesy M. Davies).

Figure 2.15b Shaded airbrush chart of the area shown in (a) of part of the Kuiper quadrangle, Mercury (from Davies & Batson 1975).

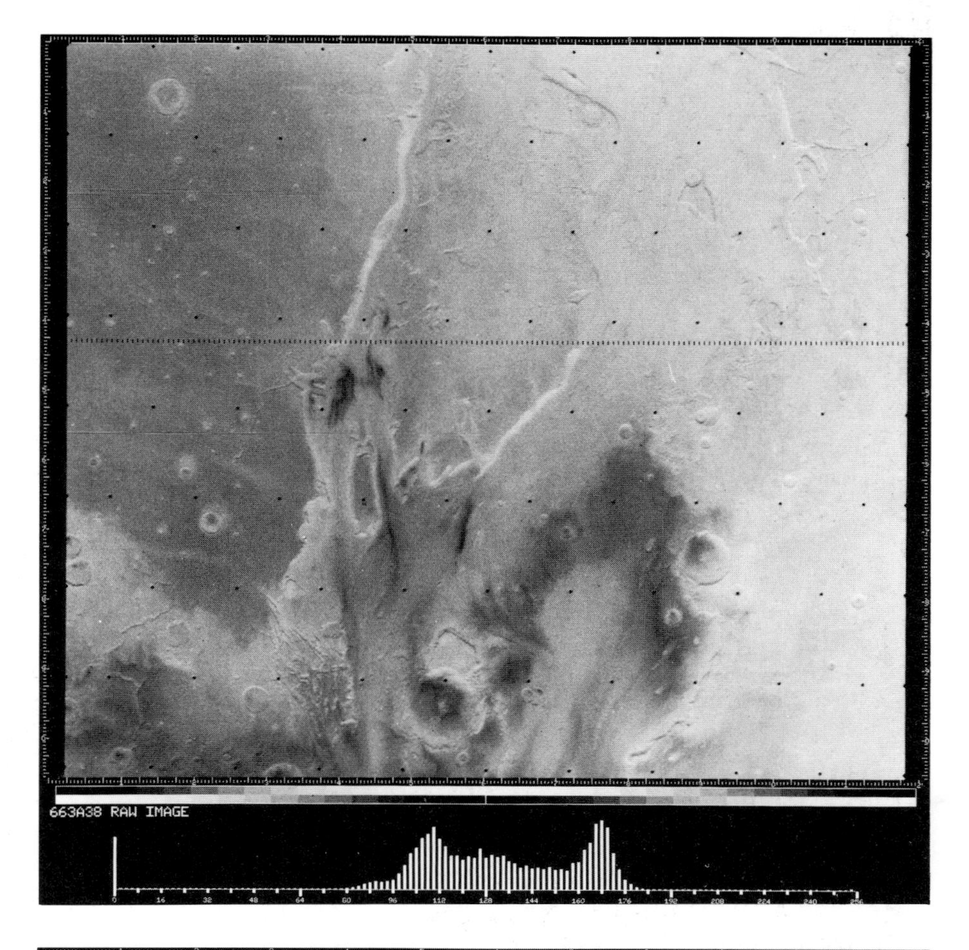

Figure 2.16a "Raw" or unprocessed Viking orbiter image containing dropped line of data across the picture and reseau marks (evenly spaced black dots used for geometric analyses). Distribution of DN values shown in histogram (Arizona State University Image Processing Facility).

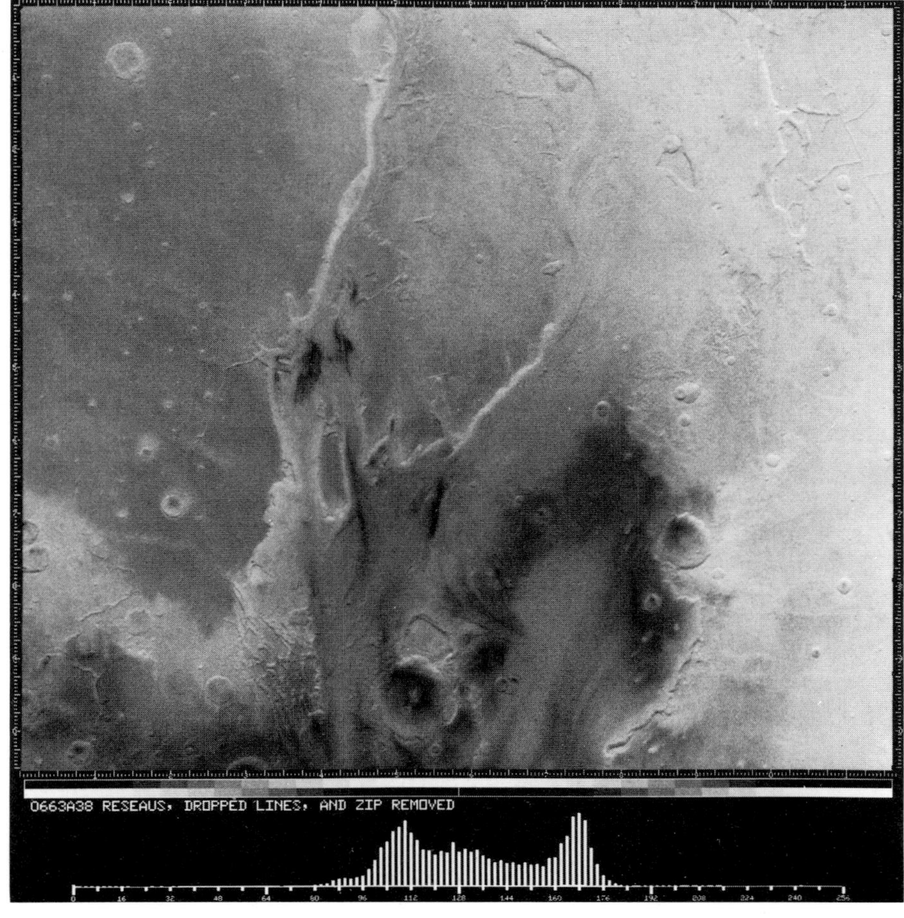

Figure 2.16b Image in which correction routines have been applied to remove the reseaus and the dropped line by filling in pixels based on averaging values of surrounding pixels (Arizona State University Image Processing Facility).

Figure 2.16c Same image which has been stretched to enhance contrast (as indicated by the histogram), in addition to being corrected and calibrated (Arizona State University Image Processing Facility).

and enhance small detail, although there is generally a loss of tonal variations (Fig. 2.17a). **Low-pass filters** enhance low-frequency detail and tend to smooth image detail, while preserving tonal quality such as albedo (Fig. 2.17b).

2.5.4 Multispectral images

Various color pictures are generated from multiple images of the same terrain taken through colored filters as part of most spacecraft camera systems. Each selected range of wavelength (determined by the appropriate filter) is registered on separate images. These multiple images are then registered geometrically to generate a color view. Except for color film returned from the Moon, all color images of the planets have been obtained in this manner and are termed **false color** images.

By ratioing the DN values of images taken through different filters of the same scene, the spectral reflectances, or color difference between rock units, are often enhanced. Thus, color ratioing is often a useful technique for photogeologic mapping.

2.5.5 Other image-processing techniques

The development of image-processing techniques is an ever-expanding field as more researchers become familiar with the general process and as new applications are discovered. Some of the more important techniques for planetary geology include **computer mosaicking** (Fig. 2.18), **picture differencing** in which images taken at different times of the same area are subtracted from one another to detect possible changes (Fig. 2.19), and combining multiple data sets to determine possible correlations (such as merging multispectral images with radar or infra-red images).

Because of the manner in which spacecraft images are acquired and processed, artifacts are frequently introduced. In the decade and a half of planetary digital image processing, techniques have been devised to reduce these artifacts; nonetheless, a great many problems exist in using digital images. Furthermore, use of images from earlier missions requires familiarity with these artifacts. Some of the more common artifacts and problems are illustrated in Figure 2.20.

Despite the difficulties encountered in digitally processed images, the benefits derived from image-

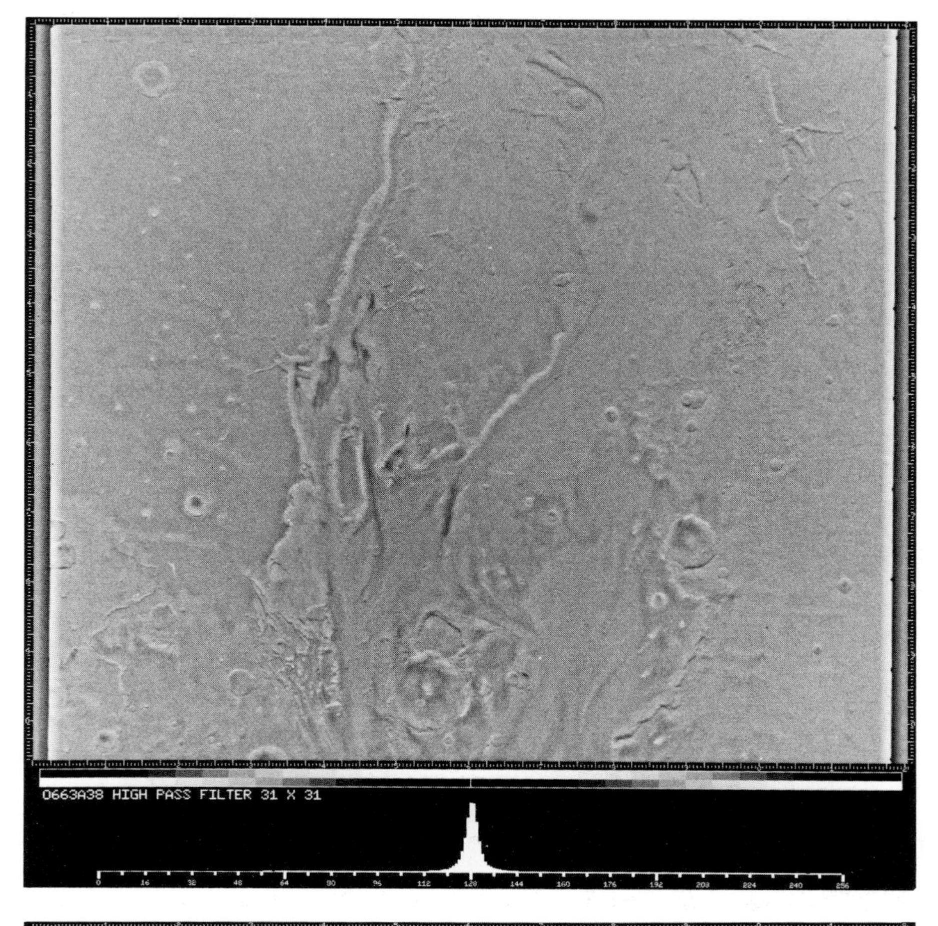

Figure 2.17a Same image as in Figure 2.16, but a *high-pass filter* has been applied to sharpen edge features such as crater walls (Arizona State University Image Processing Facility).

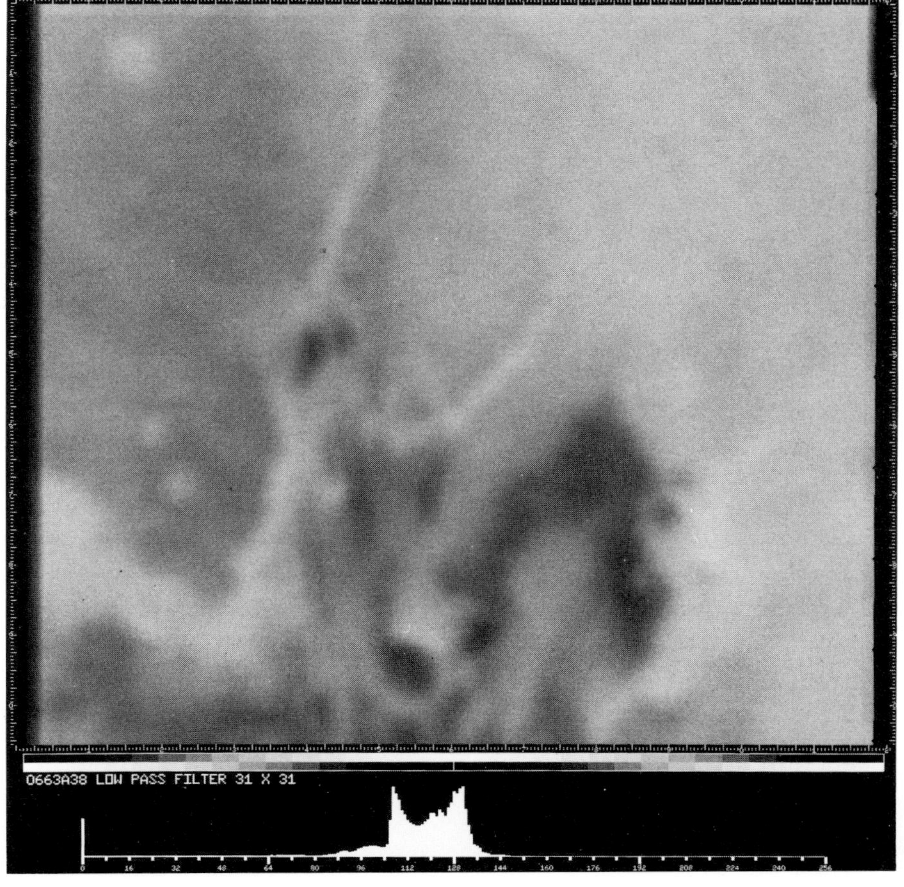

Figure 2.17b Same image in which a low-pass filter has been applied to smooth image detail while preserving tonal quality.

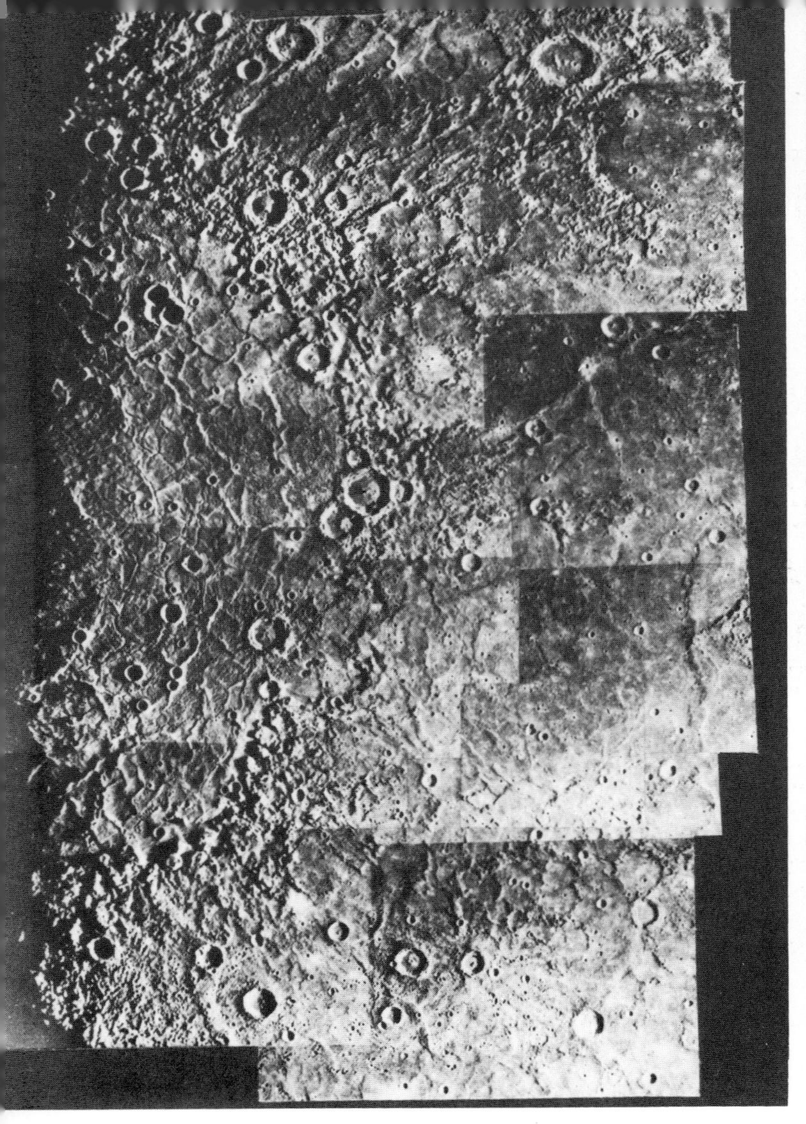

Figure 2.18a Mosaic of Mariner 10 frames assembled by hand, showing the Caloris basin of Mercury.

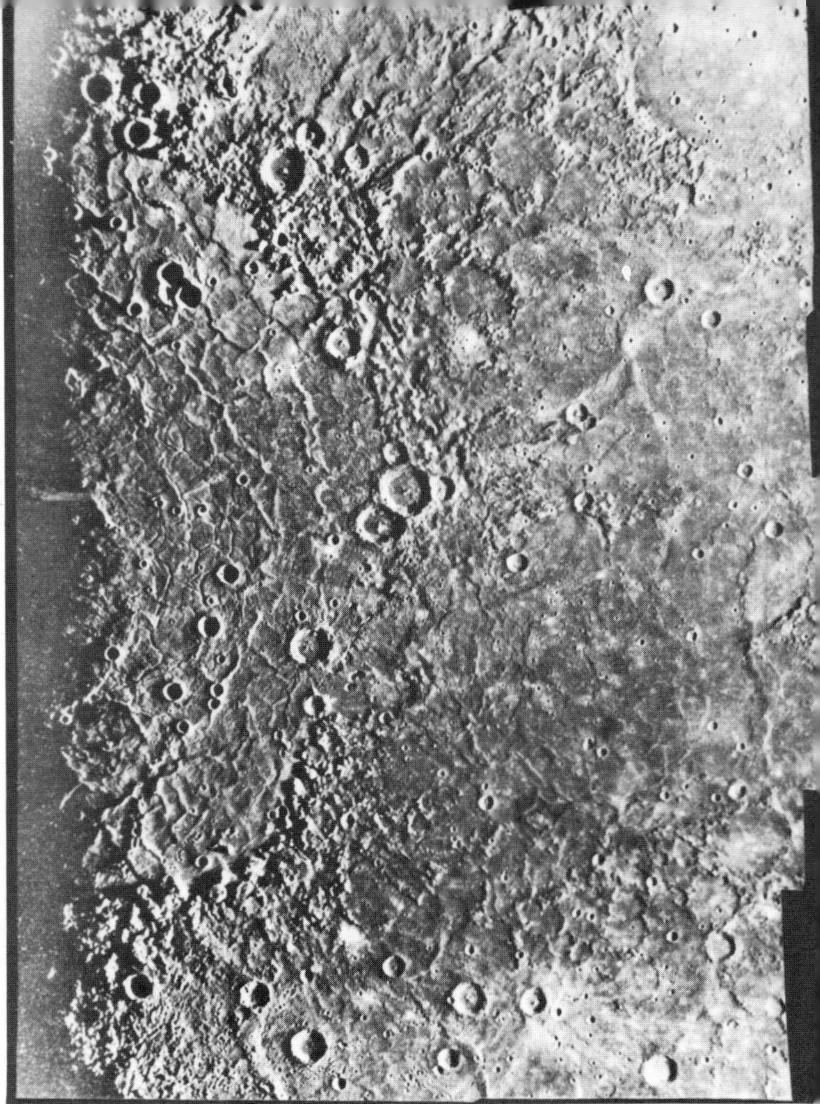

Figure 2.18b Same area as (a), but computer-generated mosaic in which frame boundaries have been "processed-out" and the entire scene is filtered and enhanced as a single image (NASA images courtesy of Jet Propulsion Laboratory).

processing techniques far outweigh the disadvantages and will continue to be an important part of planetary exploration.

2.6 Planetary cartography

Maps are an essential part of planetary exploration. They provide the base for plotting observations and give a context for the drawing of conclusions.

Beginning with Galileo's simple sketch maps of the Moon, planetary cartography has advanced to a sophisticated science. The typical sequence of generating planetary maps begins with the development of an uncontrolled mosaic of images covering a general region. Control points, such as well defined and identifiable surface features, are established within the region and are used to generate a controlled reference network, or planetary grid system. The second stage is to produce a controlled mosaic of images which geometrically fits the network. This mosaic is generated in some conventional format, such as a Mercator projection, and may have additional information superimposed, such as contour lines, to produce a photo-topographic map. Figures 2.21 and 2.22 show the layout of map quadrangles for the major planets and satellites.

Because the images composing the controlled mosaic may have been taken under a wide range of lighting conditions and may be of different spatial resolutions, it is often desirable to produce a more uniform cartographic product. The **shaded airbrush relief map** serves this purpose. Using available images for detail and the controlled mosaic as a base, skilled artists generate these renditions which then serve as uniform base maps. Shaded airbrush maps may have additional information superimposed, such as contour lines, albedo and place names, as discussed by Inge and Bridges (1976) and shown in Figure 2.23.

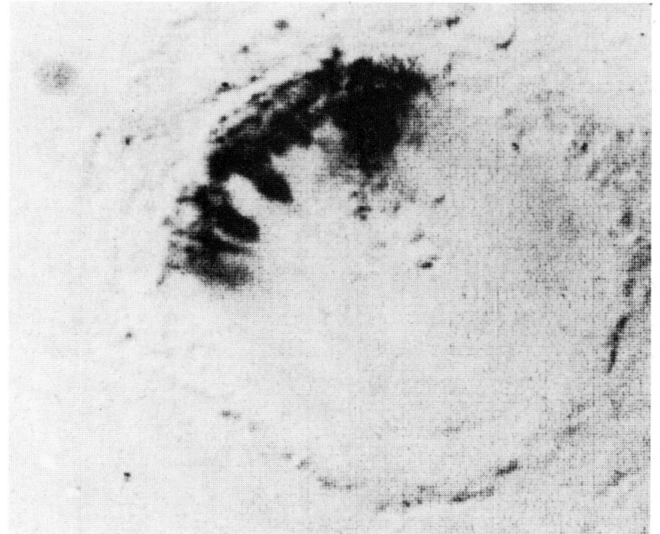

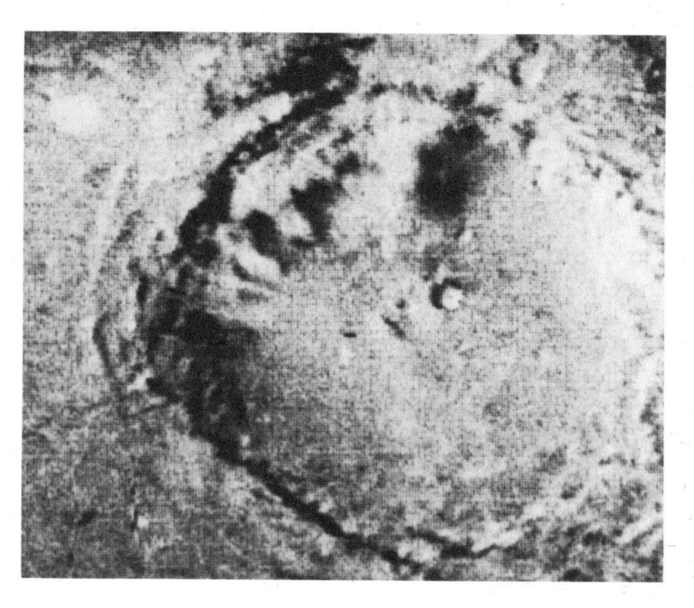

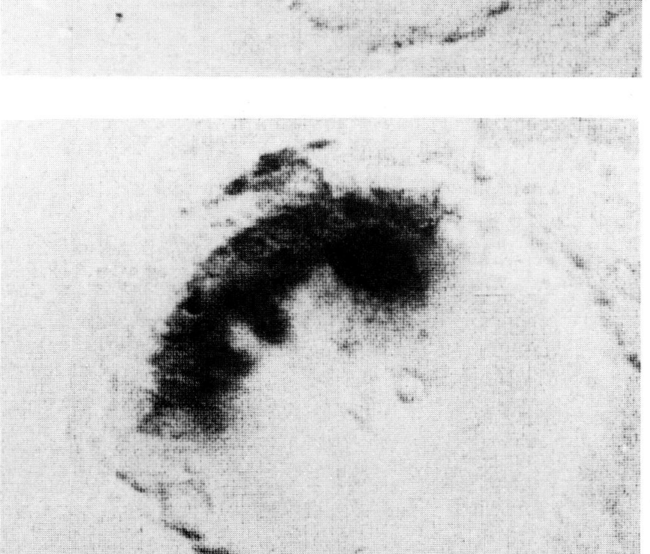

Figure 2.19 Image processing for change detection ("picture difference"). (a, top left) Mariner 9 image (DAS 08189414) showing dark patches inside a large crater, (b, left) same crater imaged later in the mission (Mariner 9 image DAS 10649609), (c, above) image of the "difference" between (a) and (b) in which the DN values of (a) were subtracted from (b) and the resultant DN values used to construct a new image. The dark areas in (c) represent newly darkened zones between the time (a) and (b) were observed.

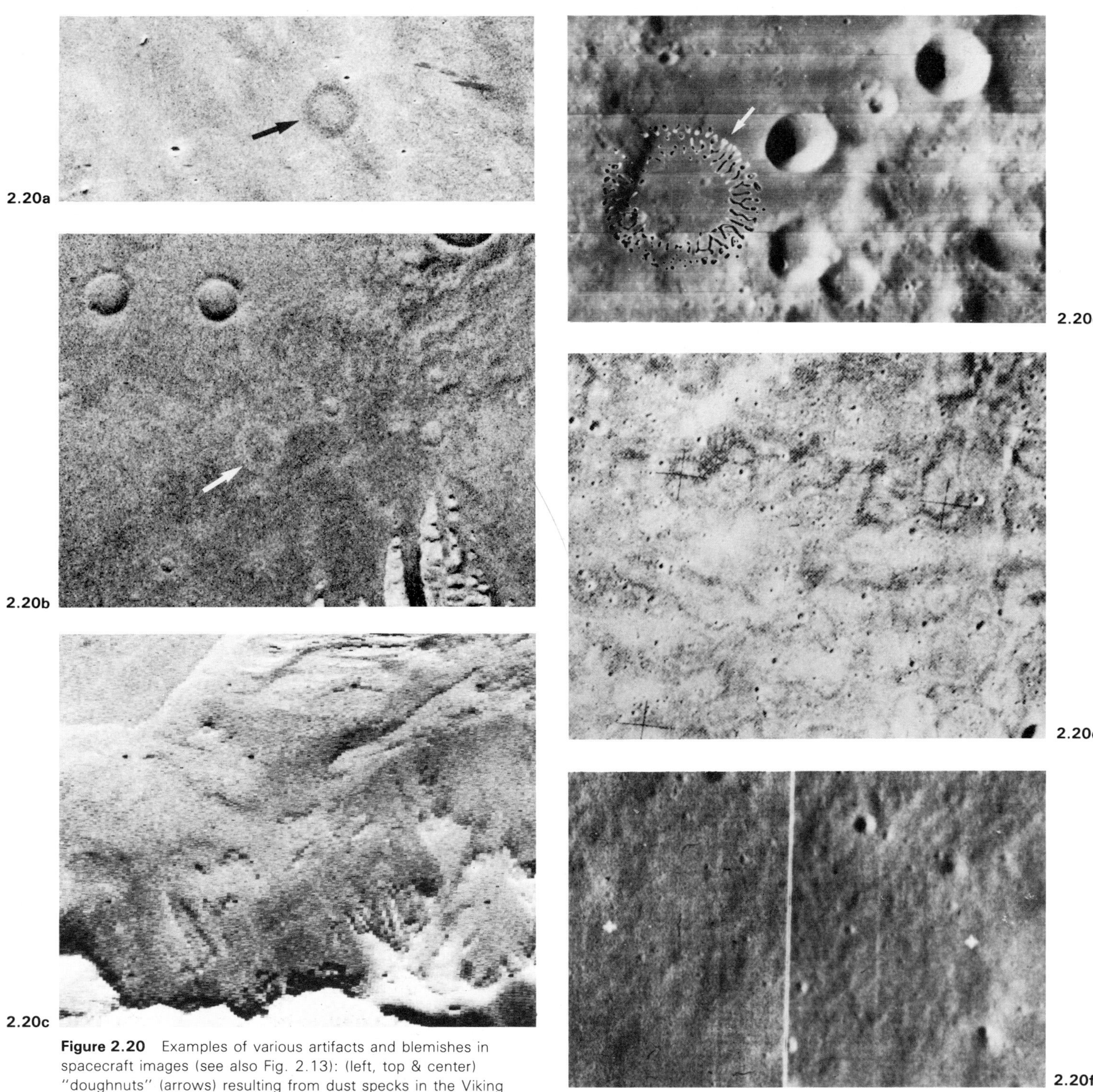

Figure 2.20 Examples of various artifacts and blemishes in spacecraft images (see also Fig. 2.13): (left, top & center) "doughnuts" (arrows) resulting from dust specks in the Viking orbiter camera system; these artifacts could be mistaken for surface patterns on Mars, (top image, VO 826A68; center image, VO 846A40); (bottom left) an oblique Viking orbiter image showing "blocky" pixels which result from problems in transmission of digital data from the spacecraft to Earth (VO 815A63); (top right) "bimat bubbles" (arrow) in Lunar orbiter photograph of the Moon result from tiny air pockets between the film layer and developer layer (part of LO IV-67H_1), (center right) "wormy" texture in Apollo metric photographs that results from processing, shows in both photographic prints and lunar photomosaics, and (bottom right) tiny white crosses are not blemishes in Lunar orbiter photographs but are reseau marks built into the camera system; some non-scientists have interpreted these as representing cultural features by intelligent life on the Moon! (Lunar Orbiter V-H67).

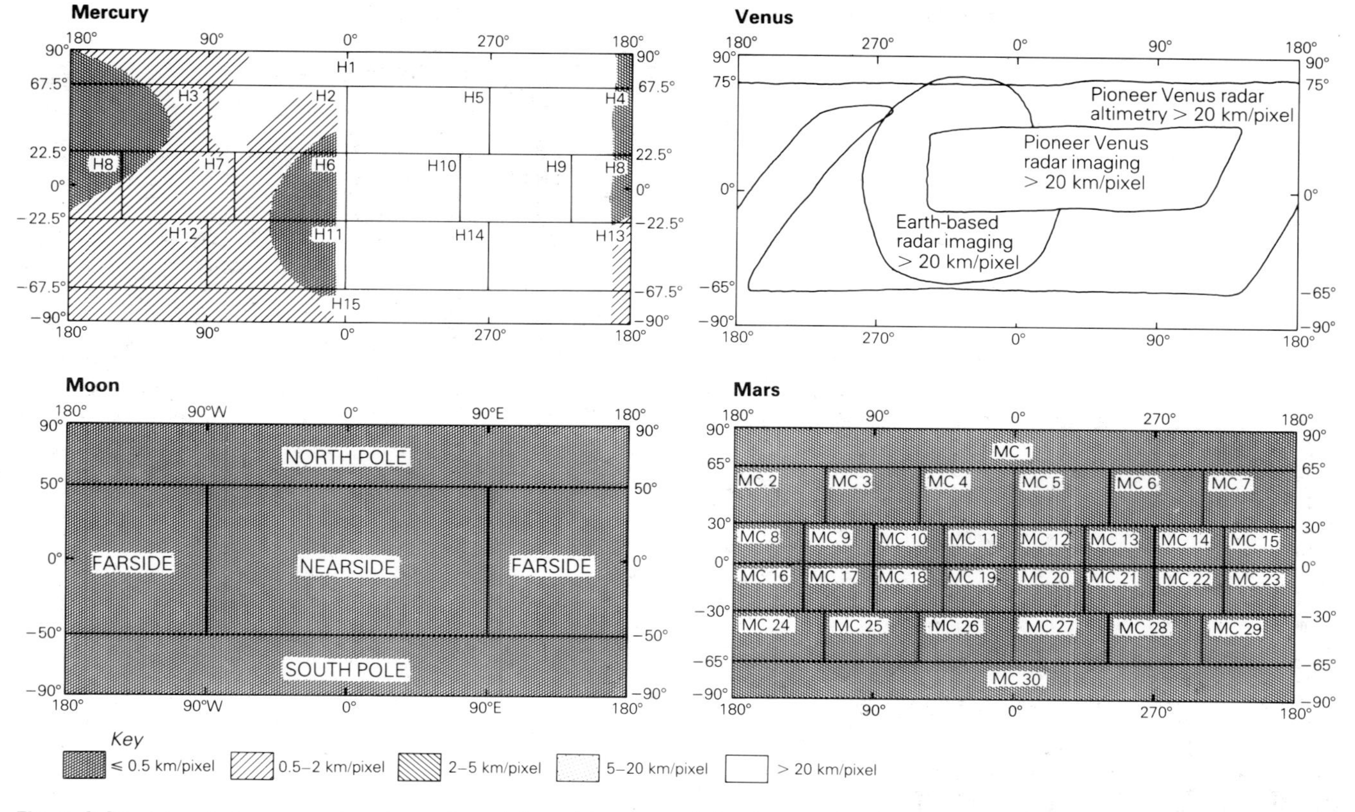

Figure 2.21 Layout of map quadrangles and resolution of images available for mapping Mercury, Venus, the Moon and Mars. As the term is used here, resolution is the size of a picture element on the surface of the planet (from Batson 1981).

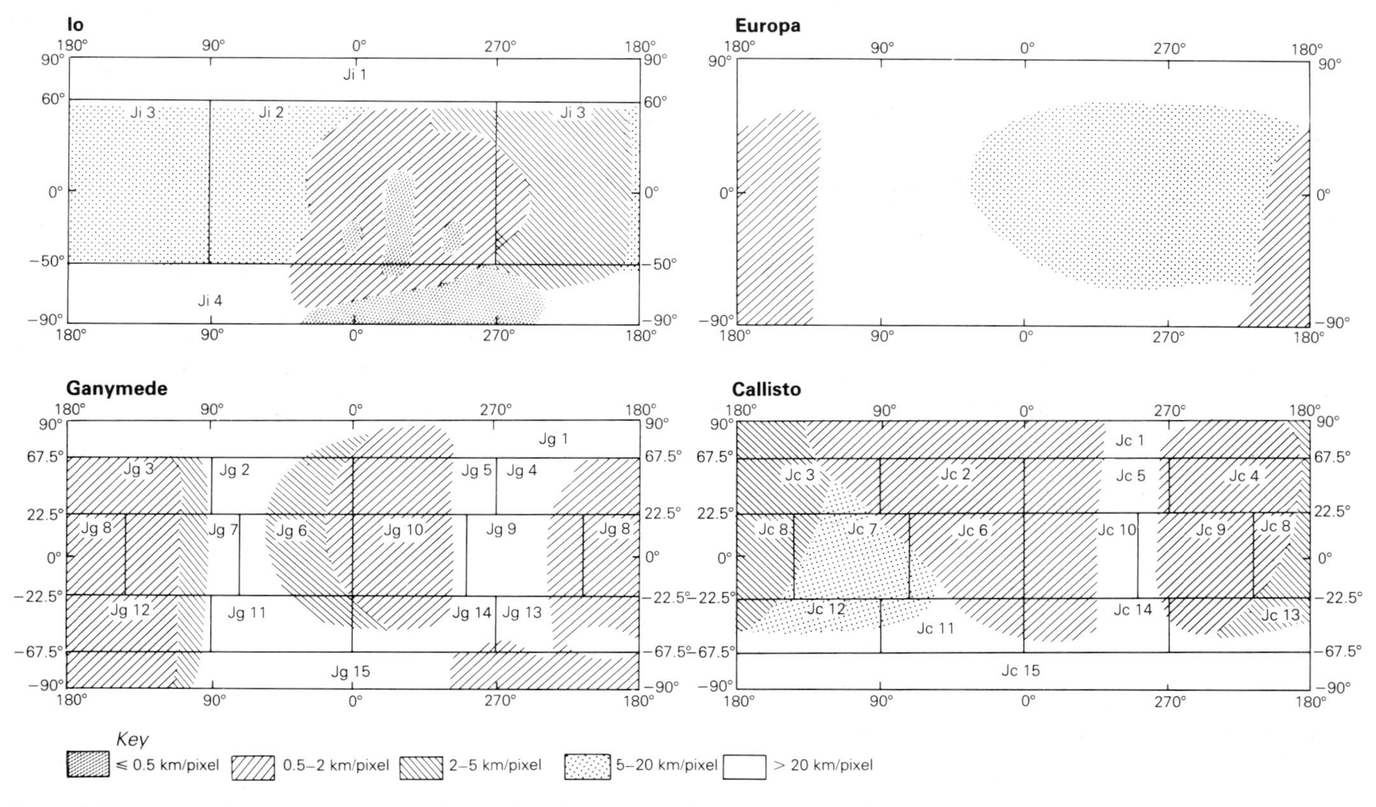

Figure 2.22 Layout of map quadrangles and resolution of images available for mapping Io, Europa, Ganymede and Callisto (from Batson 1981).

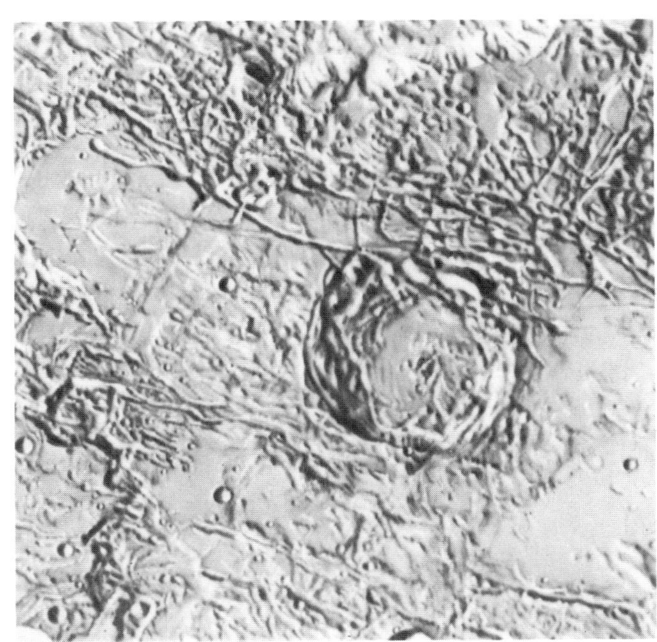

Figure 2.23 Mariner 9 image (a, left) of part of Mars and corresponding airbrush drawing (b, right) which reverses the illumination, removes camera artifacts and incorporates data from other images (courtesy P. Bridges, US Geological Survey, Flagstaff).

In the early 1970s, when it became evident that topographic nomenclature would soon be required for a large number of planetary bodies in the Solar System, the International Astronomical Union organized its Working Group for Planetary System Nomenclature. Modeled on earlier working groups for lunar and martian nomenclature, this new Working Group, divided into five task groups (Moon, Mercury, Venus, Mars and Outer Solar System), provided a formal system for naming surface features. The names used are applied to the features according to systematic plans and are drawn from many sources, including famous deceased scientists and artists, gods and goddesses from mythology, classical places and even small modern towns of Earth.

2.7 Summary

Nearly two decades of Solar System exploration have resulted in a wealth of planetary data. Most geologic information for the Moon has stemmed from the US unmanned missions of the Ranger series, the Surveyor program, the Lunar orbiters and the US manned Apollo program. Soviet contributions have come from the Zond and Luna missions. Geologic exploration of planets beyond the Moon has been accomplished by the US Mariner 10 mission to Mercury (with Venus flyby), the Soviet Venera series of orbiters and landings on Venus, the Pioneer Venus orbiter/probe, the US Mariner and Viking missions to Mars, the Soviet Mars missions, and the US Pioneer and on-going Voyager missions to the outer Solar System.

An integral part of nearly all these missions has been various imaging experiments. Planetary images provide the primary data base to determine the physiography of the planets and to interpret the processes that have shaped their surfaces.

3 Planetary morphologic processes

3.1 Introduction

Geomorphologists have long recognized that the surface of the Earth is shaped by various processes. These processes are generally placed into three primary groups: (a) **tectonic processes**, involving deformation of the lithosphere; (b) **volcanic processes**, involving eruptions of magma onto the surface; and (c) **gradational processes**, involving the erosion, transportation and deposition of surface materials by various agents, such as wind and water, to bring surfaces to a more uniform level. In the planetary context, we must add a fourth major process, **impact cratering**, which involves collision of solid objects.

Each process produces landforms, some of which are diagnostic of the processes involved in their formation. Learning to recognize these landforms is one of the primary goals of planetary geology, for it is through this recognition and the interpretation of the relevant processes that geologic histories can be derived for the planets and satellites.

In this chapter, each of the major surface-modifying processes is described and illustrated. Because most of the knowledge about these processes has been obtained from field and laboratory studies on Earth and provides the "ground truth" for extraterrestrial interpretations, most of the examples shown will be taken from Earth. Some processes, however, are understood far better on other planets and, in those cases, extraterrestrial examples will be used.

It should be noted that not all geomorphologic processes will be reviewed here; rather, only those processes that seem to be appropriate for planetary geology will be considered. The potential flaw in this selectivity is that some processes which are critical for planetary interpretations may be neglected. However, to attempt to include a complete review of geomorphology is not the intent, and the reader is referred to the numerous excellent textbooks on the geomorphology of the Earth, including King (1967), Twidale (1976), Bloom (1978) and others.

Included with the description of each process is a brief discussion of how the different planetary environments might affect the process and the resulting landforms.

3.2 Impact cratering

The American geomorphologist William Morris Davis (1926; reviewed by Roddy 1977), summarized the controversy surrounding the origin of lunar craters with the following statement:

> It has been remarked that the majority of astronomers explain the craters of the moon by volcanic eruption – that is, by an essentially geological process – while a considerable number of geologists are inclined to explain them by the impact of bodies falling upon the moon – that is, by an essentially astronomical process. This suggests that each group of scientists find the craters so difficult to explain by processes with which they are professionally familiar that they have recourse to a process belonging in another field than their own, with which they are probably imperfectly acquainted, and with which they therefore feel freer to take liberties.

Planetary research has resolved many of the misconceptions regarding impact craters and has laid the foundation for understanding this important geologic process. Several conferences have been held on impact cratering mechanics and related phenomena, with proceedings publications which are important to planetology (French & Short 1968, Roddy *et al.* 1977, Silver & Schultz 1982).

Impact cratering involves the nearly instantaneous transfer of energy from an impacting object, called the **bolide**, to the target surface. Bolides can include meteoroids, asteroids and comets. Velocities of these objects upon impact on Earth range from 5 to 40 $\mathrm{km\,s^{-1}}$. Using the simple expression for kinetic energy of $KE = 0.5\ (mv^2)$, where m is the mass of the bolide and v is the velocity, it can be seen that an average nickel–iron meteoroid 30 m in diameter traveling at 15 $\mathrm{km\,s^{-1}}$ could impart about 1.7×10^{16} J of energy onto a planetary surface, the equivalent of exploding about 4 million tons of TNT! Such an impact event is considered to be responsible for the formation of Meteor Crater in northern Arizona, in which more than 175 million tons of rock were excavated to leave a crater more than 1 km across and 200 m deep (Fig.

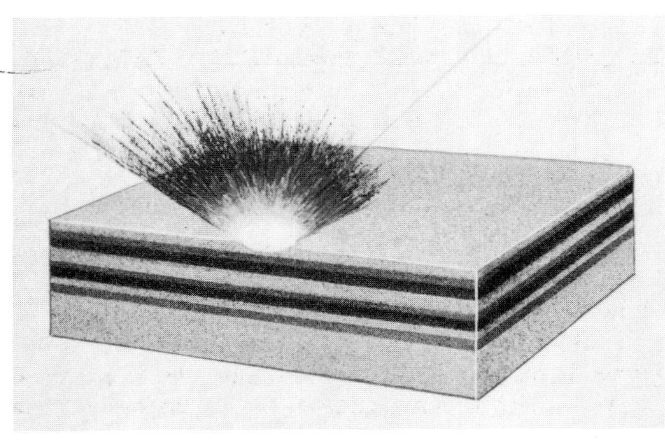

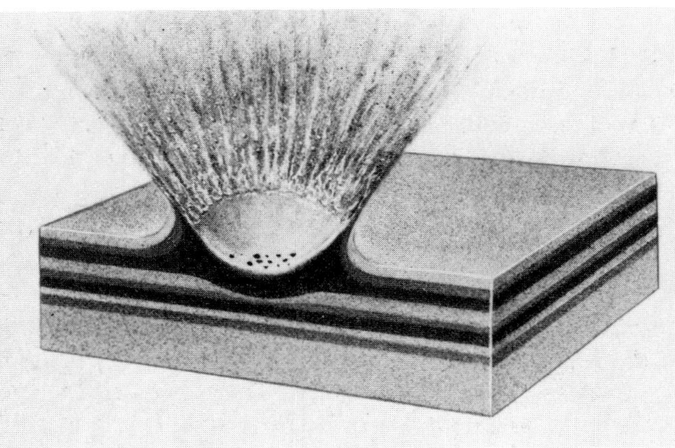

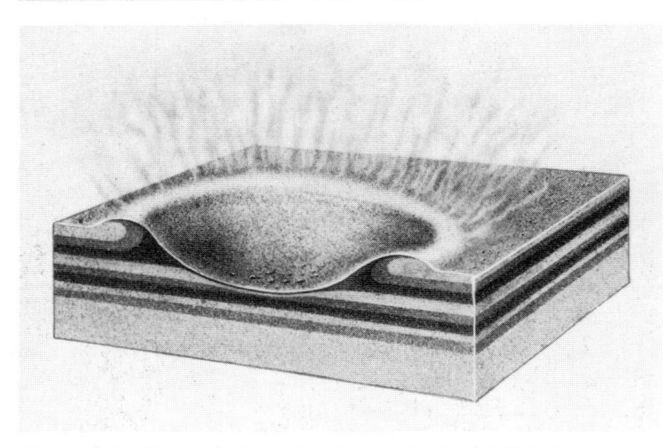

Figure 3.1 Stages in impact cratering: (a, top) initial stage involves contact of bolide with target and high-speed jetting of upper target materials; (b, center) passage of shock waves through target and subsequent generation of rarefaction waves sets target material into motion as ejecta leading to (c, bottom) formation of crater and emplacement of ejecta. Note the inversion of stratigraphy in the overturned "flap" of the crater rim. Post-impact cratering modification stage, not shown, involves slumping of walls and infill of crater (from Gault *et al.* 1968).

1.8e). Unlike most geologic processes, an individual impact cratering event is of very short duration; Meteor Crater probably formed in about one minute.

3.2.1 Impact cratering mechanics

No natural impact craters have been observed in active formation. This, together with the very short duration of a cratering event, has led most investigators into the laboratory to conduct cratering experiments where the process could be studied under controlled conditions (Fig. 1.8). From analyses of high-speed motion pictures and detailed analyses of cratered targets, Don Gault and his colleagues (1968) at the NASA Ames Research Center derived the general sequence of cratering (Fig. 3.1). First is the *compression stage*, in which the projectile contacts the target, penetrates the surface and is engulfed, resulting in high-speed jetting of material outward from the zone of contact. At the same time, intense shock waves are sent through both the target and the projectile. In this stage of the impact, shock pressures of several megabars are common, exceeding by three to four orders of magnitude the effective material strength of common rocks. It is this high shock pressure which sets impact cratering apart from other geologic processes. The impact process results in intensely crushed and broken target material, some of which is so severely shock-metamorphosed that some of the rock is melted and vaporized.

The second stage involves *excavation* of the crater as shock waves and attendant rarefaction (or decompression) waves set target material into motion. The material excavated from the crater (Fig. 3.2), termed **ejecta**, is distributed radially as a blanket of fragmental debris. **Continuous** ejecta covers the surface as an uninterrupted blanket from the crater rim outward and gives way to the **discontinuous** ejecta, which in turn grades into zones of *secondary craters*, formed by the impact of ejecta blocks and clods.

The third stage includes various *post-cratering modifications* not directly attributable to shock waves. These include slumping of the crater walls, isostatic adjustments to the floor and rim, and erosion and infill of the crater. The third stage may continue over long periods of time until the crater is eventually obliterated by processes of gradation and viscous flow.

3.2.2 Impact craters on Earth

The existence of impact craters on Earth was not readily accepted by the scientific community. Even to advocates of impact cratering, prior to 1930 fewer than ten impact structures were known on Earth. By 1966, the number had only risen to about 33, but after intensive searches and establishment of criteria for the recognition of impact craters (Table 3.1), by the 1980s nearly 200 craters and related structures had been fairly well documented as resulting from impact processes (Classen 1977).

Compared to the heavily cratered surface of the Moon, 200 craters on Earth would appear to be an anomalously low number. However, impact cratering involves chance collisions of planetary objects. Statistically, the longer a surface is exposed, the greater the likelihood that it will be struck by an extraterrestrial object. Thus, we would expect to find impact craters only on older surfaces. Because most of the Earth's crust has been recycled by crustal tectonics and modified by gradation, most impact craters on Earth have been destroyed or obliterated, in contrast to the Moon where most of the early history is preserved in its ancient crust (Fig. 1.2). Thus, to some degree the number of craters one can see on planetary surfaces provides a clue to the degree of surface evolution that the planet has experienced during its geologic history.

Meteor Crater in northern Arizona is one of the best preserved and most intensely researched impact craters on Earth. Studies by Shoemaker (1963), Roddy (1977) and others have shown that the crater was formed between 20 000 and 30 000 years ago in flat-lying sedimentary rocks. Detailed field studies and analyses of drill cores show that the rocks were highly deformed; strata in the rim of Meteor Crater were folded back as an overturned flap (Fig. 3.1), and rocks beneath the floor of the crater were severely brecciated to a depth of about 350 m.

Unlike many lunar craters of the same size, Meteor Crater is distinctly polygonal in plan view (Fig. 1.8e), demonstrating the importance that structural control can exert on crater form. The strata surrounding Meteor Crater have a distinct joint pattern which at the time of impact evidently controlled the passage of shock waves and the subsequent excavation of the fragmented rocks, leading to the polygonal outline of the crater which was further enlarged by post-excavation modifications.

Table 3.1 Criteria for the recognition of impact craters (modified from Dence 1972).

Criterion	Characteristics	Reliability
Remote sensing		
plan view	distinctly circular; may be modified by slumping, tectonic patterns, or erosion	fair, but can be attributed to other processes
rim structure	Inverted stratigraphy	definitve
central zone	floor lower than surrounding plain; may contain central uplift	fair, but can be attributed to other processes
Geophysical observations		
gravity anomaly	generally negative	supportive, but not conclusive
magnetic field	variable; may be distinct anomaly over melt rock	supportive, but not conclusive
seismic velocities	generally lower in brecciated zones	supportive, but not conclusive
Ground observations		
presence of meteorites	rare except in very young craters	definitive
shock metamorphism	features such as high pressure minerals, impact melt, planar shock features and shatter cones	definitive
brecciation	observed in ejecta, rim and floor of craters	may be attributed to other processes

Figure 3.2 The lunar crater Euler (27 km in diameter) showing principal features of a typical impact crater (Apollo 17 AS 17-2923).

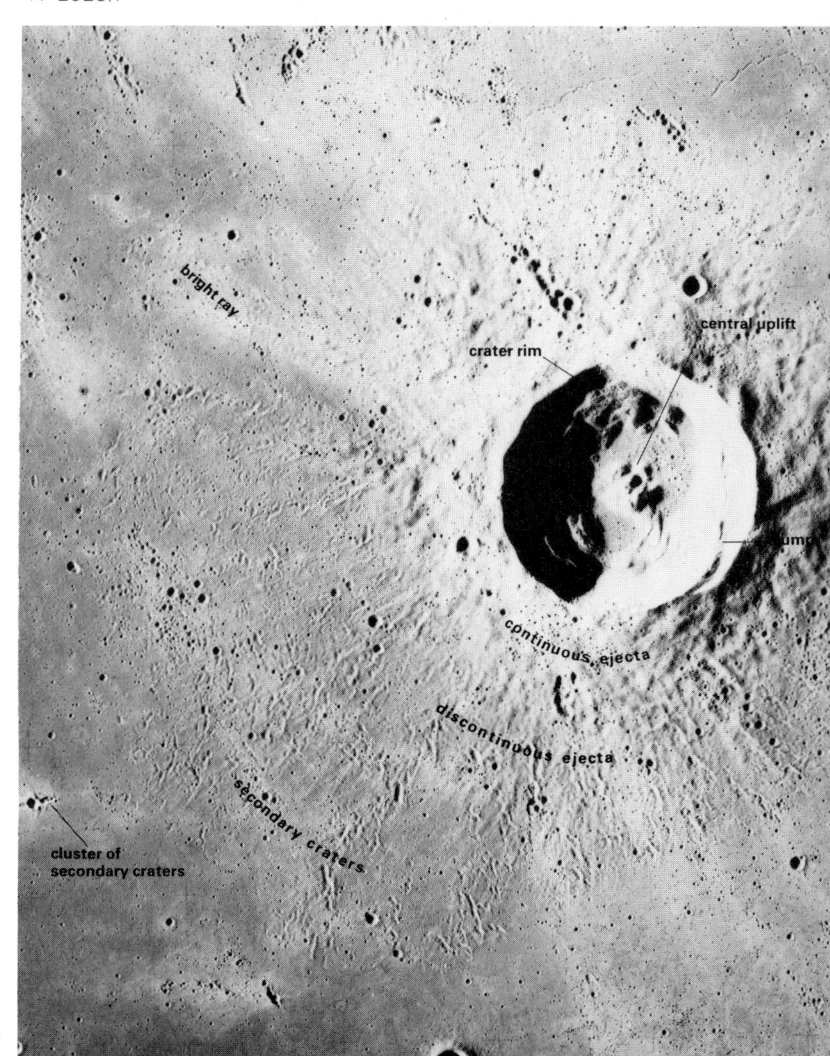

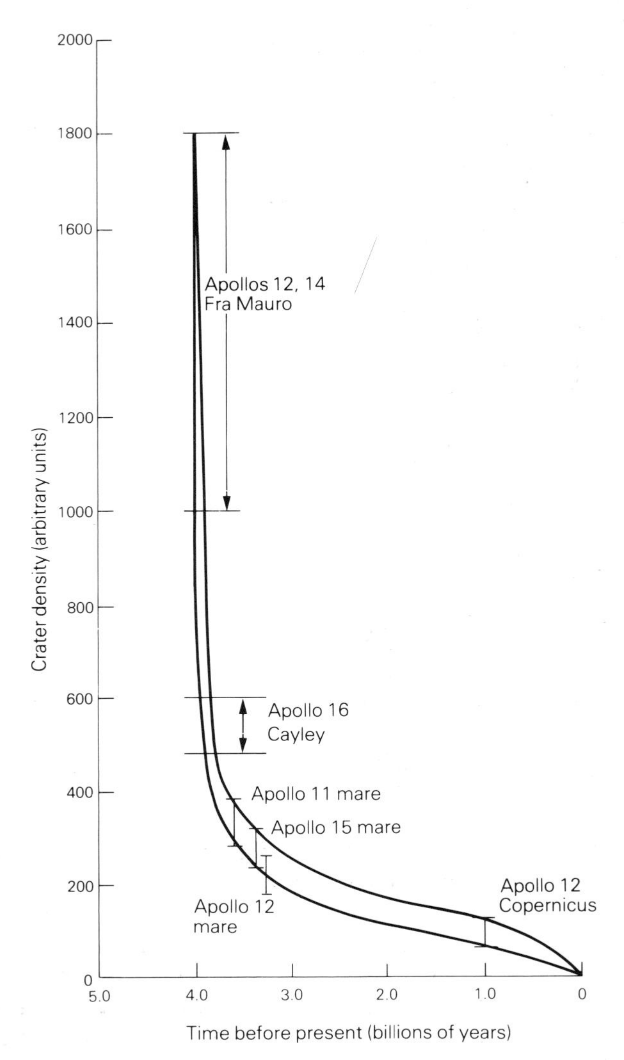

Figure 3.3 Impact crater frequency on the Moon as a function of time, showing that cratering was much more frequent in the early history of the Moon and that there has been a subsequent decay in the rate of impact (after Soderblom *et al.* 1974. Copyright © 1974 Academic Press).

Gene Shoemaker predicts that an impact of the magnitude to form Meteor Crater would occur every 50 000 to 100 000 years on Earth. Although the uncertainties in this estimate are numerous, it gives some appreciation of the frequency of impact crater formation spread over geologic time. Analyses of the crater record on the Moon, which is much better documented than the record for Earth, shows that while smaller impact events are much more frequent, there has been a general decay in the rate of cratering through time (Fig. 3.3).

Field studies of eroded impact craters have yielded important clues about the deformation of the rocks surrounding the crater. For example, the Sierra Madera structure in Texas is estimated to represent a ~16 km in diameter crater that has been deeply eroded. Detailed mapping of the rock units and reconstructions of the original position of the rocks show that the central zone of the structure was uplifted more than 1200 m (Wilshire *et al.* 1972). Such an uplift probably corresponds to the *central uplift*, or *central peak*, of many lunar craters (Fig. 3.2). However, central uplifts are observed only on craters larger than a few kilometers in diameter on Earth (there is no central uplift at Meteor Crater), and it appears that such features require a certain minimum impact energy to form. Although the causes of central uplift are not well defined, they are generally considered to be the result of rapid elastic rebound immediately following the excavation of the crater bowl.

The Ries Kessel of southern Germany has provided insight into the mechanics of ejecta emplacement. This 24 km structure has been the subject of considerable debate among European geologists and traditionally has been attributed to volcanic or crypto-explosion origins. Discovery of the high-pressure minerals *coesite* and *stishovite*, which form only by impact processes, have now convinced most investigators of the impact origin for the Ries Kessel. Extensive surface mapping and drilling to obtain subsurface core samples have provided knowledge on the extent of the ejecta deposit and its properties. The ejecta is composed largely of a brecciated mass known as the Bunte breccia. Petrologic analyses and considerations of ballistics suggest that fragmental ejecta was mixed with melt rock and volatiles as it was excavated from the **transient cavity** (original crater bowl) and thrown in a ballistic trajectory to rain upon the surrounding surface where the material churned up and mixed with local rock. Then the entire mixture continued to slide outward a short distance to settle into place. This model suggests the important role that target volatiles may play in ejecta emplacement, and may be applied in part to the interpretation of certain craters on Mars (Fig. 3.4).

Among the many other structures on Earth that have yielded important data on the morphology of impact craters and the cratering process are the Clearwater Lakes, Canada, which appear to be a double impact crater, Serra da Canghala structure, Brazil, which displays a central-ring uplift (Fig. 3.5), and the Henbury Craters, Australia, which consist of 13 or more craters, possibly reflecting the breakup of an incoming object to form multiple craters.

3.2.3 Impact crater morphology and effects of different planetary environments

The primary factors governing the size and morphology of impact craters are the impact energy (a function of the size and velocity of the bolide), various properties of

Figure 3.4 Mosaic of Viking Orbiter images of the 28 km diameter martian impact crater, Arandes, showing ejecta which is considered to have been emplaced partly as a liquid slurry (from Gault & Greeley 1978).

the target such as rock strength and the presence or absence of volatiles, and gravity. As discussed by Gault (1974), gravity affects the cratering process by influencing: (a) the dimensions of the excavation bowl, (b) the extent of the ejecta, and (c) various post-impact crater modifications. For equal-size impact events, fragmented blocks of ejecta could be lifted and excavated more easily on low-gravity planets, leading to larger craters in comparison to high-gravity environments. Furthermore, in low-gravity environments, ejecta is thrown a greater distance, as shown in Figure 3.6. Thus, we would expect to see a wider (but thinner) zone of ejecta surrounding impact craters on the Moon than on higher-gravity planets such as Mars or Mercury. In the modification stages of impact cratering, gravity plays a role by governing the rate of isostatic adjustments, influencing the degree of slumping and perhaps governing the magnitude of potential central uplifts.

Lunar impact craters show a distinctive progression in morphology with increasing size, with simple bowl shapes predominating up to about 10 km in diameter, central-peaked craters occurring at 10–150 km in diameter, clusters of central peaks in the range 100–175 km, peak-ring craters for diameters of 150–220 km, and multi-ringed craters (more commonly called **basins**) for structures greater than about 220 km in diameter. This general progression is observed on other planets as well, but there is a variation in the size ranges and there are additional categories, as will be discussed in subsequent chapters.

The shape of impact craters in map view is controlled by the angle of the incoming projectile and the tectonic fabric of the target, such as the presence of joints or other fractures, as noted in the case of Meteor Crater. Because impacts involve essentially point-source transfers of energy, both the crater and the distribution of ejecta for most impacts are concentrically symmetrical about the point of impact. Although intuition might suggest that oblique angles of impact would cause elongate craters, experiments have shown that only for very low angles ($<15^\circ$) do impact craters become noticeably asymmetrical (Gault & Wedekind 1978).

Figure 3.5 Oblique aerial photograph of Serra da Canghala in Brazil showing an uplifted central ring and pit, possibly resulting from impact into water-saturated sediments. Outer diameter is 12 km (from Greeley *et al.* 1982; photograph courtesy of John McHone).

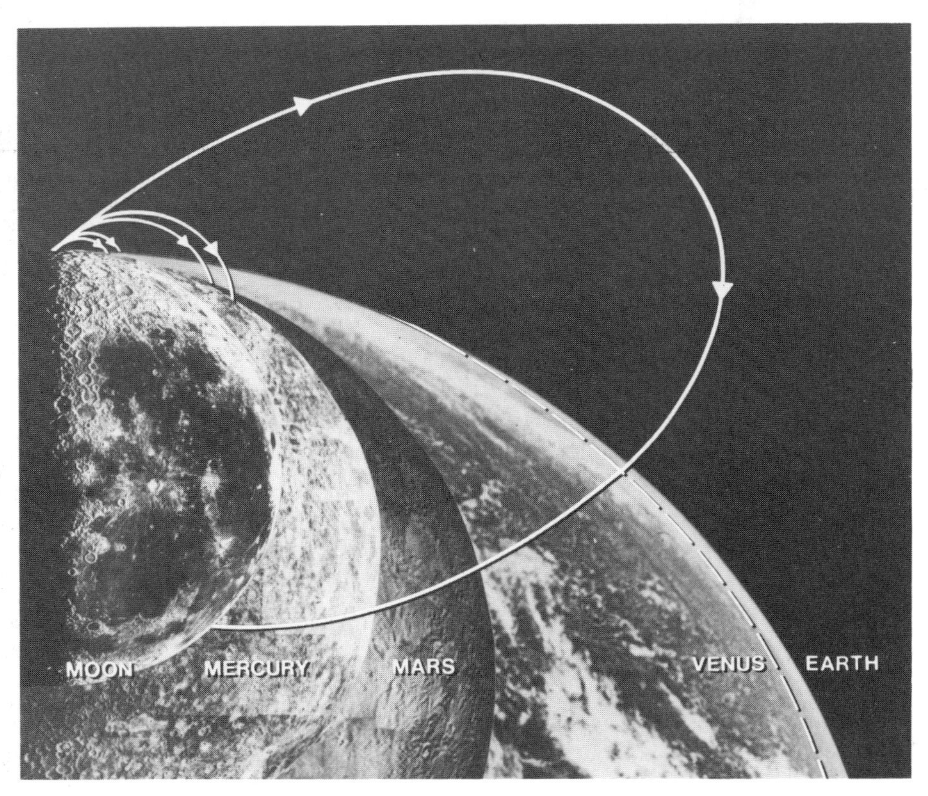

Figure 3.6 Comparison of range of heights for ejecta traveling 2 km s^{-1} on the Moon, Mercury, Mars, Venus and the Earth. Note that the high planetary density of Mercury (Table 2.1) permits the ejecta to travel only a little farther on its surface than ejecta on Mars, despite the fact that Mars has a larger diameter than Mercury (from Schultz 1976b).

3.2.4 Crater counting as a technique for age determinations

The longer a planetary surface is exposed to impact bombardment, the more craters it should display. Thus, the more craters observed on any given surface (Fig. 1.7), the older it should be, and by counting the number of impact craters on various surfaces, it should be possible to place each surface in a relative age sequence (Fig. 3.7). In principle, if the flux of incoming bolides is known through geologic time, then it should be possible to assign "absolute" ages for cratered surfaces by dividing the total number of craters observed on a surface by the number of craters formed each year (Hartmann *et al.* 1981).

In practice, however, a great many difficulties arise in using crater statistics for age determinations. Among other problems, these include:

(a) *Non-impact* craters, such as those formed by volcanic, karst, thermokarst or other processes, may be indistinguishable from impact craters, and if they are present in significant numbers, the surface would appear anomalously old.

(b) *Secondary craters* add to the total crater population and must be taken into account by various models which predict how many secondary craters would form as a function of primary crater size, target material properties, etc. Unfortunately, such models are imperfect, and it is very difficult to determine the presence and number of secondary craters.

(c) *Variations in target properties* could cause variations in crater sizes. For example, experiments show that craters formed in targets containing fluids are larger than craters formed in "dry" targets. This difference could cause the size-frequency distribution for the volatile-containing target to be interpreted as representing an older surface.

(d) *Crater equilibrium* studies by Gault (1970) have shown that with time, cratered surfaces reach a stage in which craters of a given size are obliterated by impact erosion at the same rate of formation, as shown in Figure 1.7. Thus, only surfaces that have not yet reached equilibrium for the crater sizes being considered can be analyzed.

Despite these difficulties and uncertainties, impact crater statistics are commonly used as a means for obtaining relative dates of formation for different planetary surfaces. Ages derived from crater counts have been compared with (and calibrated against) radiometric dates obtained from lunar samples and demonstrate the validity of the technique, at least on the Moon where surface-modifying processes are minimal. This result suggests that crater counts may be used to

Figure 3.7 Apollo 15 view of the Moon showing two distinctly different surfaces, the smooth, relatively sparsely cratered mare and heavily cratered uplands. Differences in crater frequency can be used for relative age-dating of surfaces. *Sinuous rilles*, the channel-like depressions, are of volcanic origin and were channels for the emplacement of the mare lavas in this area of east central Oceanus Procellarum. Area shown is about 160 km wide (Apollo 15 AS 15-2483).

obtain dates for surfaces on other "airless" bodies, such as Mercury. Great caution must be exercised, however, in using crater counts on planets where differences in gradation may occur as a function of latitude, or where significant differences in target properties may alter the crater morphology.

The recognition of impact craters and the understanding of the mechanics of impact cratering as a process remain among the most important areas of investigation in planetary science. As more diverse objects are explored in the Solar System, knowledge of crater form as functions of planetary size and target composition becomes ever more important for the interpretation of surface history.

3.3 Tectonic processes

Deformation of Earth's crust by tectonic processes is easily demonstrated by features such as faults, folds and fractures, as shown in Figures 3.8 to 3.12. Geologists recognize that these local features can be related, in part, to the style of deformation, such as tension or compression. However, insight into crustal deformation on larger scales was not gained until the unifying concept of global plate tectonics was derived in the 1960s. This insight has been important in understanding the evolution of the crust on Earth and is critical in the interpretation of other planets. More recently, analyses of the styles and timing of tectonism on the terrestrial planets has enabled the derivation of general geophysical models of planetary evolution and thermal history as reviewed by Head and Solomon (1981).

3.3.1 Earth's interior

The interior of the Earth is divided into distinct layers, based on seismologic characteristics and assumptions of composition. The **crust** is a thin zone composed of relatively low-density rocks extending from the surface to a depth of 5–50 km. The boundary between the crust and the underlying **mantle** is marked by an abrupt increase in seismic wave velocities and is called the **Mohorovicic Discontinuity** (the "Moho"), after the Yugoslavian scientist who discovered it. Making up less than 0.1 percent of the total volume of the Earth, the crust occurs in two forms, **sima** and **sial**, terms which are derived from their predominant compositions (Si = silicon, Ma = magnesium, Al = aluminum).

Figure 3.8 High-altitude oblique aerial photograph of the eastern San Francisco volcanic field, Arizona, showing prominent grabens (two parallel faults bounding a down-dropped block), cinder cones (left side) and lava flows, some of which have flowed into the grabens (photograph courtesy US Geological Survey, Flagstaff).

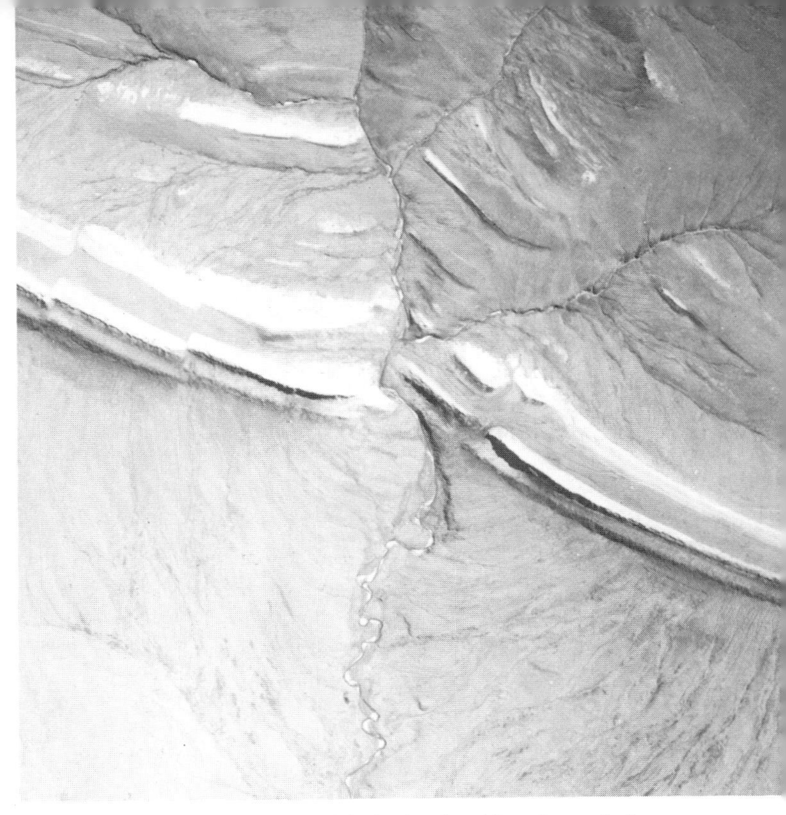

Figure 3.9 Aerial photograph of faulted strike ridges formed of steeply dipping strata (Lookout Ridge, Alaska), cut by high-angle cross-faults. The surrounding plains are being eroded by solifluction (US Geological Survey photograph BAR 44-47).

Figure 3.10 Oblique aerial view northwestward along the San Andreas fault in California. The fault marks the major boundary between the Pacific and the North American plates. Sheared rocks along the fault zone are preferentially eroded (photograph by Robert Wallace, US Geological Survey).

Oceanic crust is composed of sima and is relatively thin (~5 km), whereas continental crust is composed mostly of less dense sial overlying a zone of sima and may exceed 50 km in thickness beneath major mountain chains.

The mantle extends to a depth of nearly 3000 km and is composed of materials rich in iron and magnesium. From seismic wave properties, Earth's **core** has been shown to consist of an outer liquid part and an inner solid part, both composed predominantly of iron.

3.3.2 Plate tectonics

In order to understand global tectonic processes, it is necessary to take a slightly different view of the subdivision of the outer few hundred kilometers of the Earth. The crust–mantle–core scheme described above is based principally on compositional differences.

Figure 3.11 Anticlinal fold in Silurian sandstones and shales, Washington County, Maryland (US Geological Survey photograph by I. C. Russell).

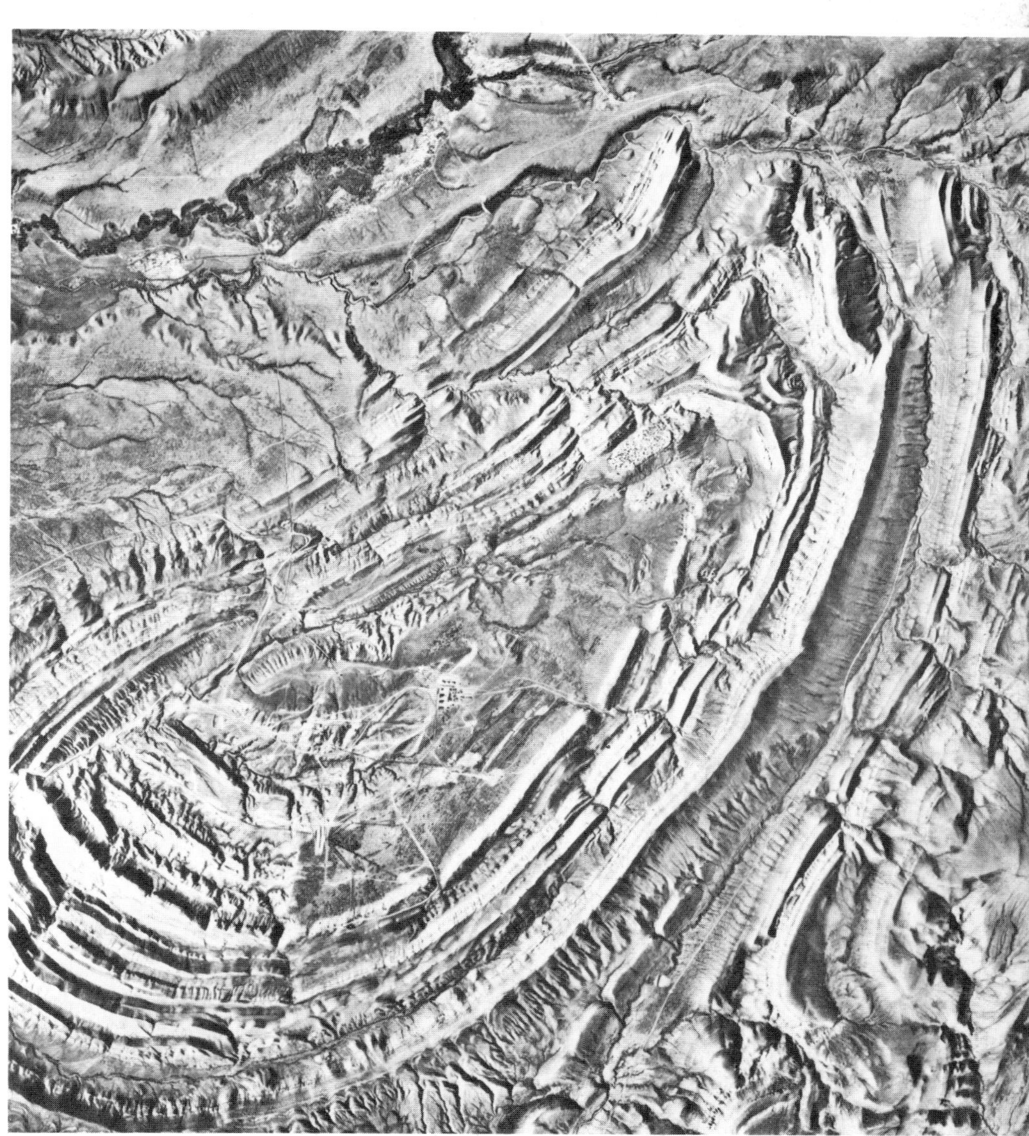

Figure 3.12 Vertical aerial photograph of Circle Ridge Dome, Fremont County, Wyoming. This eroded dome has a central core underlain by Triassic shales and sandstones surrounded by outward-dipping Jurassic sediments (US Geological Survey photograph GS IE6-32, 1948).

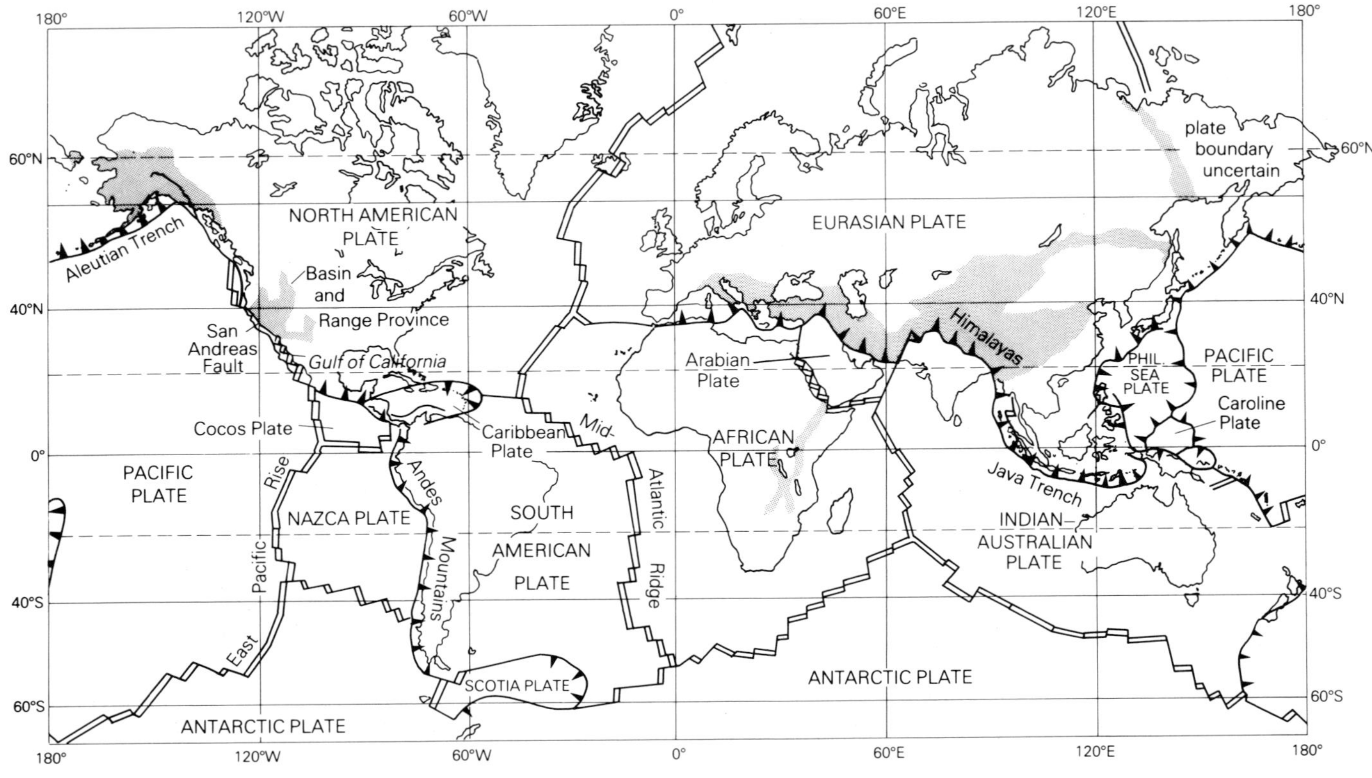

Figure 3.13 Map showing principal crustal plates on Earth (courtesy US Geological Survey).

Analyses of the physical properties of these zones led to the recognition of the **lithosphere** and **asthenosphere**. The lithosphere consists of the crust and the upper part of the mantle, which together behave as a relatively rigid, solid shell. The lithosphere rests on, and is subject to movements by, the underlying semi-molten and mechanically plastic layer of the upper mantle, termed the asthenosphere.

Within the mantle, heat sources (derived principally from decay of radioactive material) generate convection cells and plastic flow in the asthenosphere. Upward-converging convection cells may result in upward arching of the lithosphere and zonal concentration of heat at the surface. These zones may fracture, pull apart and become sites of volcanism, which introduces new rock to the surface. Lateral flow of the asthenosphere may, in turn, drag segments of the lithosphere outward from these zones of rifting. Because heat sources apparently are not evenly distributed within the mantle, nor are they of equal magnitude, the sizes and geometries of the convection cells are variable. This results in non-uniform tectonic patterns on a global scale, in which individual segments of the lithosphere, or **plates** (Fig. 3.13), are of different sizes and are moving at different rates.

In general, zones of upward-converging convection are sites of **mafic** (magnesium- and iron-rich magmas) volcanism, which reflect the iron-rich sources of magma derived from the mantle. Such volcanism leads to the generation of new crust and commonly occurs in oceanic settings. Up-arching of the crust and accumulation of lava form a symmetric mid-ocean ridge and central rift. Lateral flow away from the central rift, termed **sea-floor spreading**, has been measured to a maximum of about 16 cm yr^{-1} along the East Pacific Rise.

Downward-converging convection cells result in drag of lithospheric plates toward one another into collision. Any one of several styles of plate collision may occur, depending upon the composition of the crustal segments that are involved and the angle of collision (head-on, orthogonal, etc.). Downward dragging of slabs of crust, or **subduction**, generates seismic disturbances (earthquakes) and possible remelting of the crust, leading to volcanism. Because the magma is generated at least partly from sial (either from continental crustal materials or from oceanic sediments), volcanism in subduction zones tends to be more **silicic** (silica rich) in comparison to zones of up-welling convection.

The global tectonic patterns, related to plate motion on Earth, provide an important model for comparing the geophysical characteristics of all solid-surface planets. The morphologic expression of these tectonic patterns provides the key to the interpretation of internal processes on all planets, including the Earth.

3.3.3 Surface morphology of tectonic features

Local tectonic features – more commonly referred to as **structures** – include various faults, joints and folds, shown in Figure 3.14. Faults involve shearing of rocks and occur in three principal types: (a) **normal faults** which result from tensional stresses; (b) **reverse faults** which result from compressional forces (thrust faults are reverse faults involving very low angle fault planes); both normal and reverse faults involve primarily vertical displacements of rocks along the fault plane; and (c) **strike-slip faults** in which displacement is principally in horizontal directions.

Joints, like faults, also involve fractures, but unlike faulting in which rocks shear along the fracture, jointing only involves separation of rocks away from the fracture. Because the faulted and jointed rocks are fractured, often they are more easily eroded and the fractures themselves may become widened by preferential erosion (Fig. 3.10).

Folds also result from crustal deformation, but rather than fracturing, as occurs in faulting and jointing, the rocks yield by bending. Folds of several geometries may occur, some of which produce topographic expressions that directly reflect the form of the fold. Thus, in Figure 3.15 the mountains coincide with **anticlines**. More commonly, however, there is an inversion of topography: during the process of folding, rocks along the axes of anticlines are subjected to tension which may cause them to be jointed, or in effect to "open-up"; conversely, rocks along the axes of **synclines** are subjected to compression, which may cause pore space and other voids to close. Thus, weathering and erosion are enhanced along the axes of anticlines and are retarded along the axes of synclines. With time, net erosion is more rapid over the anticlines, leading to topographic inversion in which valleys develop along the axes of the anticlines and the synclines remain as ridges (Fig. 3.16).

Although individual faults, joints and folds may range in size up to a few tens of kilometers and may be related to localized tectonic or igneous events, complex systems of multiple structures can extend hundreds of kilometers as part of tectonic deformation associated with crustal plate motion (Fig. 3.17). From orbital views, only the larger individual structures and structural systems are visible, and often it is difficult to identify the specific type of structure, such as normal versus reverse faults, or normal versus inverted topography.

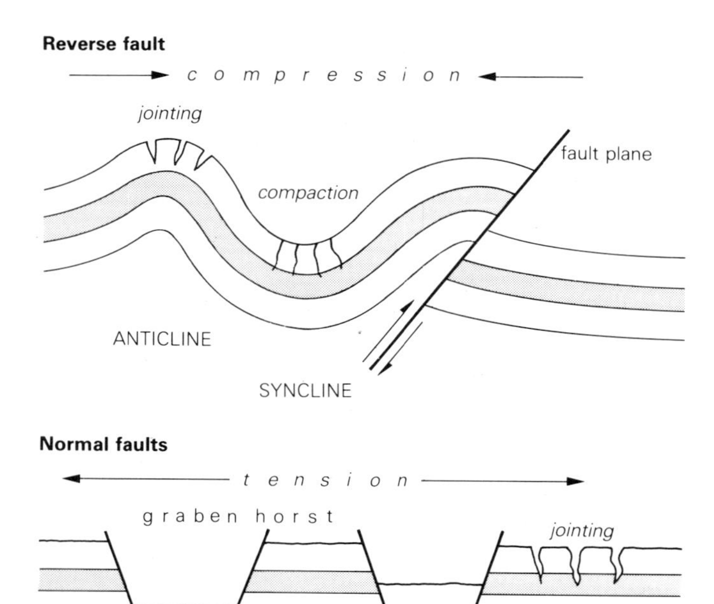

Figure 3.14 Diagrams showing common styles of tectonic deformation into faults, folds and joints and the formation of horsts and grabens.

3.3.4 Comparative planetary tectonism

As reviewed by Head and Solomon (1981), the types of landforms exhibited on the terrestrial planets and the timing of their formation are intimately linked to thermal evolution and the subsequent planetary interior characteristics. The smaller terrestrial planets – Mercury, Moon and Mars – have thick lithospheres which show little evidence of destruction or renewal over the last 80 percent of their history, in contrast to Earth and possibly Venus. These so-called *one-plate* planets are dominated by early-formed crusts which preserve the period of heavy bombardment and have histories that suggest ever-thickening lithospheres. Tectonism on these planets is expressed primarily by vertical movements, forming features such as grabens.

In contrast to one-plate planets, the lithosphere of Earth is highly mobile and exhibits extensive lateral movement, as discussed in previous sections. Venus, similar in size to Earth, has landforms unlike the smaller planets, but may also show evidence suggestive of large impact craters. However, until high-resolution images of Venus become available, its tectonic history remains open to considerable debate, as reviewed in Chapter 6.

3.4 Volcanic processes

Volcanic processes involve the generation of magma and magma-related materials and their eruption onto the surface. Thus, volcanic structures provide direct clues to the thermal evolution and interior characteristics of planetary objects. On Earth, magma appears to be generated in the lower crust and upper part of the mantle as a result of heating from a variety of sources, including: (a) heat generated by the differentiation of

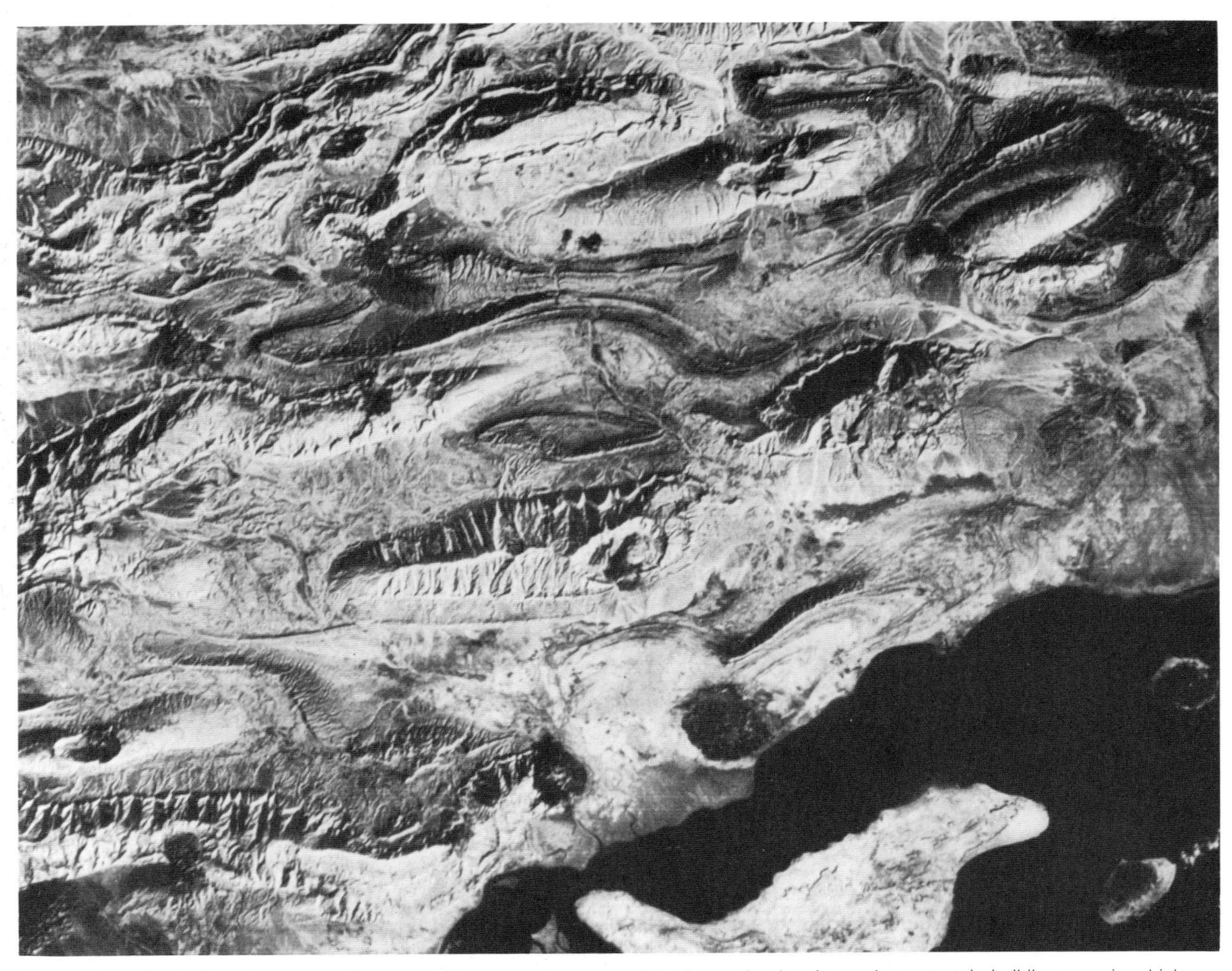

Figure 3.15 Apollo 7 photograph of the Zagros Mountains, Iran viewed to the north, showing active mountain-building zone in which topography mirrors the structure; most of the mountains are anticlines; dark circular areas are intrusive salt domes (Apollo 7 AS7-15-1615; after Lowman 1981).

Figure 3.16 Landsat mosaic prepared by US Soil Conservation Service showing eastern USA, centered on 40°N, west of the Atlantic coast. Present ridges and valleys of Appalachians were formed by differential erosion after regional uplift, not directly by tectonism as in the Zagros Mountains shown in Figure 3.15 (after Lowman 1981).

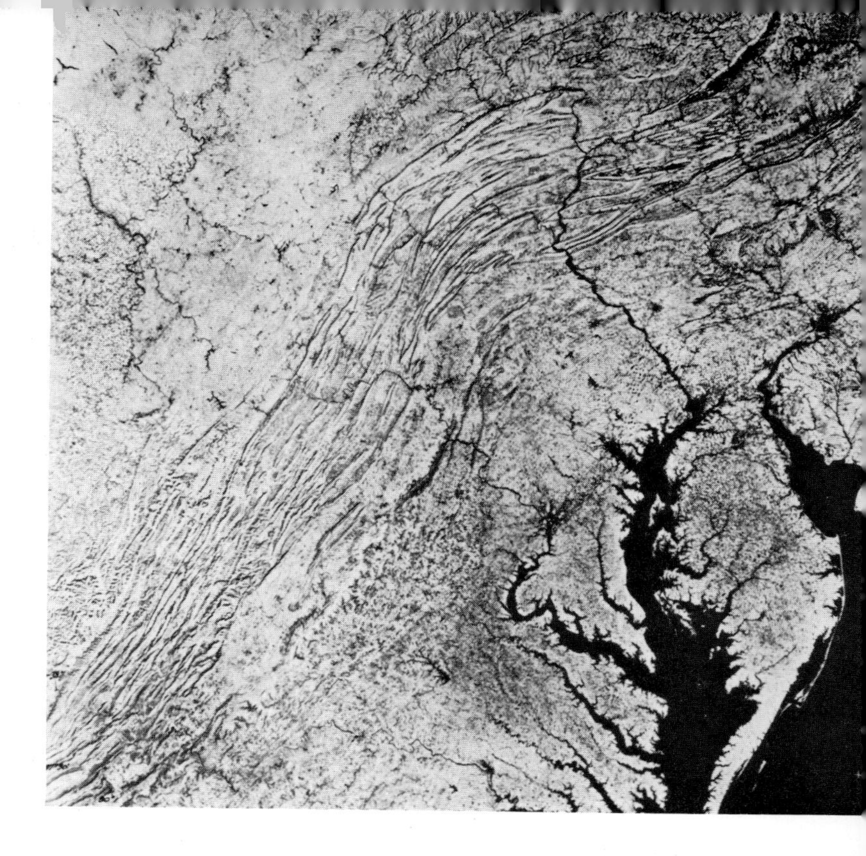

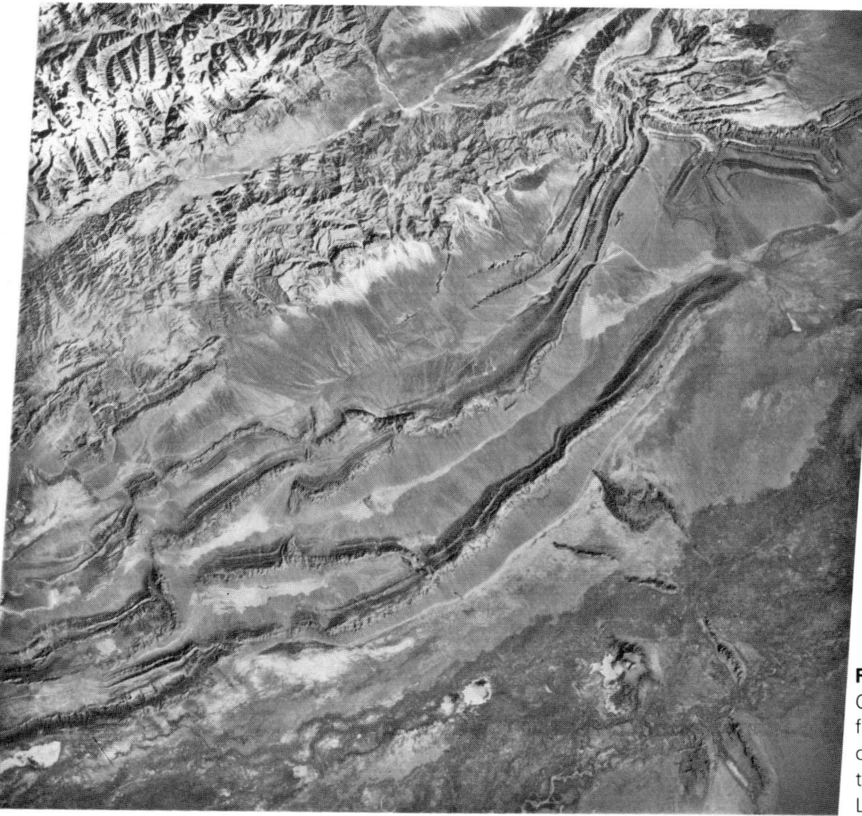

Figure 3.17 Landsat view of Tien Shan, China, centered near 78°E, 40°N, showing folds and overthrusts; note prominent overthrust at lower left, bounded by left-lateral tear faults (Landsat image 1206-05000; after Lowman 1981).

3.18a Landsat image of Cerro Panizos ignimbrite shield, Bolivia/Argentina border (from Francis & Wood 1982), showing radial drainage gullies cut into ash flows and domes in the central area. Area shown is 100 km across, centered at about 23°S, 66.5°W (Landsat 1008-13531 and 2256-13411, courtesy P. Francis).

Figure 3.18b Mount St Helens volcano in the Cascade Range, USA, prior to the eruptions of 1980. This composite, or strato-volcano, grew by intermittent eruptions of lavas which ranged in composition from basalt to andesite and dacite. The volcano is 10 km in diameter and reaches 2000 m in elevation; view is to the east (photograph 76-A, R. Greeley).

Figure 3.18c Oblique aerial view of Mauna Loa shield volcano, Hawaii (left side), composed predominantly of basaltic lava flows, and Mauna Kea (right side) a more "evolved" volcano, the summit of which is marked by cinder cones (US Navy photograph, 0066, Nov. 1954).

the mantle and core; (b) frictional heat generated by tectonic and body-tidal processes; and (c) heat derived from radionuclides. An examination of the location of most active volcanoes on Earth shows a more than coincidental correlation with tectonic plate boundaries. As discussed in Section 3.3, spreading zones are typically marked by mafic volcanism, involving ferromagnesium-rich silicate magmas which typically produce basaltic rocks.

Subduction zones are also sites of volcanism and typically involve silica-rich magmas which commonly produce andesitic, dacitic and rhyolitic lavas and ash deposits (Fig. 3.18a & b). As the subducted plates are carried to greater depth, they begin to melt, and because the crustal plates are composed of lower-density materials than the mantle, the melt tends to rise. Depending upon the geometry of subduction, additional crustal materials may be incorporated in the melt during its ascent to the surface.

Volcanism may also occur in mid-plate zones, both in oceanic environments, as typified by the Hawaiian Islands, and in continental environments, represented by the Tibesti Highlands of north-central Africa. Some of these mid-plate volcanic zones are postulated to result from a rising thermal "plume" in the mantle which upwells in a fixed, centralized location. As the overlying crustal plate slides across the plume, magma periodically erupts through the plates to the surface (Fig. 3.18c). Thus, chains of volcanoes may be generated on the plate in an assembly line fashion in which the volcanoes are progressively older in the chain away from the source. Close correspondence of the rate of plate motion and of the ages of the volcanic rocks in the Hawaiian Emperor chain on the Pacific plate lends credence to this hypothesis (Dalrymple *et al.* 1973).

Basaltic volcanism is an extremely important process on the terrestrial planets. Basalts form the floors of the oceanic basins and have erupted on Earth throughout its known geologic history. The dark mare areas of the Moon are basaltic lava flows, constituting at least one-fifth of the lunar surface; perhaps 50 percent or more of the martian surface is covered with basaltic materials, as may be substantial parts of the plains on Mercury. Several lines of evidence suggest that the asteroid Vesta may be basaltic, and some meteorites appear to be fragments of bodies that experienced basaltic volcanism. Although speculative, some of the mountains and plains of Venus may also be the result of basaltic volcanism. The importance of basaltic volcanism in the inner Solar System prompted the formation of a project involving more than 100 scientists to consider all aspects of the topic in the planetary context. Begun in 1976, the project culminated with the publication of a key reference entitled *Basaltic volcanism on the terrestrial planets* (BVSP 1981).

Numerous classifications of volcanic eruptions have been derived. Most classifications for the *styles of volcanism* are based on the characteristics of the products erupted and on the degree of eruption explosivity, as given in simplified form in Table 3.2.

3.4.1 Volcanic morphology

The forms of volcanoes and volcanic terrains have been studied for many years. Cotton (1952) analyzed volcanic geomorphology as related to various processes including styles of eruption. Wentworth and Macdonald (1953), Macdonald (1967, 1972) and Green and Short (1971) also describe volcanic features. More recently, Head *et al.* (1981b) and Whitford-Stark (1982) have discussed factors involved in the morphology of volcanic landforms, especially as related to planetary geomorphology. Most of these references are focused on basaltic volcanism but, unfortunately, much less attention has been given to silicic volcanism.

Table 3.2 Styles of volcanism.

Style	Products	Activity	Typical composition	Vent types
flood eruption	fluid lavas	high rate of effusion, little explosivity	basaltic	fissure
plains volcanism	fluid lavas	moderate rates of effusion	basaltic	aligned central vents and fissures
Hawaiian	fluid lavas, modest pyroclastics	moderate rates of effusion, sporadic eruptions	basaltic	predominantly central vents, some fissures
Strombolian	pyroclastics, some lavas	low to moderate rates of effusion, sporadic	basaltic, andesitic	central vent
explosive	pyroclastics (mostly ash); modest, viscous flows	moderate rate, but highly energetic	andesitic, dacitic	central vent
rhyolitic flood	ash flows	very high rate of eruption	rhyolitic	presumably fissure

The ultimate forms of volcanoes and related terrains are the result of many complex, often interrelated, parameters. Whitford-Stark (1982) notes that these factors fall into three groups (Table 3.3): (a) planetary variables, (b) magma properties controlling rheology and (c) intrinsic properties of eruptions. Planetary variables include those factors that are characteristic for the particular body in question. For example, the height of an ejection plume in an explosive eruption would be governed by such considerations as the presence or absence of an atmosphere, the gravitational acceleration and the escape velocity. In turn, these factors influence the shape of the volcano; in an airless, low-gravity environment such as the Moon, pyroclastic deposits would be widespread, in contrast to Earth where the ejection distance would be retarded by the atmosphere and higher gravity (McGetchin & Head 1973), which could lead to the formation of cinder cones (Fig. 3.19).

Magma rheology is an extremely important parameter in volcanic morphology. Fluid lavas spread more easily, leading to the emplacement of volcanic **plains** often "fed" by **lava channels** (Fig. 3.20) and **lava tubes** (Fig. 3.21), in contrast to viscous lavas which typically form short, stubby flows accumulating as **domes** (Fig. 3.22). Many properties of the magma control the rheology of lavas, as shown in Table 3.3.

Table 3.3 Factors governing the morphology of volcanic landforms (from Whitford-Stark 1982).

Planetary variables	Magma properties controlling rheology	Properties of eruption
gravity	viscosity	eruption rate
lithostatic pressure	temperature	eruption volume
atmospheric properties	density	eruption duration
surface/subsurface liquids	composition	vent characteristics
planetary radius	volatiles	topography
planetary composition	amount of solids	ejection velocity
temperature	yield strength	
	shear strength	

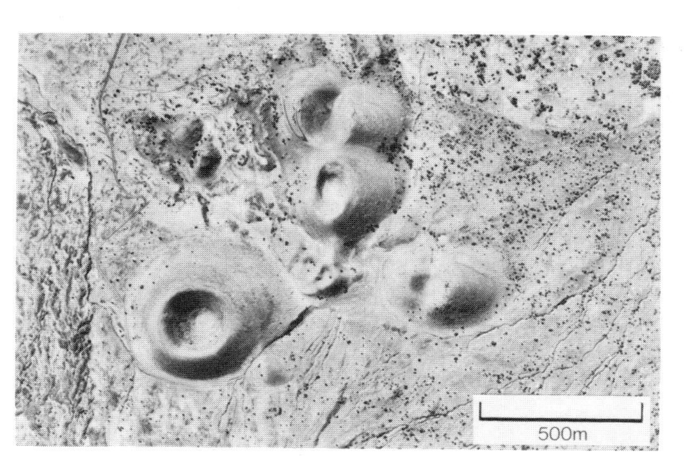

Figure 3.19 Cluster of small cinder cones in Hawaii; morphology of cinder cones may be highly variable as to planimetric shape and profile; some cones lack summit crater.

Figure 3.20 Lava channels up to 10 m wide on the southwest rift zone of Mauna Loa, Hawaii; the rift zone runs diagonally from upper right and is marked by a series of cinder-and-spatter cones which were the vents for flows in which the channels developed (NASA-Ames photograph by R. Greeley). ▷

Figure 3.21 Vertical aerial photograph of the southeast rift zone of Hualalai Volcano, Hawaii, showing collapsed lava tube that originated in flows from the cinder cone on the left (US Dept of Agriculture photograph, 1965).

Figure 3.22 Oblique aerial photograph of Mono Craters, eastern California. These volcanic domes are composed predominantly of rhyolitic obsidian. On the left are two short, stubby lava flows originating from a large dome. Panum Crater, on the right, shows a central dome surrounded by a ring of pyroclastic deposits. View is to the southwest, with Mono Lake in the foreground and the Sierra Nevada mountains in the distance (US Geological Survey photograph by C. D. Miller).

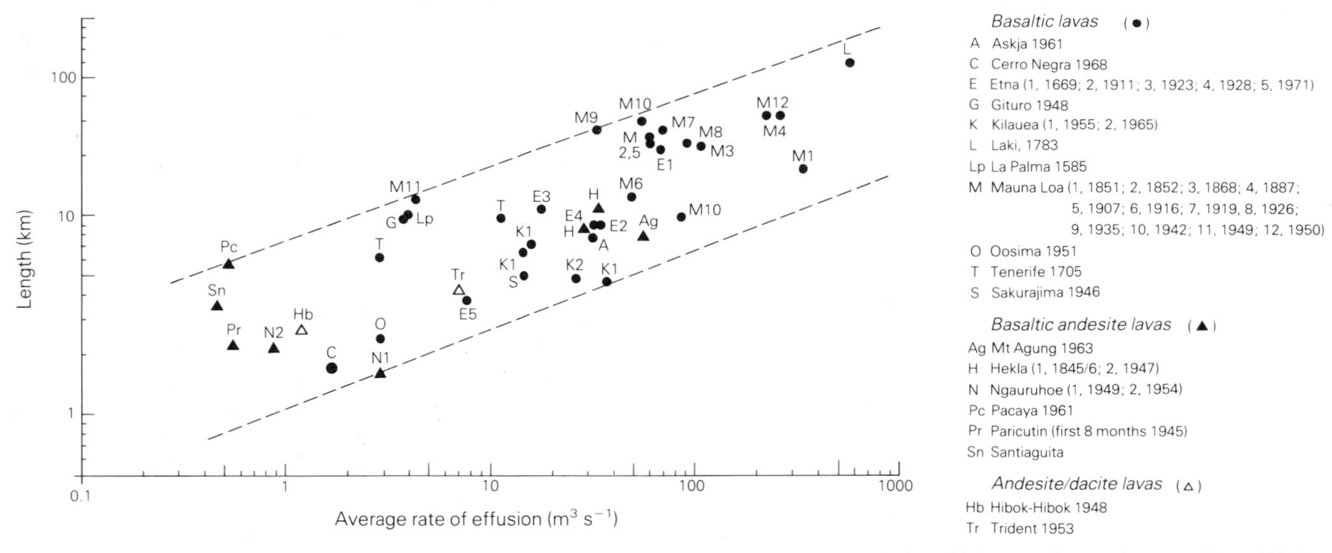

Figure 3.23 Plot of lava flow length against average effusion rate for lava eruptions (mostly basaltic) on various volcanoes (from Walker 1973; reprinted with permission of the Royal Society).

Various characteristics of the eruption constitute the third main group of factors. Figure 3.23 illustrates how one of these factors, rate of effusion, may influence volcanic morphology. Low rates of effusion produce relatively short flows, whereas high rates of effusion produce long flows, as documented by Walker (1973). Short flows tend to accumulate close to the vent, forming lava domes and cones, in contrast to long flows which produce lava plains.

3.4.2 Volcanic craters

Volcanic craters can form through a variety of processes. Craters larger than about 2 km are termed **calderas** and can form by collapse, explosive eruptions, erosion or a combination of these processes. Typically, calderas involve multiple eruptions and as a consequence show multiple vents and complex histories (Fig. 3.24). Smaller (⩽1 km) collapse features, termed **pit craters**, commonly form in basaltic lavas and, although frequently are vents, can form without associated eruptions. Still smaller craters, termed **collapse depressions** (Fig. 3.25), form on basalt flows and are not generally related to vent activity.

Maars are volcanic explosion craters which result from phreatomagmatic eruptions in which rising magma encounters water (either surface or subsurface water). Because these are "point-source" explosions, in many ways maars resemble impact craters (Fig. 3.26).

Figure 3.27 is an attempt to classify common volcanic features and to relate them to some of the more important factors involved in volcanic eruptions. Thus, for planetary purposes it is possible to approximate the styles of eruption based on the morphology of the volcanic features. Unfortunately, because so many of the factors involved in volcanic morphology are interrelated and impossible to disentangle by commonly available remote-sensing methods, the details of the volcanic processes involved in their formation usually remain unknown.

3.4.3 Intrusive structures

Not all magma reaches the surface to produce volcanoes. Some magma intrudes crustal rocks, cools and crystallizes, and may later be exposed through weathering and erosion. Frequently, these intrusive structures are more resistant to erosion than the intruded host rock and stand out as topographic features. Figure 3.28 shows Green Mountain in Wyoming, a dome of sedimentary rocks underlain by an igneous intrusion. Figure 3.29 shows several intersecting ridges composed of igneous rocks which intruded as vertical sheets termed **dikes**.

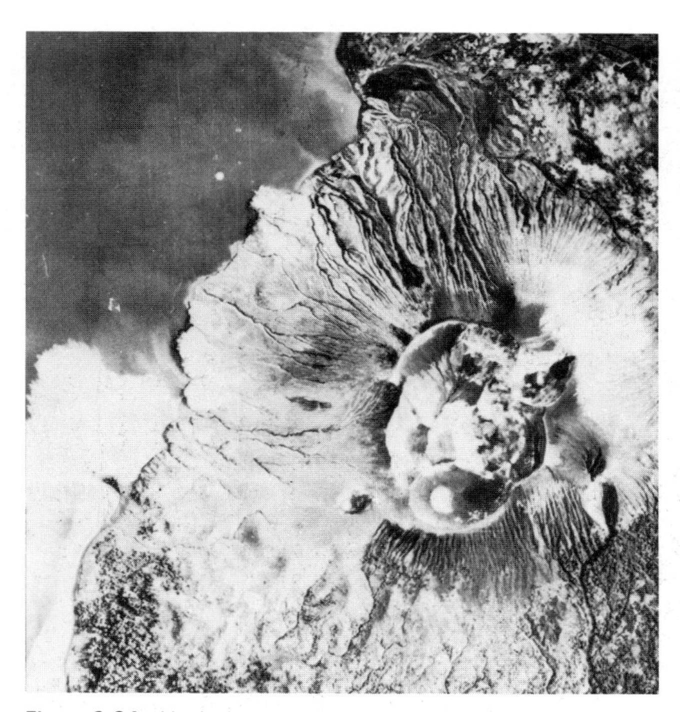

Figure 3.24 Vertical aerial photograph of Mount Tavurur, New Britain, showing complex caldera reflecting multiple eruptions (stereogram no. 102 of the University of Illinois Committee on Aerial Photography).

3.5 Gradation

Gradation is a complex process that begins with weathering and erosion, continues with transport of the weathered debris, and ends with deposition of the material. Thus, gradation is the "leveling off" process in which topographically high areas are worn away by erosion and low areas are filled by deposition. The driving force of gradation is gravity. Through gravity, material is moved by **mass wasting** – such as landslides – by running water, by frozen water (glaciers) or by wind.

3.5.1 Mass wasting

Mass wasting is the downslope movement of rock and debris under the influence of gravity (Sharpe 1968) and, thus, is a universal geologic process. Figure 3.30 classifies various forms of mass wasting. In general, mass wasting is subdivided on the basis of rate of movement, the types of material that are involved and water content. Water acts in several ways to enhance mass wasting: (a) a film of water, acting as a lubricant, can destroy the cohesion between particles; (b) in many materials, particularly the clay minerals, water may enter the crystal structure, causing swelling and disruption of the strength of the material; (c) water adds weight to the potential landslide masses and thus helps

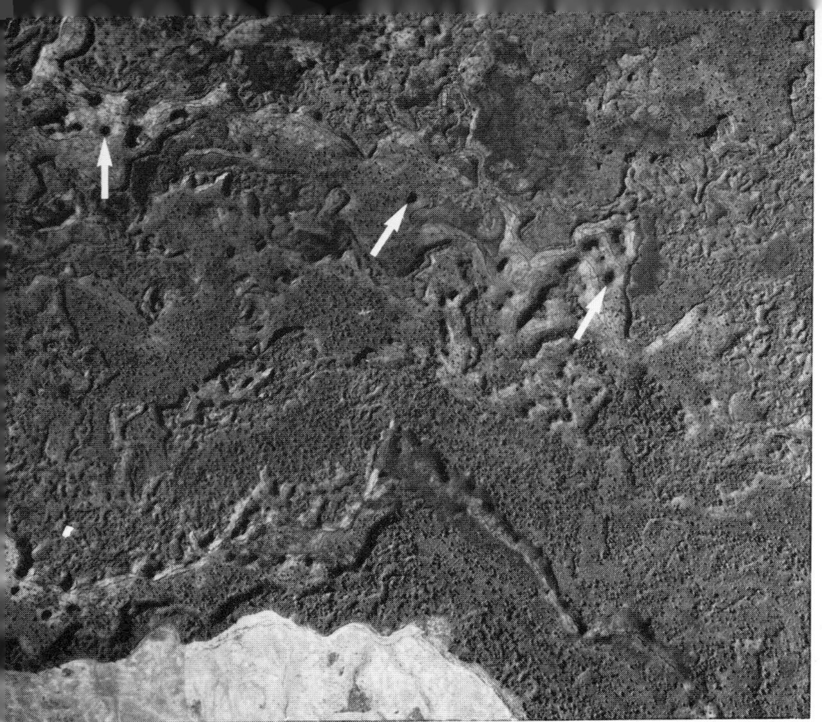

Figure 3.25b Oblique aerial photograph showing collapse depressions; most collapse depressions average 10 m in diameter (from Greeley & Gault 1979).

◁ **Figure 3.25a** Aerial photograph showing collapse depressions (arrows) on the Wapi lava field (basalt), Idaho; also shown is a pressure ridge (lower right corner); area of photograph is about 1.5 by 1.3 km (NASA-Ames photograph 878 5-1).

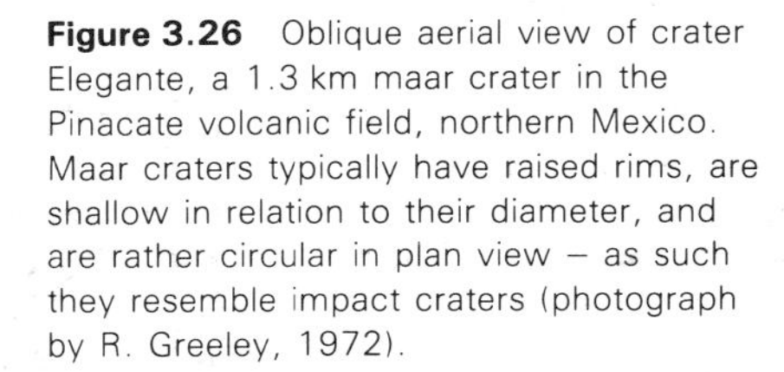

Figure 3.26 Oblique aerial view of crater Elegante, a 1.3 km maar crater in the Pinacate volcanic field, northern Mexico. Maar craters typically have raised rims, are shallow in relation to their diameter, and are rather circular in plan view – as such they resemble impact craters (photograph by R. Greeley, 1972).

Figure 3.27 Classification of volcanic features based on style of eruption and properties of the magma (modified from Rittmann 1962; reproduced with permission of Ferdinand Enke).

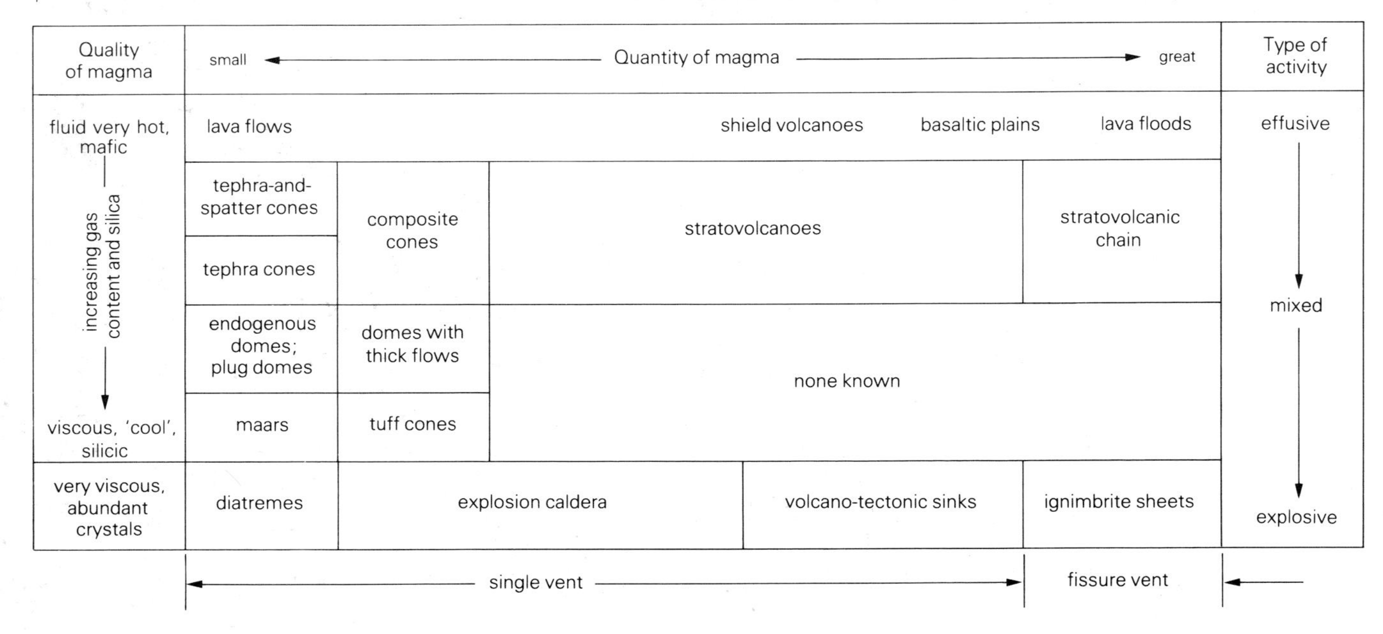

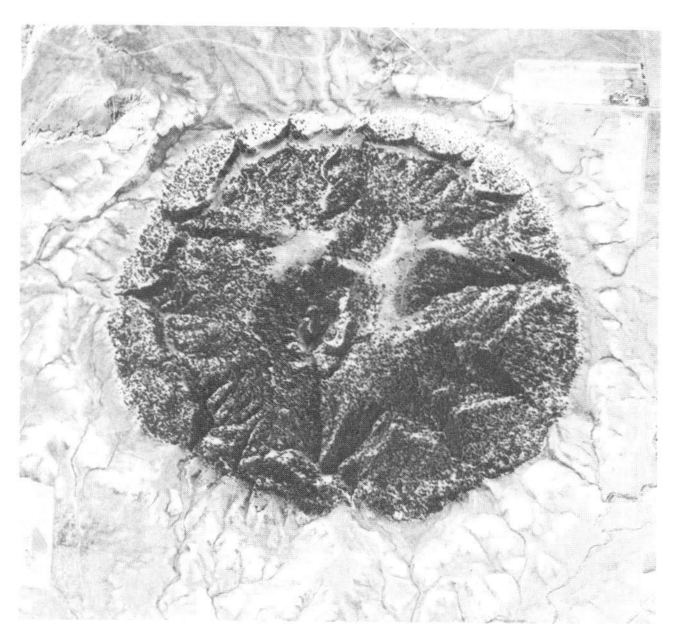

Figure 3.28 Vertical aerial photograph of Green Mountain (Crook County, Wyoming), a dome of outward-dipping sedimentary rocks underlain by a laccolith (a plutonic intrusion of igneous rocks). The dome is about 1.8 km across (US Dept of Agriculture photograph BBU-29-78).

Figure 3.29 Vertical aerial photograph of intersecting dikes, Spanish Peaks area, Colorado. Dikes of several ages are shown, indicated by cross-cutting relationships; area shown is about 3.1 by 4.5 km (US Geological Survey photograph CL 34-71).

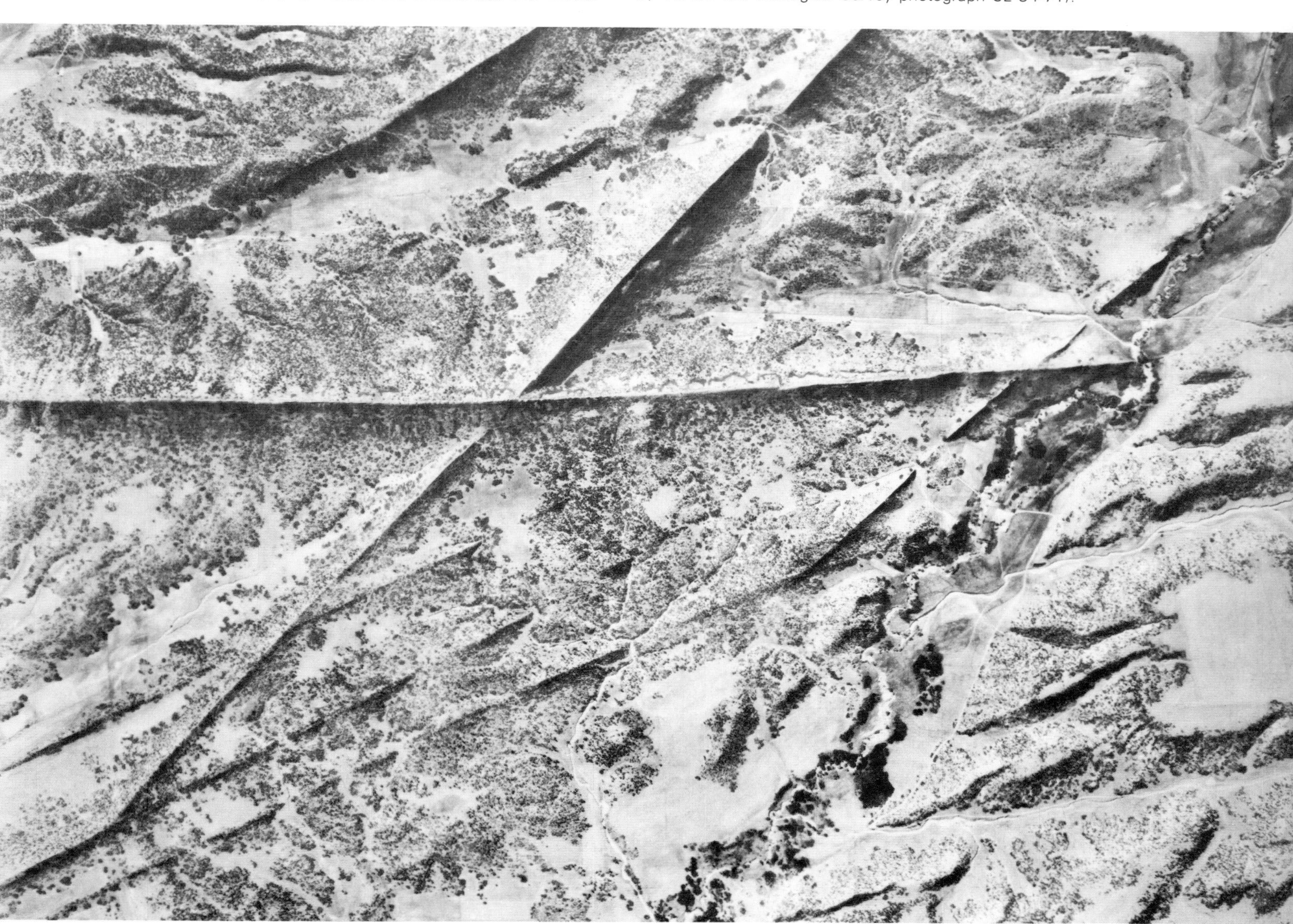

Type of movement		Type of material			
		Bedrock		Soils	
falls		rockfall		soilfall	
		rotational	planar	planar	rotational
slides	few units	slump	block guide	block guide	block slump
	many units		rockslide	debris slide	failure by lateral spreading
		All unconsolidated			
		rock fragments	sand or silt	mixed	mostly plastic
flows	dry	rock fragment flow	sand run; loess flow		
			rapid earthflow	debris avalanche	slow earthflow
	wet		sand or silt flow	debris flow	mudflow
complex landslides		combinations of materials or type of movement			

Figure 3.30 Classification of landslides (modified from Varnes 1978).

to "push" the mass down hill; and (d) fluid pore pressure can reduce the amount of energy necessary to initiate movement in both faults and landslides. Figure 3.31 shows various styles of mass wasting on Earth.

Features related to mass wasting have been observed on Mars, where water may have been a contributing factor, and on the Moon and Mercury, both of which lack water.

3.5.2 Processes associated with the hydrologic cycle

The **hydrologic cycle** defines the circulation of water among surface reservoirs (such as oceans), the atmosphere and groundwater systems. On Earth, water is a dominant geologic agent and the hydrologic cycle figures prominently in surface processes. Little can be said about the possible hydrologic cycles on other planets. Water does not exist on the Moon, nor has it ever existed according to models based on analyses of lunar samples, and it is highly unlikely that water has ever existed on Mercury, because of its close proximity to the Sun. Although water exists on Mars and many of the outer planet satellites, only on Earth are conditions favorable for liquid surface water. However, the presence of ancient channels and valley networks on Mars suggests that climatic conditions in the geologic past were different and that liquid water probably existed on the surface. Moreover, liquid water may exist at depth in certain regions of Mars and some of the outer planet satellites.

Venus, too, may have had liquid surface water, although present conditions far exceed the boiling point of water (Table 2.1). Interpretations based on measurements of certain isotopes in the atmosphere have led some investigators to propose oceans of water in the past (Donahue *et al.* 1982). When high-resolution radar images are obtained of the venusian surface, the search for ancient river beds and other water-related landforms will be a high priority.

With channels observed on Mars and the possibility of water having existed on Venus, it is important to assess the morphologic features associated with the hydrologic cycle, at least as seen on Earth, as a basis for comparison. Numerous texts on geomorphology and hydrology have been written about streams (Leopold *et al.* 1964 and others), but here we consider only the major features which appear to be important for planetary comparisons.

River and stream patterns provide clues to the structure and characteristics of the underlying rocks and topography. Howard (1967) provides an exhaustive classification of drainage patterns for Earth and gives possible geologic significance for each category. His classification takes into account the whole system (valleys, gullies, channels) on a regional scale and is, therefore, particularly appropriate for planetary comparisons which deal with large areas.

Table 3.4 summarizes Howard's scheme and is supplemented by diagrams shown in Figure 3.32. This classification is largely empirical and patterns may grade from one form into another. Nonetheless, the system provides a basis of comparison that is extremely useful. Streams and some associated features are shown in Figure 3.33.

One part of the hydrologic cycle involves groundwater. Erosion produced by groundwater dissolving certain rocks (typically limestones, rock salt or gypsum) leads to a terrain termed **karst topography**. Depending upon the stage of evolution, karst topography may

Figure 3.31 Various forms of mass wasting. (a, previous page, top) View of slumping in shale (Oahe quadrangle, South Dakota; US Geological Survey photograph by D. R. Crandell). (b, previous page, bottom) Talus cones resulting from rock falls and stream wash; South Stinking Water Canyon, Park County, Wyoming (US Geological Survey photograph by T. A. Jaggar, 1893). (c, above left) Rock glacier on McCarthy Creek, Copper River region, Alaska, showing source supply in talus cones and flow lobes into valley. Rock glaciers are masses of poorly sorted rocks and fine material held together by interstitial ice (US Geological Survey photograph by F. H. Moffit). (d, above right) Earthquake-induced landslide (dark flow) on the Sherman Glacier. Slide material is estimated to be 10^7 m^3 and covers an area of about 8 km^2. Tashuna district, Copper River region, Alaska (US Geological Survey photograph by Austin S. Post, 1965).

display only a few **sinkholes** (collapse pits), or numerous sinkholes plus **solution valleys** (collapse drain network), or highly eroded karst in which only **haystacks** (Fig. 3.34), **pinnacles** and **spires** remain as erosional remnants.

Landforms with the imprint of former lakes, swamps and oceans are highly diverse. Typically, these are sites of former sedimentary deposition and, with the removal of water, leave flat, broad plains, typified by intermontane **playas**. Shoreline processes may lead to features such as **terraces** (both erosional and depositional, which may reflect former shorelines), **sea cliffs** or **beaches**. Except for some craters and canyons on Mars, which may have contained ponded water in the past (Fig. 3.35), and the possibility of former oceans on Venus, only Earth displays landforms associated with large bodies of water.

3.5.3 Aeolian processes

Any planet or satellite having a dynamic atmosphere and a solid surface has the potential for **aeolian** (wind) processes (Greeley & Iversen 1985). Most deserts, coastal areas and glacial plains and many semi-arid regions on Earth are subject to aeolian processes. Seasonal dust storms sweep across Mars where aeolian activity appears to be the dominant active process. Measurements of wind speeds on Venus and analyses of images from its surface suggest the possibility of aeolian processes. Discovery by the Voyager mission of the predominantly nitrogen atmosphere on Titan raises the possibility of aeolian activity on this satellite of Saturn.

As outlined in the classic reference by Bagnold (1941), winds transport sediments via three modes: **suspension** (mostly silt and clay particles, i.e. smaller than about 60 μm), **saltation** (mostly sand-size particles, 60–2000 μm

Table 3.4 Classification of drainage patterns (modified from Howard 1967).

Pattern	Significance
dendritic	Horizontal sediments or uniformly resistant crystalline rocks. Gentle regional slope at present, or at time of drainage inception.
parallel	Moderate to steep slopes; also in areas of parallel, elongate landforms.
trellis	Dipping or folded sedimentary, volcanic or low-grade metasedimentary rocks; areas of parallel fractures.
rectangular	Joints and/or faults at right angles. Streams and divides lack regional continuity.
radial	Volcanoes, domes and residual erosion features.
annular	Structural domes and basins, diatremes and possibly stocks.

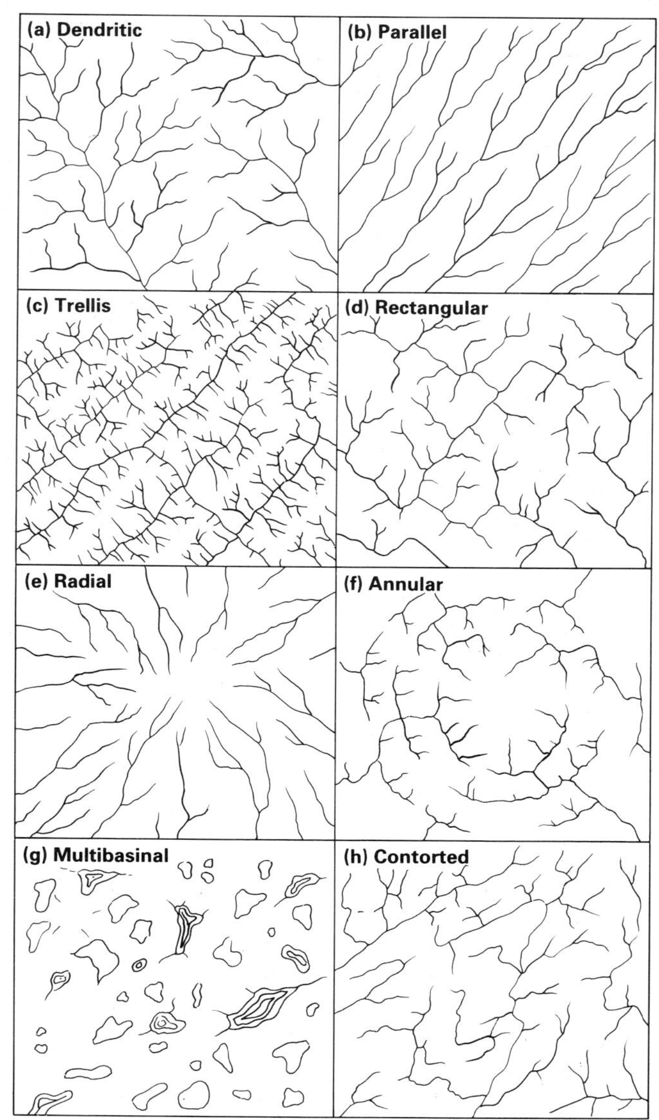

Figure 3.32 Diagram showing basic stream patterns (from Howard 1967, with permission of American Association of Petroleum Geologists).

in diameter) and **surface creep** (particles larger than about 2000 μm in diameter). Wind threshold curves (Fig. 3.36) define the minimum wind speeds required to initiate movement of different particles for given planetary environments. The ability of wind to attain threshold is a function primarily of atmospheric density, viscosity, composition and temperature. Thus, the very low-density atmosphere on Mars (Table 2.1) requires wind speeds that are about an order of magnitude stronger than on Earth.

Aeolian processes are capable of redistributing enormous quantities of sediment over planetary surfaces, resulting in the formation of landforms large enough to be seen from orbit and deposition of windblown sediments that can be hundreds of meters thick. Because aeolian processes involve the interaction of the atmosphere and lithosphere, an understanding of aeolian activity sheds light on meteorological problems. Aeolian activity can be considered in terms of large-scale and small-scale modifications.

Large-scale modifications involve features that can be observed from distances of orbiting spacecraft. One of the most useful types of features for interpretation of surface processes is the **dune**, a depositional landform (Fig. 3.37). Both the planimetric shape and cross-sectional profile of dunes can reflect the prevailing winds in a given area (Fig. 3.38). Thus, if certain dune shapes or slopes can be determined from orbital data, local wind patterns can be determined. Repetitive viewing of the same dunes as a function of season may reveal seasonal wind patterns.

On Earth great quantities of silt and clay are transported in dust storms and eventually deposited as **loess**. Thick loess deposits are found throughout the geologic column. Even where relatively young and well-exposed on the surface, loess deposits are nearly impossible to identify as such by remote-sensing methods. Yet identification of such deposits could be very important in understanding planetary surfaces.

Large-scale aeolian erosional features include pits and hollows (called **blowouts**) that form by deflation (the removal of loose particles) and wind-sculptured hills called **yardangs** (Fig. 3.39).

Observations of active aeolian features provide direct information on the atmosphere. For example, **variable features** on Mars are surface patterns which form as a result of aeolian activity. The patterns are visible as contrasts in albedo, or surface reflectivities. Repetitive imaging shows that many of them disappear, reappear or change their size, shape or position with time. Mapping the orientations of variable features has been used to derive patterns of near-surface atmospheric circulation.

3.5.4 Glacial and periglacial processes

This section deals with planetary surface features and processes associated with cold regions and water ice. **Periglacial** refers to processes, conditions, areas, climate and topographic features in cold regions or in any environment where frost action is important.

A review of the various environments in the Solar System shows that all planets and satellites except Venus experience temperatures below freezing. Subsurface ice probably exists on Mars and ice is a major constituent of many of the outer planet satellites.

Surface features on an icy or ice-rich body result largely from processes of flow and fracture. Although ice is often modeled as a Newtonian viscous fluid, experiments indicate that it can be considered a "pseudo-

Figure 3.33a Streams and stream-related features. Headward erosion of streams, San Bernardino Mountains, California (US Geological Survey photograph by J. R. Balsley, 1949).

Figure 3.33b Oblique aerial photograph looking south toward the junction of the Yukon and Koyukuk Rivers, Alaska, showing meander loops, former channels and flood plain deposits (US Geological Survey photograph by US Army Air Corps, 1949).

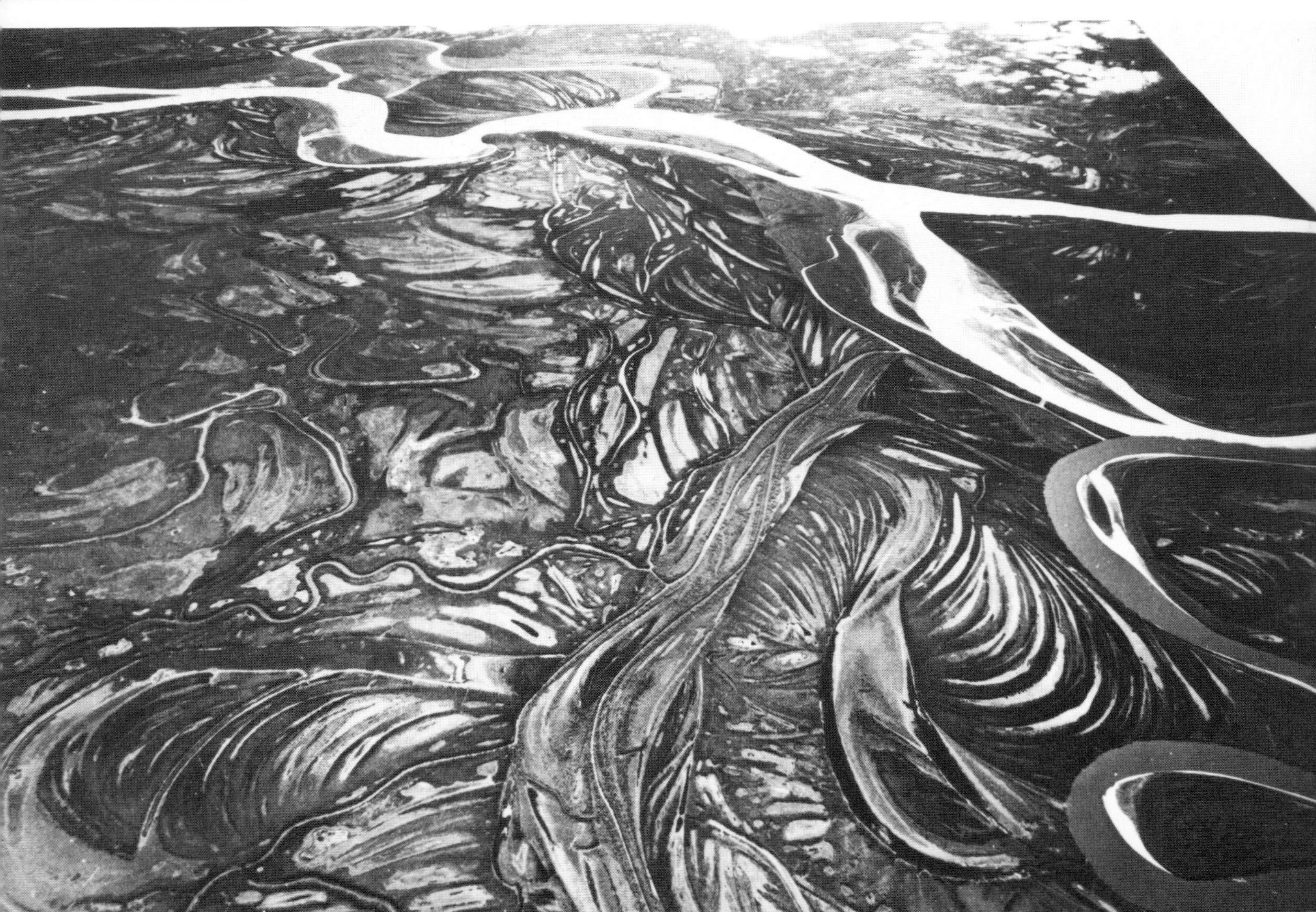

Figure 3.33c Alluvial fans spreading into intermountain basins, Mojave Desert, California (US Geological Survey photograph by J. R. Balsley, 1949).

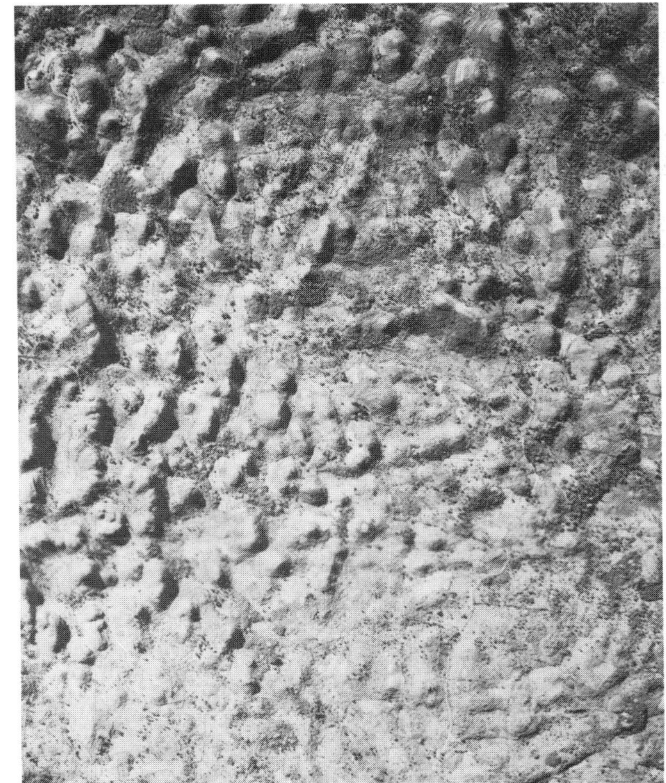

Figure 3.34 Vertical aerial photograph of karst terrain in Puerto Rico showing haystack remnants resulting from solution of limestone (stereogram no. 170 of the University of Illinois Committee on Aerial Photography).

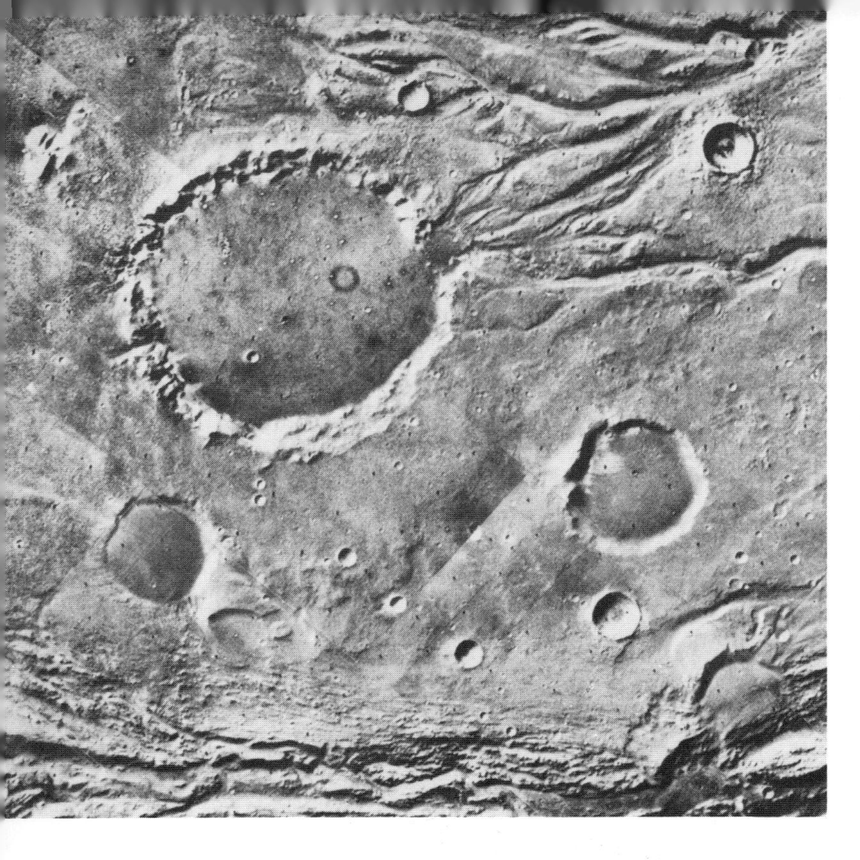

Figure 3.35 Mosaic of Viking orbiter pictures of the Lunae Planum region of Mars showing flat, dark floor of a 30 km diameter crater that appears to have been flooded with deposits associated with channels, as described by Baker and Kochel (1979) and Lucchitta and Ferguson (1983) (part of US Geological Survey subquadrangle MC-10 EC).

plastic fluid" which deforms by creep (Glen 1974). In a Newtonian fluid, the rate of strain is linearly proportional to the applied stress and the viscosity is the ratio of strain proportional to the stress raised to some power. Thus, as the stress level is increased, the material deforms more and more rapidly. The result is that ice appears to become less viscous at higher rates of strain. However, under very rapid strain rates, such as during an impact event, ice behaves more like a brittle elastic material than a fluid.

On Earth, glaciers are classified as either **valley glaciers** (Fig. 3.40), or as **ice sheets** (also called continental glaciers, or ice caps) if they are too large to be contained by valleys. All glaciers move downslope or outward, leaving distinctive terrains (Figs 3.41 & 3.42), and a "retreating" glacier simply means that the melting and ablation exceed the rate of forward movement by the glacier.

On Earth, precipitation of snow in the **area of accumulation** (headward part of glacier) forms a deposit that is about 20 percent ice and 80 percent air. Melting and refreezing plus compaction converts the snow to spherical ice particles called **firn**. As the firn accumulates, further compaction causes recrystallization to form the main ice mass, typically having less than 10 percent air.

Most glaciers also incorporate rocky materials within the ice mass. This material can include dust and other airborne particles and chunks of rock gouged by the ice as it moves across a surface or from the accumulation of debris derived from valley walls. Surface material often coalesces into various **moraines** which are linear deposits (Fig. 3.41). On Earth, material carried by the ice eventually reaches the front of the glacier. It may be deposited *in situ* or carried away by melt water. The finer material is often transported by the strong winds which are generated along ice margins. Coarse glacial deposits, termed **drift**, may assume a variety of geometries which provide clues to the form and position of glaciers after the ice mass has "retreated".

On Earth, glaciers have effected extensive changes in the landscape. U-shaped valleys (Fig. 3.42), grooves and striations parallel to the flow of ice, and amphitheater-

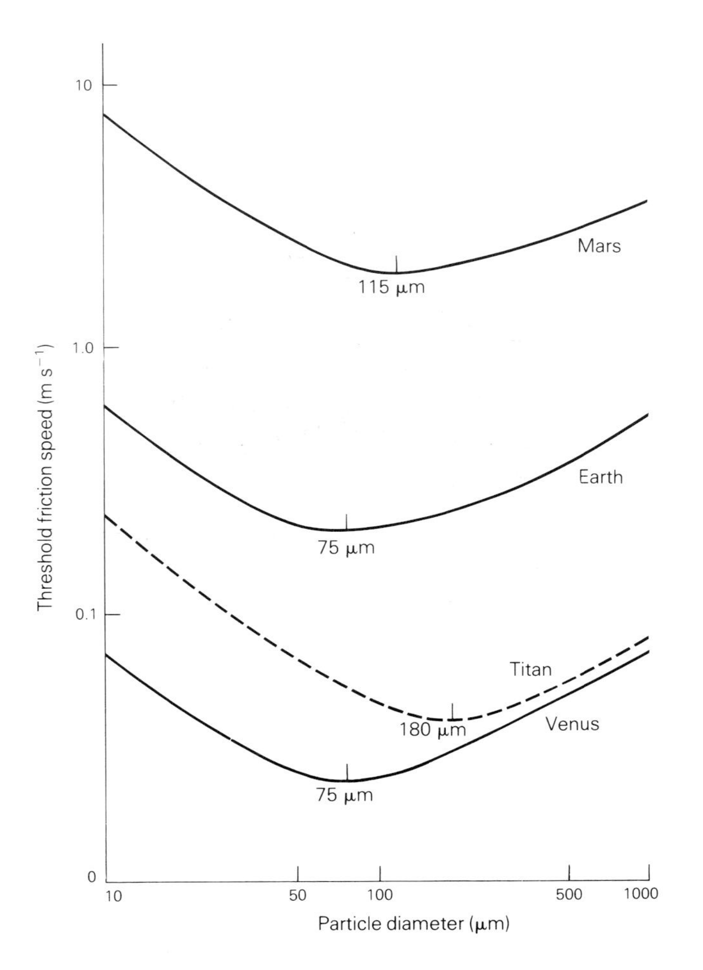

Figure 3.36 Diagram showing the minimum threshold friction speed (a function of wind speed) required to move particles of different sizes on Mars, Earth, Titan and Venus; note that as the atmospheric density decreases from Venus to Mars, minimum winds needed to set particles into motion increases.

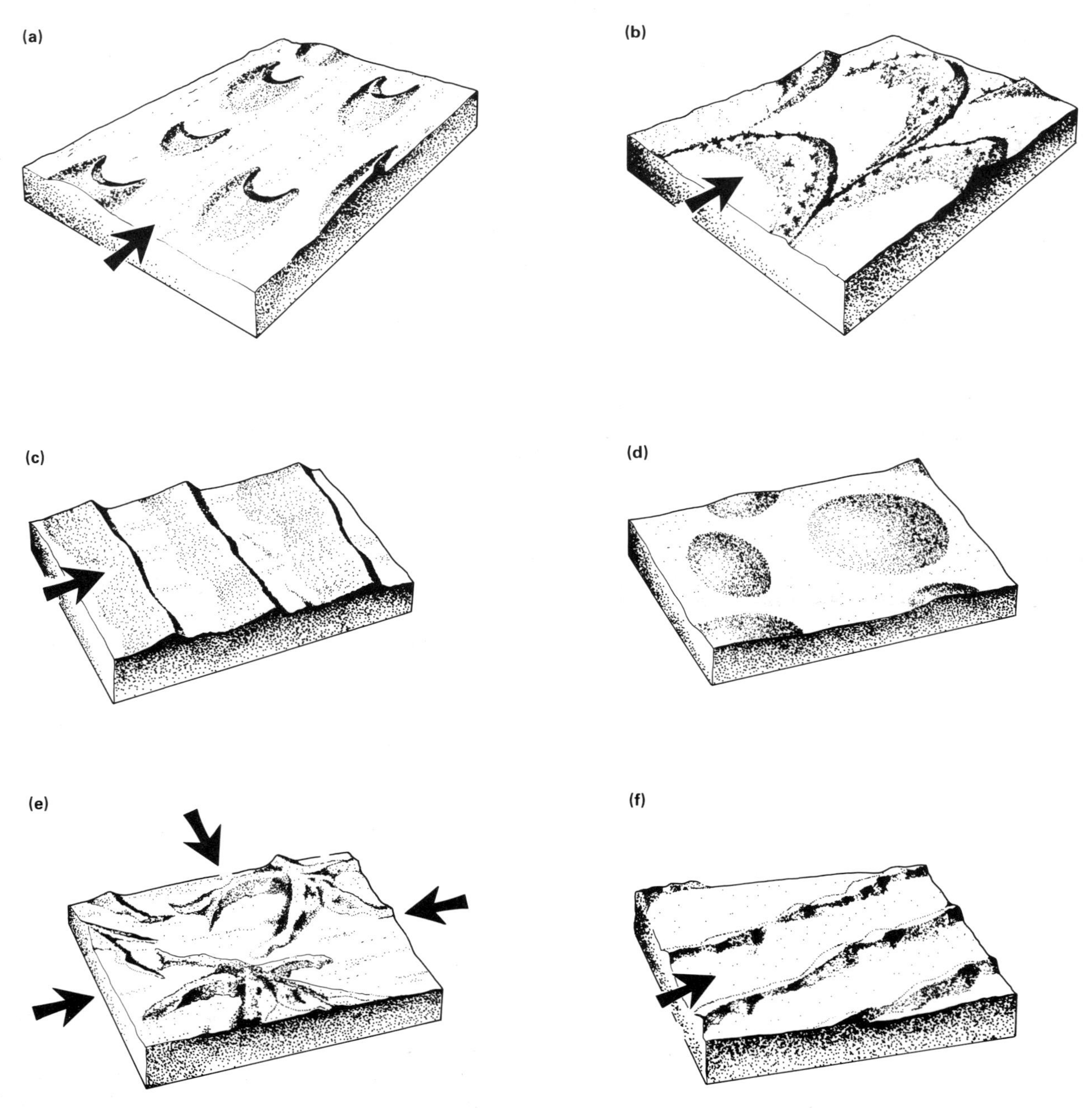

Figure 3.37 Diagrams illustrating principal types of sand dunes and the wind direction(s) (arrows) responsible for their formation: (a) barchan, (b) parabolic, (c) transverse, (d) dome, (e) star, (f) linear (from McKee 1979).

Figure 3.38 Sand dunes on Earth. (opposite) Oblique aerial view of barchan dunes in Peru; prevailing wind is from the upper left; prominent *slip faces* are on the downwind side of the dunes, toward the camera (US Geological Survey photograph by E. C. Morris). (below) Complex transverse dunes in the Algodones dune field, southern California; prevailing wind is from the left.

Figure 3.39 Wind-sculpted hills, termed yardangs, several hundred meters long, Peru (US Geological Survey photograph by J. McCauley).

Figure 3.40 Oblique aerial photograph of the Cook Inlet Region, Alaska, showing valley glaciers; dark stripes represent rocks and debris carried along with the ice (US Geological Survey photograph by A. Post, 1970).

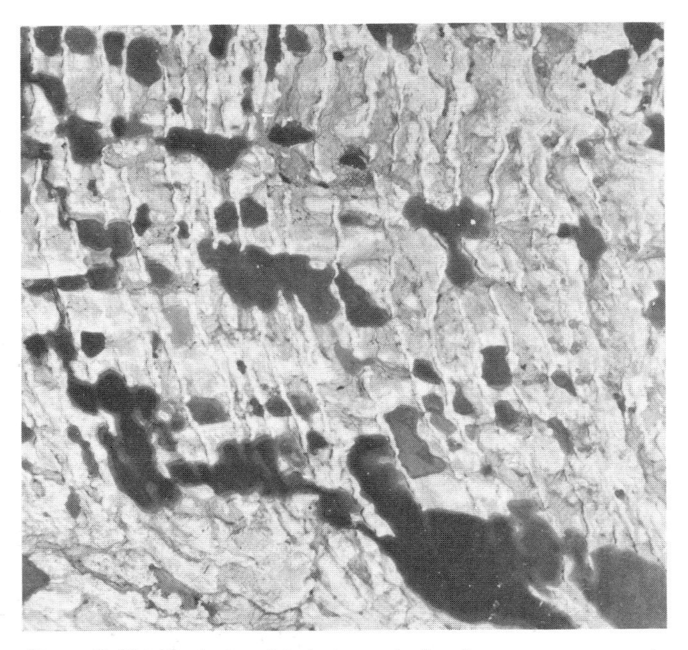

Figure 3.41 Vertical aerial photograph showing numerous, nearly equally spaced ridges (moraines) in Quebec; each ridge is 5–7 m high and is considered to result from debris pushed up near the front of advancing ice (from Royal Canadian Air Force photograph, prepared by the University of Illinois Committee on Aerial Photography, Stereogram No. 535).

shaped **cirques** in the headward parts of valleys are indicative of glacial erosion and can be seen on images obtained from orbit.

The term **periglacial** refers to a specific climatic zone in which the processes of **solifluction** (Fig. 3.43; the slow, viscous, downslope flow of water-saturated, unconsolidated materials), **gelifluction** (the flow of ice-saturated materials) and **nivation** (the erosion of rock or soil by snow and ice, by frost action and by chemical weathering) are characteristic and within which such geomorphic features as permanently frozen ground **(permafrost)**, **patterned ground** (Fig. 3.44), **pingos** (Fig. 3.45) and **thermokarst** topography are developed. The occurrence of a periglacial region is not genetically related to the proximity of glaciers or continental ice sheets, contrary to what is implied by its etymology. However, the presence of water is essential for most periglacial processes to occur. This broader definition is useful in that it allows us to consider the possible operation of periglacial-type processes on the surfaces of other objects in the Solar System.

Figure 3.42 View of Deadman Canyon, Tulare County, California, showing typical U-shaped glaciated valley (US Geological Survey photograph by F. E. Mathes, 1925).

Extensive periglacially modified plains have been inferred for Mars, particularly in the northern latitudes (Carr & Schaber 1977). This deduction is based on observations and interpretations of mass wasting, some types of polygonally patterned ground and the radially striated, apparently fluidized ejecta blankets surrounding many craters.

Voyager provided a wealth of information about the outer planet satellites. With the exception of Io and possibly Amalthea, all are considered to have complex crusts consisting of mixtures of water ice and silicates. Thus, the possibility exists for mass wasting and processes of surface modification similar to terrestrial glacial and periglacial terrain. These processes may occur in association with other ices such as methane, ammonia and their clathrates. For example, the surface of Titan may consist, at least in part, of frozen methane. On Mars, deposits of carbon dioxide frost occur in the annual polar caps. By 1986, Voyager 2 will reach Uranus and image its satellites, whose surfaces may consist of water ice and other materials.

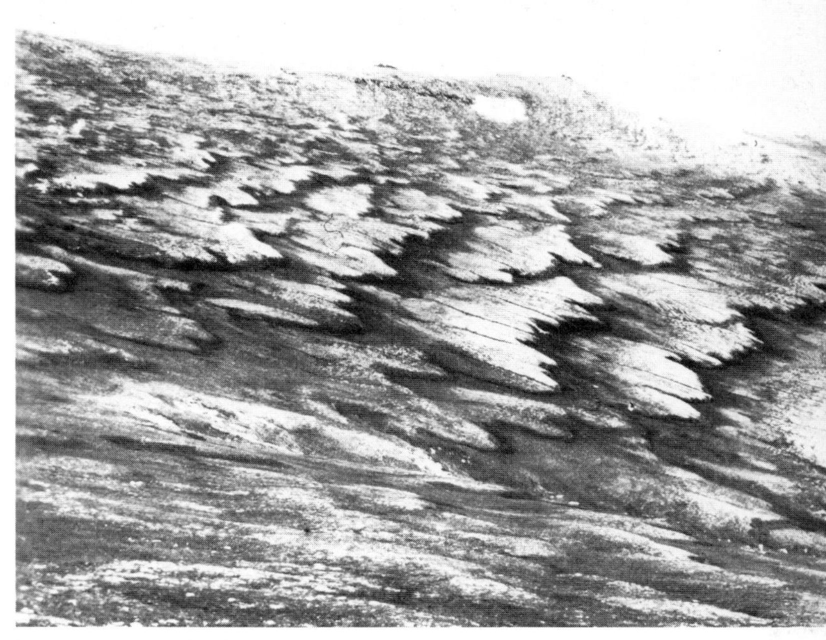

Figure 3.43 Solifluction lobes in the Seward Peninsula region, Alaska (US Geological Survey photograph).

Figure 3.44 Oblique aerial view of raised-edge ice wedge polygons on the Alaskan sea coast near Barrow; polygons are 7–15 m across (US Geological Survey photograph by R. I. Lewellen).

Figure 3.45 View of collapsed *pingo* (soil-covered ice mound) on Mackenzie Delta near Tuktoyaktuk in northwestern Canada. Long dimension of pingo is about 110 m (photograph courtesy of T. L. Péwé, Arizona State University).

3.6 Summary

Planetary surfaces are shaped and modified by four principal processes: (a) tectonism, (b) igneous activity, (c) gradation and (d) impact cratering. Each of these processes produces distinctive landforms on Earth where most of these processes have been studied in detail. One of the goals of planetary geology is to determine how these landforms might be different in extraterrestrial environments.

Views of the Earth obtained from orbit show that while some processes can be identified by remote-sensing techniques, others cannot, thus introducing uncertainties into interpretations of planetary histories. Furthermore, with increased knowledge of the outer planet satellites, planetary geologists must assess the validity of applying Earth analogs to bodies composed mostly of ice and having markedly different environments.

4 The Moon

4.1 Introduction

The Moon has long been an object of fascination to everyone from casual observers to scientists. Aside from Earth, it is the planet about which we know most. The Moon has been photographed, probed and analyzed from orbit to determine its surface composition. Nearly a half ton of it has been returned to Earth for analyses by every conceivable method. More missions have been flown to the Moon than to any other object (Table 4.1). As discussed in Chapters 1 and 2, studies of planetary geomorphology began with the Moon. Thus, the Moon served as a "training ground" for learning how to study the geology of extraterrestrial worlds. It is fortunate that the Moon played this role because, in many respects, it is a relatively simple object for analysis. Lacking an atmosphere, none of the complicating "imprints" of wind or running water are superimposed on its surface. Consequently, the Moon was a far easier subject for developing methods to analyze planetary objects than if the first planet studied had been a more complicated body such as Mars.

Furthermore, the Apollo and Luna landings provided the opportunity to field-check some of the more critical relationships and interpretations that were based on remote sensing. The information from these landings enabled not only a better understanding of the Moon, but also an assessment of the techniques for deriving geologic data by remote sensing. Moore *et al.* (1980) provide an excellent review of remote sensing applied to the Moon.

The mid-1960s to mid-1970s saw the greatest efforts in lunar exploration through the acquisition of high-resolution images and other remote-sensing data, along with samples and results from the lander missions as reviewed by El-Baz (1979) and others. From this enormous data set, it has been possible to synthesize the general geologic history of the Moon, especially as related to the evolution of the surface. However, because of the orbital constraints placed on the Apollo missions, the equatorial regions are far better documented than the higher latitudes, which must await future missions before the knowledge for the Moon is globally uniform.

4.2 General physiography

Even with the naked eye, it is readily apparent that the surface of the Moon consists of different terrains (Fig. 4.1). The original two-fold division into the dark (albedo 5–8%), flat **maria** and the light-toned (albedo ~9–12%), rugged highlands or **terrae**, first described by lunar observers in the 16th and 17th centuries, is still valid, but these terrains can be further divided. As shown in Figure 4.2, five major terrains can be recognized on the Moon: (a) heavily cratered terrains, (b) moderately cratered terrains, (c) terrains associated with older impact basins, (d) terrains associated with the younger impact basins and (e) mare regions. In addition to these major units and the surface features that characterize them, impact craters are ubiquitous and important features on the Moon. Ranging in size from the microcraters observed on lunar samples to impact basins exceeding 1000 km in diameter, impact craters and related features dominate all regions of the Moon.

4.2.1 Cratered terrains

The highlands consist mostly of cratered terrains, which can be subdivided into two units based on the frequency of impact craters. In general, cratered terrain is characterized by rugged relief and high albedo and consists primarily of impact ejecta deposits and remnants of crater rims (Fig. 1.2). Models of the formation and evolution of the Moon (Taylor 1982) suggest that cratered terrain consists of highly brecciated rock masses derived from a chemically differentiated, anorthositic (rich in calcium feldspar) lunar crust. Most lunar investigators consider this differentiation to have occurred between 4.3 and 4.2 aeons. During this phase of lunar history, heavy impact bombardment continuously churned the evolving lunar crust, leading to the rugged terrain seen today.

4.2.2 Basins and basin-related terrains

Basins on the Moon are defined as impact structures larger than 220 km in diameter and typically displaying concentric rings (Fig. 2.5). Studies of the oldest surfaces on the Moon (CLHC 1979) suggest that basins formed over an extended period prior to 3.8 aeons and that the earlier basins have been mostly obliterated (Carr 1983).

Table 4.1 Lunar missions.

Spacecraft	Encounter date	Mission	Encounter characteristics
Luna 2*	12 Sept. 1959	hard lander	Impacted lunar surface.
Luna 3*	10 Oct. 1959	flyby	First (indistinct) photos of farside of Moon.
Ranger 7	31 July 1964	hard lander	4300 high-resolution images with about 2000 times better definition than Earth-based photography; impacted in Mare Cognitum.
Ranger 8	20 Feb. 1965	hard lander	7100 images obtained; impacted in Mare Tranquillitatus.
Ranger 9	24 Mar. 1965	hard lander	5800 images obtained with resolution up to 1 m. Impacted eastern floor of crater Alphonsus.
Zond 3*	20 July 1965	flyby	System test: took 28 pictures during flyby of Moon, then flew as far as orbital path of Mars.
Luna 9*	3 Feb. 1966	lander	Soft landing in western Oceanus Procellarum; returned pictures.
Luna 10*	3 Apr. 1966	orbiter	First object to orbit Moon; measured lunar magnetism and radiation.
Surveyor 1	2 June 1966	lander	Soft landing in Oceanus Procellarum; transmitted 11 240 TV images.
Lunar Orbiter 1	14 Aug. 1966	orbiter	Obtained 216 images, including 11 of the lunar farside. Medium-resolution pictures good, high-resolution smeared.
Luna 12*	25 Oct. 1966	orbiter	Transmitted ~15 m resolution pictures.
Lunar Orbiter 2	10 Nov. 1966	orbiter	Obtained 409 images.
Luna 13*	24 Dec. 1966	lander	Soft landing in western Oceanus Procellarum; transmitted pictures.
Lunar Orbiter 3	8 Feb. 1967	orbiter	Obtained 290 images.
Surveyor 3	20 Apr. 1967	lander	Soft landing in Oceanus Procellarum; returned 6300 pictures; analysis of surface.
Lunar Orbiter 4	8 May 1967	orbiter	Obtained 546 images, covering all of nearside and 95% of farside of Moon.
Lunar Orbiter 5	5 Aug. 1967	orbiter	418 images; completed high altitude farside photographic coverage.

Table 4.1 *Continued.*

Spacecraft	Encounter date	Mission	Encounter characteristics
Surveyor 5	11 Sept. 1967	lander	Soft landing in Mare Tranquillitatus; alpha backscatter device indicated basaltic character of that mare surface; transmitted 19 120 TV pictures.
Surveyor 6	10 Nov. 1967	lander	Soft landing in Sinus Medii; alpha backscatter data indicated basaltic composition of surface; transmitted 29 950 TV pictures.
Surveyor 7	10 Jan. 1968	lander	Last Surveyor; soft landing on Tycho ejecta blanket; alpha backscatter data; 21 040 TV pictures transmitted.
Luna 14*	10 Apr. 1968	orbiter	Analysis of gravitational field.
Zond 5*	18 Sept. 1968	flyby	Circumlunar; returned to Earth and was recovered in Indian Ocean; precursor to manned mission(?).
Zond 6*	13 Nov. 1968	flyby	Circumlunar; returned to Earth and was recovered; precursor to manned mission(?).
Apollo 8	24 Dec. 1968	orbiter	First manned Apollo flight to the Moon; system test of 10 orbits around the Moon; 864 photographs.
Apollo 10	21 May 1969	orbiter	Apollo Lunar Module system test: Lunar Module descended to within 15 km of surface; 1319 photographs, color TV transmission.
Apollo 11	20 July 1969	lander	First manned lunar landing: soft landing in Mare Tranquillitatus; 1359 photographs, 22 kg of samples returned.
Zond 7*	11 Aug. 1969	flyby	Circumlunar; returned to Earth and was recovered; precursor to manned mission(?); 33 photographs.
Luna 16*	20 Sept. 1970	lander	Soft landing in Mare Fecunditatis; 100 g of sample returned to Earth.
Zond 8*	24 Oct. 1970	flyby	Circumlunar; returned to Earth and was recovered; precursor to manned mission(?); 108 photographs.
Luna 17*	17 Nov. 1970	lander	Soft landing in western Mare Imbrium; Lunokhod I roving surface vehicle traversed 20 km.

Table 4.1 *Continued.*

Spacecraft	Encounter date	Mission	Encounter characteristics
Apollo 12	19 Nov. 1969	lander	Soft landing in Oceanus Procellarum; 1.4 km traverse; investigated the Surveyor 3 spacecraft, deployed scientific experiments; 1577 photographs; 34 kg of samples returned.
Apollo 14	5 Feb. 1971	lander	Soft landing near Fra Mauro crater; 3.5 km traverse; deployed scientific experiments; 1324 photographs; 43 kg of samples returned.
Apollo 15	30 July 1971	lander	Soft landing at Hadley Rille-Apennine Mountains; 28 km (lunar rover) traverse; deployed scientific experiments; 5099 photographs plus 3375 mapping photographs (from orbit); 77 kg of samples returned.
Luna 19*	1 Oct. 1971	orbiter	Orbiter only. Returned pictures.
Luna 20*	18 Feb. 1972	lander	Soft landing in Apollonius Highlands; 30 g of samples returned to Earth.
Apollo 16	21 Apr. 1972	lander	Soft landing near Descartes crater; 27 km (lunar rover) traverse; deployed scientific experiments; 4250 photographs plus 3480 mapping photographs (from orbit); 95 kg of samples returned.
Apollo 17	11 Dec. 1972	lander	Soft landing in Taurus-Littrow Valley; 30 km (lunar rover) traverse; deployed scientific experiments; 5807 photographs plus 4710 mapping photographs (from orbit); 111 kg of samples returned.
Luna 21*	16 Jan. 1973	lander	Soft landing in eastern Mare Serenitatis; Lunokhod II roving surface vehicle traversed 30 km.
Luna 23*	2 Nov. 1974	lander	Soft landing in Mare Crisium; sample return malfunctioned.
Luna 24*	19 Aug. 1976	lander	Soft landing in Mare Crisium; 160 cm core sample returned.

* Soviet missions.

Basins and basin-related terrains are subdivided into two units on the basis of age: old areas that pre-date the formation of the Nectaris basin and younger, post-Nectaris areas and basin-related terrains. Basin geology dominates the lunar surface in several ways, as reviewed by Howard *et al.* (1974): (a) all of the named mountain ranges on the Moon are segments of basin rims; (b) basins appear to have provided the structural focus for the emplacement of the mare lavas; and (c) basin ejecta deposits drape across much of the pre-basin surface on the Moon. Because of the important role of basins in lunar geology, many of the Apollo and Luna missions were specifically designed and targeted to address questions related to basin formation and evolution.

The Imbrium basin, first described by G. K. Gilbert in 1893 as an impact structure, is particularly important because it dominates much of the northern half of the lunar nearside. Gilbert recognized the distinctive radial grooves and furrows, which he termed Imbrium sculpture (Fig. 2.2), and attributed the markings to impact processes. The Imbrium basin consists of at least three rings and, although various authors have used different criteria to define these rings, Dence (1976) and Spudis (1982) suggest that the main ring is composed of the Apennine Mountains (Figs 4.3 & 4.4), the Alpes, part of the Sinus Iridum rim, and smaller, isolated mountains which collectively define a ring about 1140 km across. Samples from the Apennines collected during Apollo 15 show that this ring consists predominantly of breccias composed of fractionated igneous rocks – mostly norites and noritic melt rocks. The intermediate ring is defined by isolated massifs such as Mons La Hire and Montes Archimedes and has a diameter of about 850 km. The inner ring is about 570 km across and is identified only by a series of ridges on the mare surface which are interpreted to reflect an underlying mountainous ring now completely buried by lavas.

Ejecta deposits originating from the Imbrium basin are spread over much of the lunar surface (Fig. 4.5). Mapped as the Fra Mauro Formation, the ejecta may be 1 km thick as far as 600 km from the basin, as determined at Julius Caesar, an impact crater that has been partly filled by ejecta from Imbrium. Because it is so widespread and can be recognized on photographs, the Fra Mauro Formation serves as a critical index formation, or datum plane, in lunar photogeologic mapping. Samples returned from the Apollo 14 mission show that the Fra Mauro Formation consists of highly brecciated rocks. Radiometric dates obtained from these samples place the age of formation for the Imbrium basin at 3.85 aeons.

Detailed geologic mapping of the Imbrium basin and analysis of the orbital geochemical data have been combined with results obtained from lunar samples to

Figure 4.1 Shaded airbrush chart of the lunar nearside (a, above) and farside (b, right) showing prominent physiography and named features (base maps courtesy US Geological Survey).

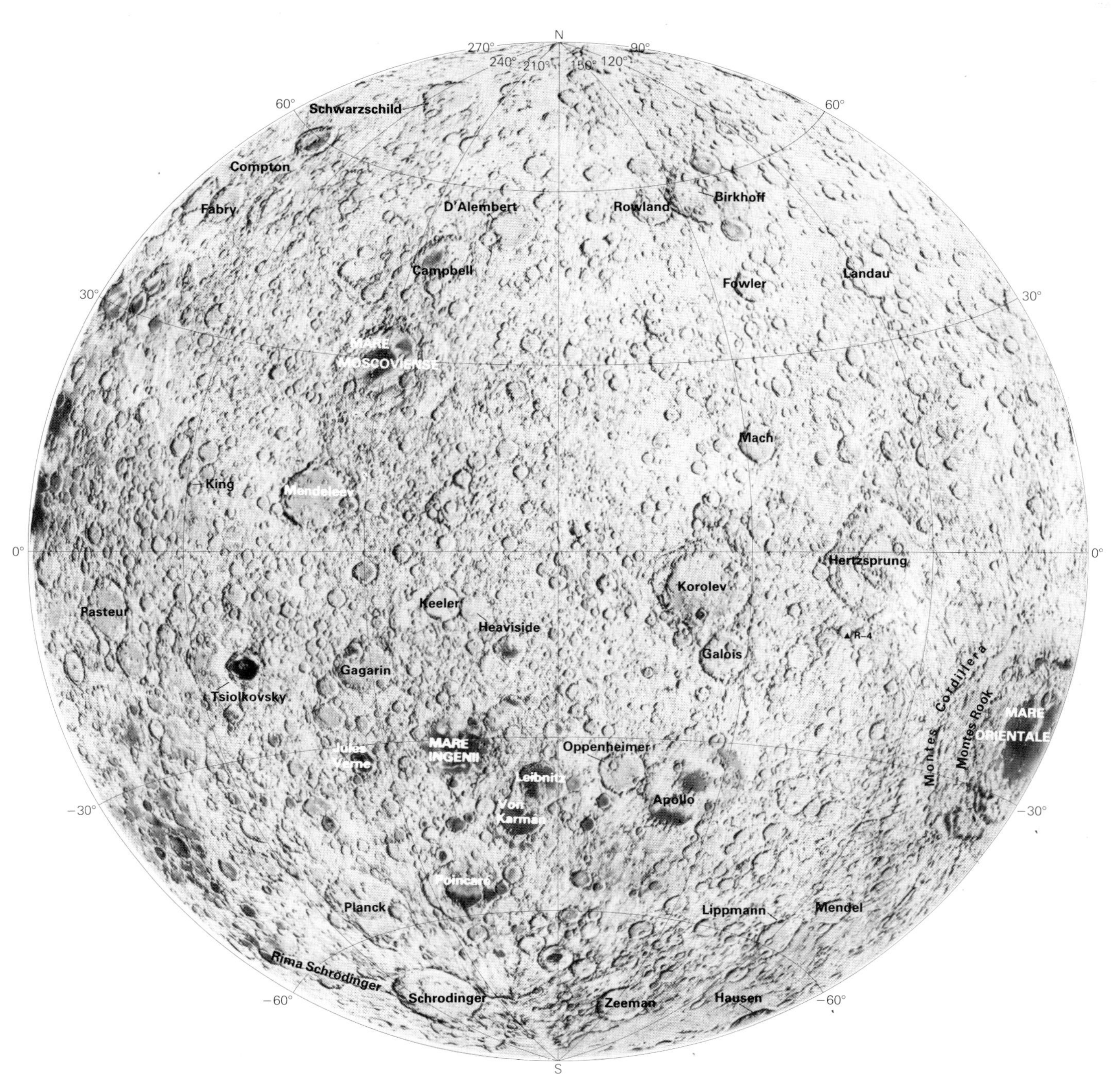
N
270°
240°
210°
150°
120°
90°
60°
60°
Schwarzschild
Compton
Fabry
D'Alembert
Rowland
Birkhoff
Campbell
Fowler
Landau
30°
30°
MARE
MOSCOVIENSE
Mach
King
Mendeleev
0°
0°
Hertzsprung
Korolev
Pasteur
Keeler
Heaviside
R-4
Galois
Gagarin
Tsiolkovsky
Montes Cordillera
Montes Rook
MARE
ORIENTALE
MARE
INGENII
Jules
Verne
Oppenheimer
Leibnitz
Apollo
Von
Karman
−30°
−30°
Poincaré
Planck
Lippmann
Mendel
Rima Schrodinger
Schrodinger
Zeeman
Hausen
−60°
−60°
S

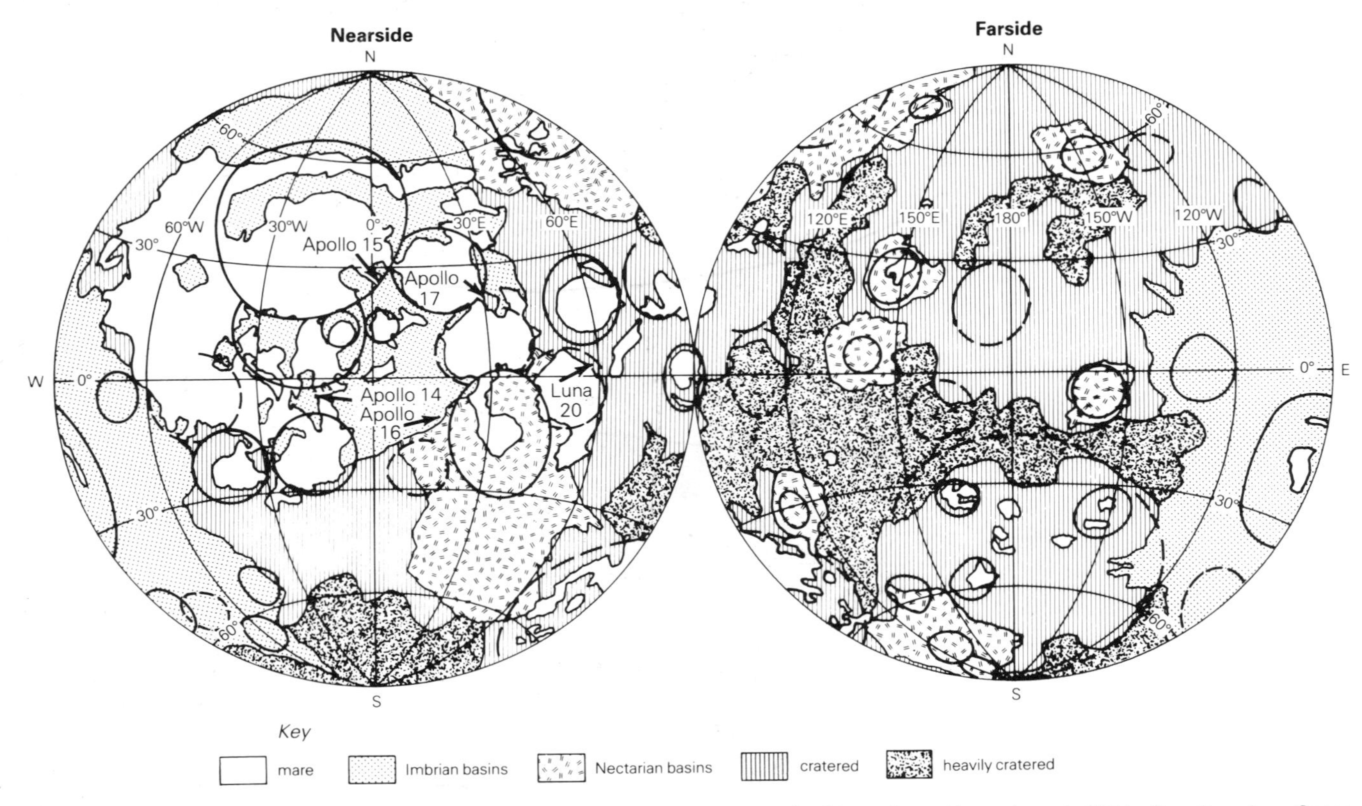

Figure 4.2 Diagram showing the distribution of the five main terrain types on the Moon (from Howard *et al.* 1974; *Rev. Geophys. Space Phys.*, copyright the American Geophysical Union).

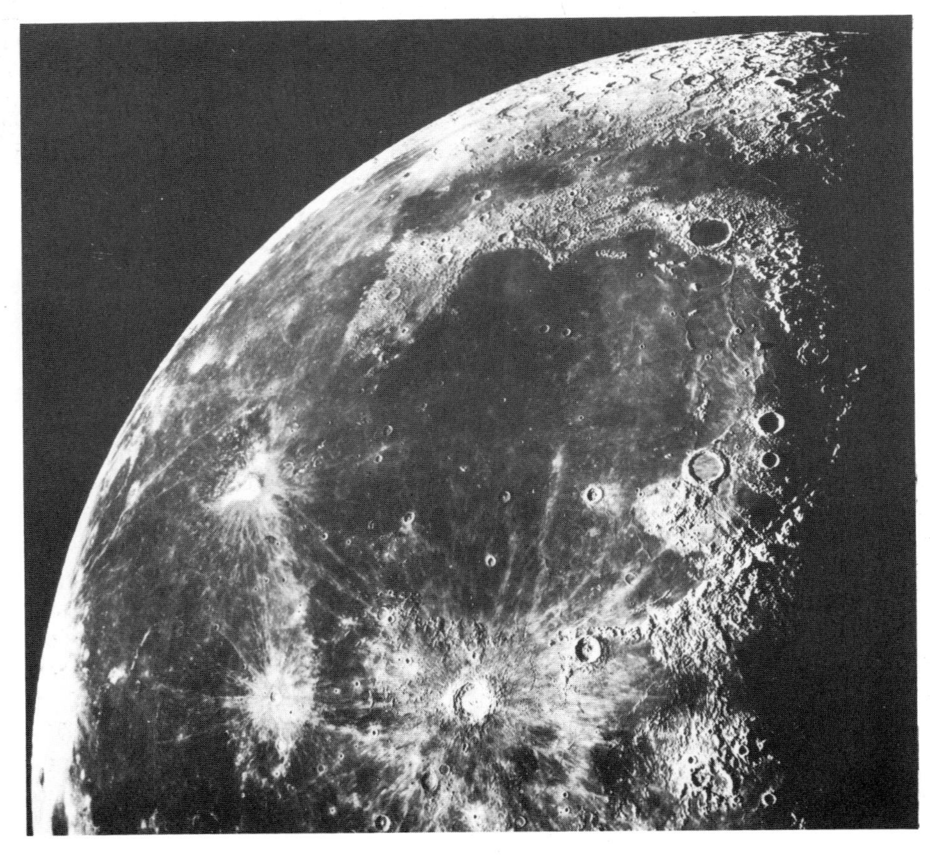

Figure 4.3 Earth-based telescopic view of the northwest quadrant of the Moon showing the Imbrium basin (Lick Observatory photograph).

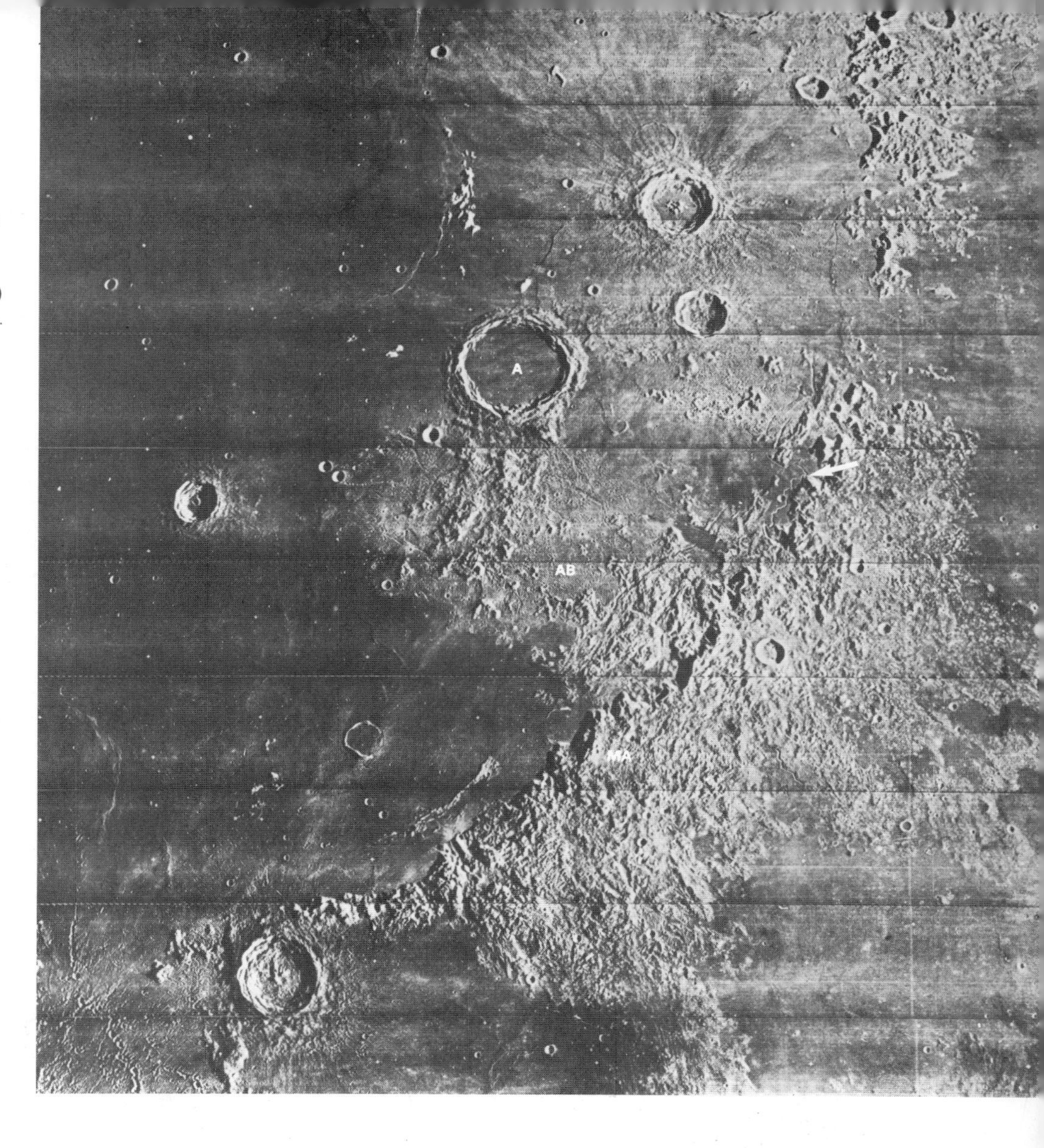

Figure 4.4 Lunar orbiter photograph showing the southeast portion of the Imbrium basin. The Apennine Bench is a structural high near the crater Archimedes (A). The light plains (AB) constitute the Apennine Bench Formation, a volcanic basin-fill unit. The Montes Apennines (MA) comprise the main basin rim in this region. Apollo 15 landing site shown by arrow. Illumination from the right (from Spudis 1982; Lunar Orbiter IV-M109).

synthesize its general geology. From this analysis, Spudis (1982) has derived a model of basin formation (Fig. 4.6) in which the transient crater cavity was a feature considerably smaller than the main ring, enlarged by slumping. Fracturing associated with the impact as well as various post-impact crustal adjustments was extensive and appears to have aided the eruption of early-stage lavas following the excavation of the basin. For example, Spudis (1978) has suggested that the Apennine Bench Formation (Fig. 4.4) consists of lavas emplaced as a result of these adjustments prior to the main eruptions of mare lavas that flooded the basin interior.

The Orientale basin (Fig. 2.5) is the youngest large structure on the Moon. Superposition of its ejecta, termed the Hevelius Formation (Fig. 4.7), over the Fra Mauro Formation shows that it post-dates the Imbrium basin. The Orientale basin consists of at least four rings, the Montes Cordillera which forms a ring about 900 km across, the outer Montes Rook, the inner Montes Rook and an inner bench (unnamed). The basin is only partly flooded by mare lavas (Fig. 4.8) and hence many of the basin's interior structures are visible and can be analyzed.

Geologic mapping of Orientale (McCauley 1977) divides the pre-mare basin units into several formations (Fig. 4.9). The Maunder Formation lies between the inner and outer Rook Mountains and is a fissured, high-albedo unit interpreted to be impact melt draped over fractured blocks of basin floor (Fig. 4.10). The Montes Rook Formation lies primarily between the outer Rook Mountains and the Cordillera Mountains, although some parts of the formation extend beyond the Cordilleras. It consists of widely spaced 10 km knobs set in a background of undulating terrain and is also considered to be impact melt, but may include non-melted ejecta emplaced late in the cratering process (Scott *et al.* 1978). Most researchers favor the outer Rook Mountains defining the Orientale impact transient cavity. This conclusion is based partly on the observation that Orientale

Figure 4.5 Lunar orbiter image of the Fra Mauro Formation (F) gradational with the Apennines material (AP). Framelet width is 12 km (from Spudis 1982; Lunar Orbiter IV-H109).

(1) Impact

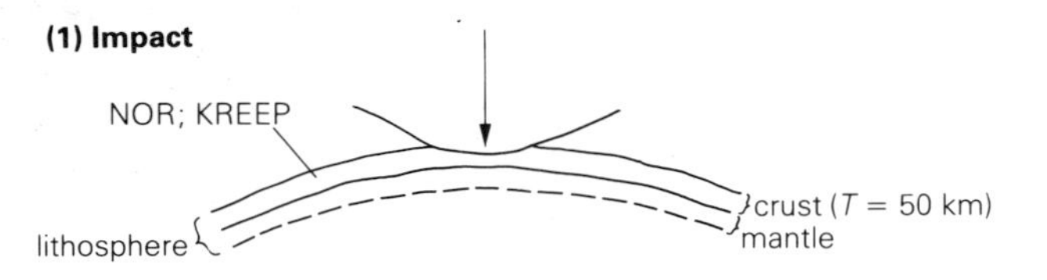

(2) Maximum cavity growth

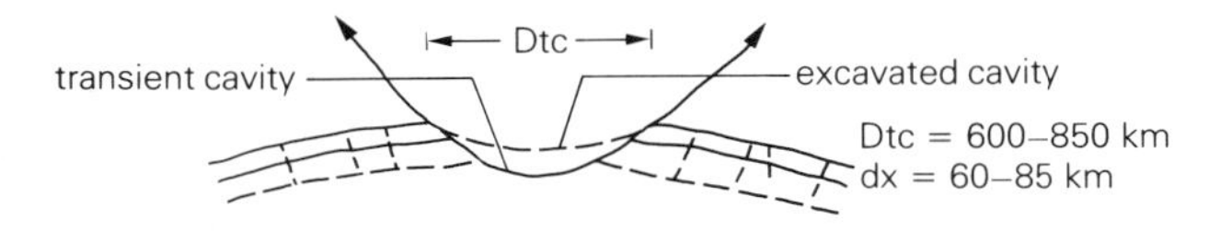

(3) Post-impact modification mare and KREEP flooding at 3.9 ae

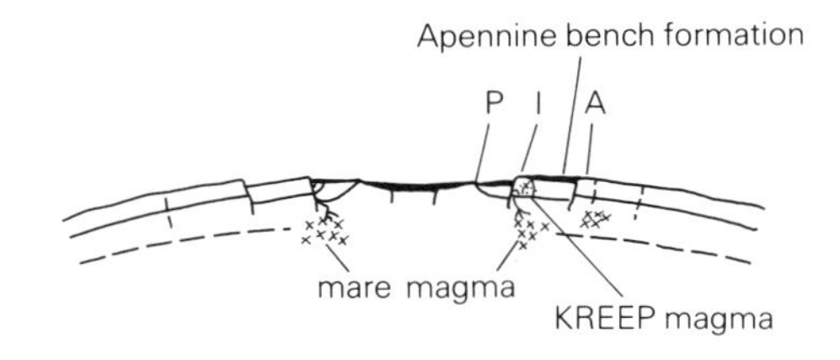

(4) Platform development and mare flooding

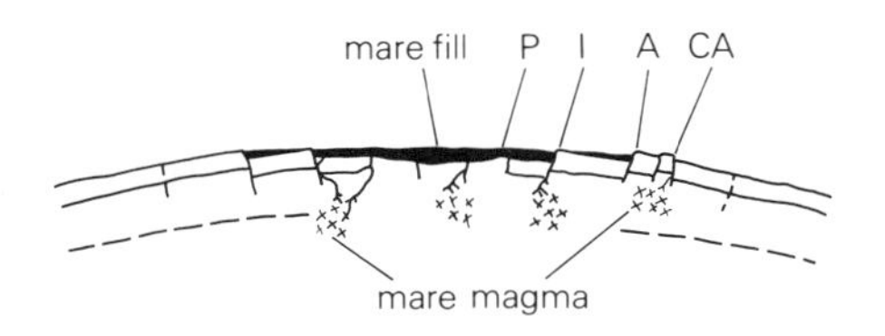

Figure 4.6 Sequential diagram illustrating the formation of the Imbrium basin (from Spudis 1982).

ejecta-related features occur exterior to this scarp. The Cordillera Mountains are taken to represent a "mega-terrace" by this theory. The mega-terrace is thought to have formed by the inward slump of material as the transient cavity grew, then collapsed, resulting in an enormous ring fault. Alternatively, Hodges and Wilhelms (1978) proposed that the Cordilleras represent the transient cavity. Morphologic support for this model is provided by the knobby texture of the Montes Rook Formation, which mostly occurs within the Cordilleran scarp and strongly resembles the knobby texture of the floors of medium to large (~50–100 km diameter) craters. The presence of ejecta-related features on this surface is explained by the assumption that high-angle ejecta falls back onto this location after the floor units are in place. Mountain rings and scarps interior to the transient crater rim, regardless of location, are thought to result from isostatic rebounding and rebound induced by impact energy.

Most lunar geologists have noted that there are significant differences among the morphologic and structural features of lunar basins (CMRB 1980). A detailed study of five basins on the Moon which span a wide range of ages suggests to Spudis (1982) that many of the observed morphologic differences can be attributed to impacts which occurred in an ever-thickening lithosphere with time. Drawing upon models of thermal evolution and estimates of lithospheric thickness based on geophysical and geochemical considerations, Spudis shows how the differences in the morphology among various lunar basins can be accounted for, as depicted in Figure 4.11.

4.2.3 Highland plains

A widespread unit found in many highland areas consists of a high-albedo, relatively smooth, flat plains unit (Fig. 4.12). First mapped by Dick Eggleton of the US Geological Survey as the Cayley Formation, this unit was considered to be ejecta from one of the late-stage impact basins. The Cayley plains were later reinterpreted by some investigators as representing eruptions of volcanic ash or flows, possibly of silicic compositions. Because the unit is widespread over much of the

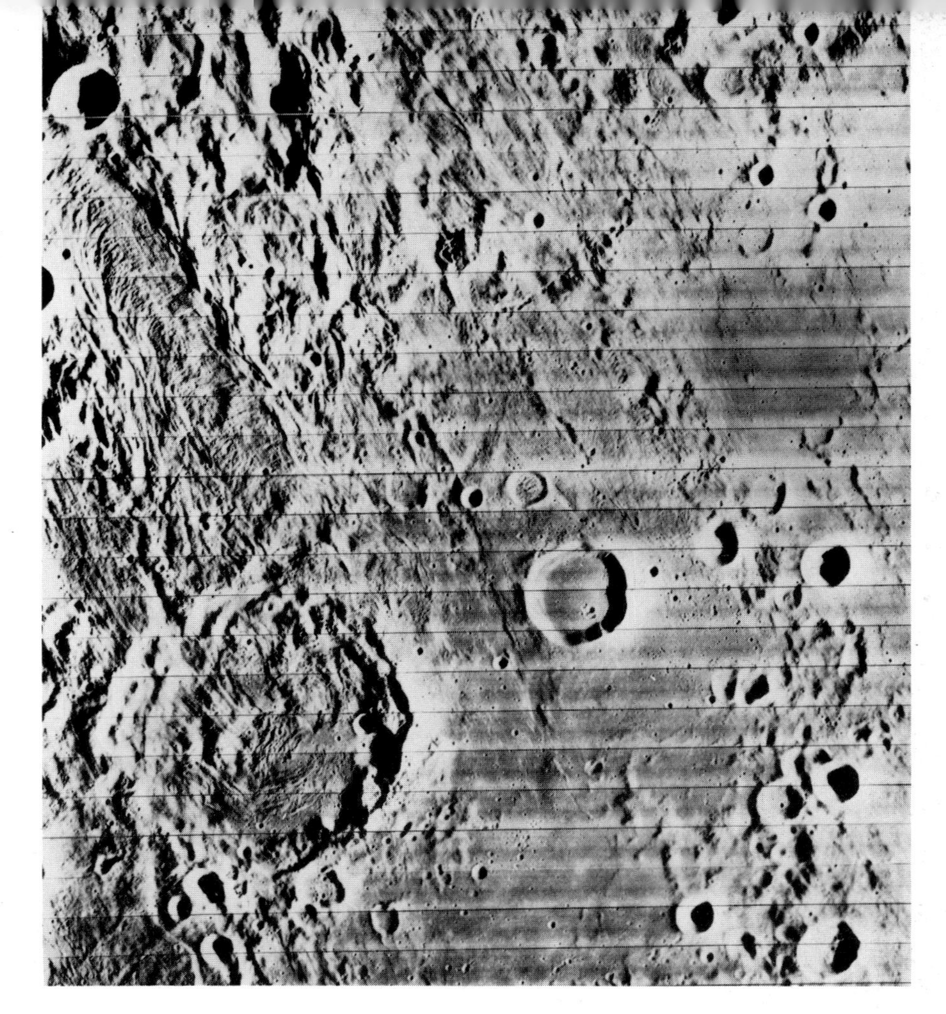

Figure 4.7 Orientale ejecta comprising the Hevelius Formation is shown here overriding craters and leaving coarsely festooned deposits. The floor of the crater in the lower left is buried by "deceleration dunes" where the ejecta appears to have encountered a topographic obstacle (the far crater wall) and "stalled". This area is 1100 km southeast of the center of the Orientale basin and ejecta movement was from the upper left to lower right. Inghirami, the crater in the lower left, is 91 km in diameter (Lunar Orbiter IV-H172).

Moon and because of its possible volcanic origin, it was given high priority as an Apollo landing site.

Samples of the Cayley Formation returned by the Apollo 16 astronauts were found to consist predominantly of impact-generated breccias, confirming its impact origin at that site. The unit is now widely regarded as ejecta deposits emplaced either as material thrown out during major basin-forming events or as local material ejected by the secondary craters of basin-forming impacts. The smooth texture of the Cayley and other Imbrium-age light plains units is attributed to the liquid-like behavior of the unconsolidated ejecta deposits when subjected to seismic shaking or acoustic fluidization.

More recent studies, however, have generated new interest in the possibility of volcanism during the early history of the Moon. Ryder and Taylor (1976) and Ryder and Spudis (1980) have synthesized data based on lunar samples and concluded mare volcanism was active prior to the formation of the Imbrium basin, possibly earlier than 4.2 aeons. Studies by Schultz and Spudis (1979) of the distribution of dark halo craters >1 km in the highlands (primarily around Mare Humorum and in the eastern hemisphere) also suggest early-stage volcanism. They propose that these impact craters excavated early-stage mare-type lavas that are presently buried by ejecta deposits from large craters and basins.

4.2.4 Maria

The maria have generally received much more attention than the highlands, despite the fact that mare areas constitute much less of the lunar surface than the highlands. It is important to note that **mare** and **basin** are not the same (see Stuart-Alexander & Howard 1970 for a review); while it is true that most maria occupy the low-lying terrain of basins, the dark maria are lava flows that were erupted generally long after the formation of most of the impact basins. Mare materials also flood some irregular depressions, such as Tranquillitatis and Procellarum, and occur in isolated patches within the highlands and on some crater floors.

Head (1976) estimates that mare lavas cover about 6.3×10^6 km^2, or about 17 percent of the lunar surface. Although thicknesses of mare deposits are difficult to determine, estimates by DeHon (1979) and Hörz (1978) based on the degree of flooding of impact craters indicate that mare lavas may be 4 km thick in some areas.

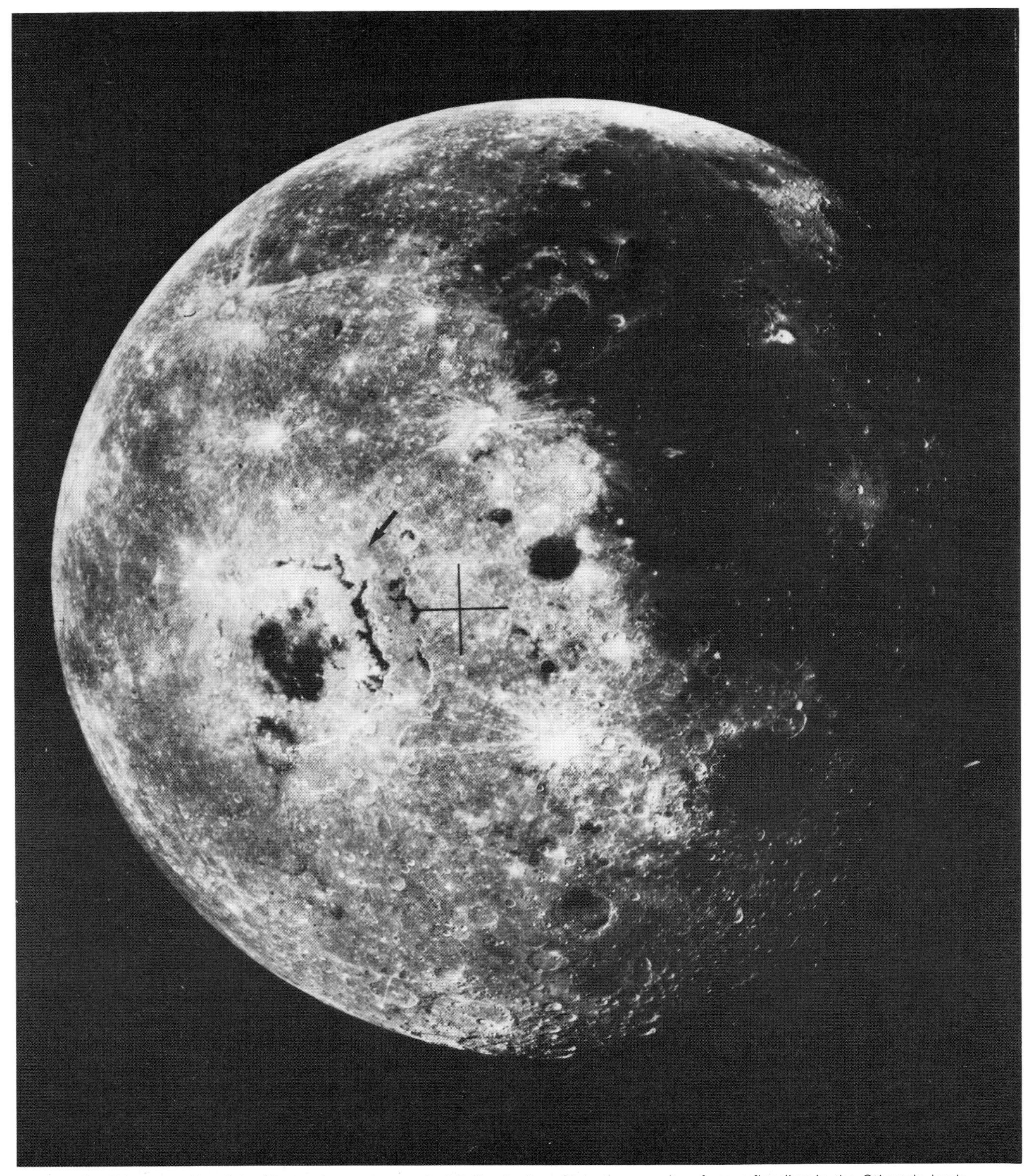

Figure 4.8 The west side of the Moon seen under nearly full illumination. Note the sparsity of mare flooding in the Orientale basin (arrow). The Rook and Cordillera scarps appear as bright rings under the high sun (USSR Zond 8 photo 12-306, 2525 H).

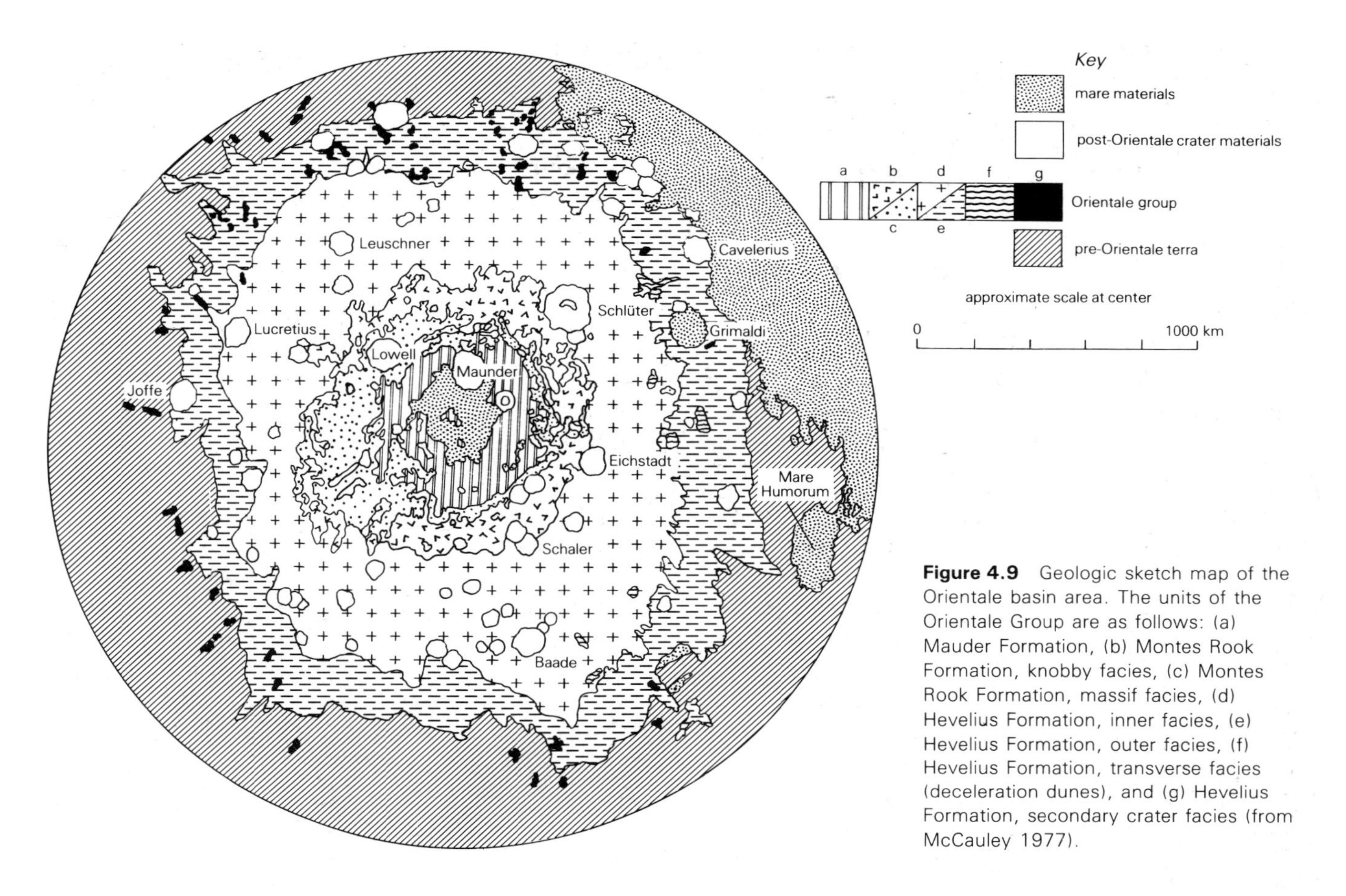

Figure 4.9 Geologic sketch map of the Orientale basin area. The units of the Orientale Group are as follows: (a) Mauder Formation, (b) Montes Rook Formation, knobby facies, (c) Montes Rook Formation, massif facies, (d) Hevelius Formation, inner facies, (e) Hevelius Formation, outer facies, (f) Hevelius Formation, transverse facies (deceleration dunes), and (g) Hevelius Formation, secondary crater facies (from McCauley 1977).

Combining these estimates with their total surface area shows that mare units comprise less than 1 percent of the total volume of the lunar crust.

Analysis of mare samples returned from the Apollo and Luna missions shows that lunar lavas are very similar to terrestrial basalts but lack hydrous minerals and have slightly higher abundances of iron, magnesium and titanium. The lack of hydrous minerals is an indication that the Moon has never experienced liquid water on the surface, nor has had an appreciable atmosphere. Analysis of the samples also indicates that most of the lunar lavas were derived from deep within the mantle, probably at depths of 150 to as much as 450 km.

Crystallization dates for the cooling lunar lavas show that most of the mare lavas were erupted during the interval of 3.9 to 3.1 aeons. Thus, for a period of more than 800 million years, the Moon experienced extensive volcanism. While this interval of time is less than one-eighth of the total lunar history, it nonetheless represents a span equivalent to the Phanerozoic Aeon on Earth. Furthermore, analysis by Schultz and Spudis (1983) indicates that both early-stage (>4.0 aeons) and very late (~1 aeon) volcanism probably occurred, extending the range of lunar volcanic activity over a 3×10^9 year period.

Laboratory studies of simulated lunar lavas provide insight into their physical properties. For example, estimates of the viscosities yield values of about 10 poise, or the equivalent to motor oil at room temperature, while studies of their thermal conductivities show that heat loss would have been minimal. Thus, the lunar lavas were extremely fluid and were able to flow long distances on relatively gentle gradients.

Photogeologic mapping has been combined with terrestrial analog studies and compositional distributions based on remote sensing to derive general models for the emplacement of the mare lavas. At least three eruptive phases can be recognized. Early-stage eruptions involved lavas high in titanium recognized by their comparatively blue spectra, which erupted at high rates to form thick, massive cooling units which flooded low-lying areas such as basin centers. These units formed vast lava lakes, seen today as flat plains which generally lack surface features other than mare ridges (described below). The early-stage lavas probably required hundreds or possibly thousands of years to cool and solidify (BVSP 1981).

The first phase was followed by the eruptions of lavas having a lower titanium content, as indicated by their relatively redder spectral characteristics. These lavas

Figure 4.10 Lunar orbiter photograph of the inner part of the Orientale basin; smooth plains (S) and fissured deposits (F) comprise the Maunder Formation considered to be impact melt, which grades into the knobby Montes Rook Formation (R). Mare lavas are visible at the top of the photograph; framelet width is 12 km (Lunar Orbiter IV-H195).

Figure 4.11 Series of diagrams illustrating the formation of five major basins on the Moon. In this model Spudis considers the differences in morphology to result from an ever-thickening lithosphere with time; thus, the formation of Crisium penetrated a relatively thin lithosphere with rapid isostatic adjustment. The sequence progresses through to the formation of the Orientale basin which impacted a relatively thick lithosphere and the attendant adjustment was followed by little extrusion of lavas in the interior (from Spudis 1982).

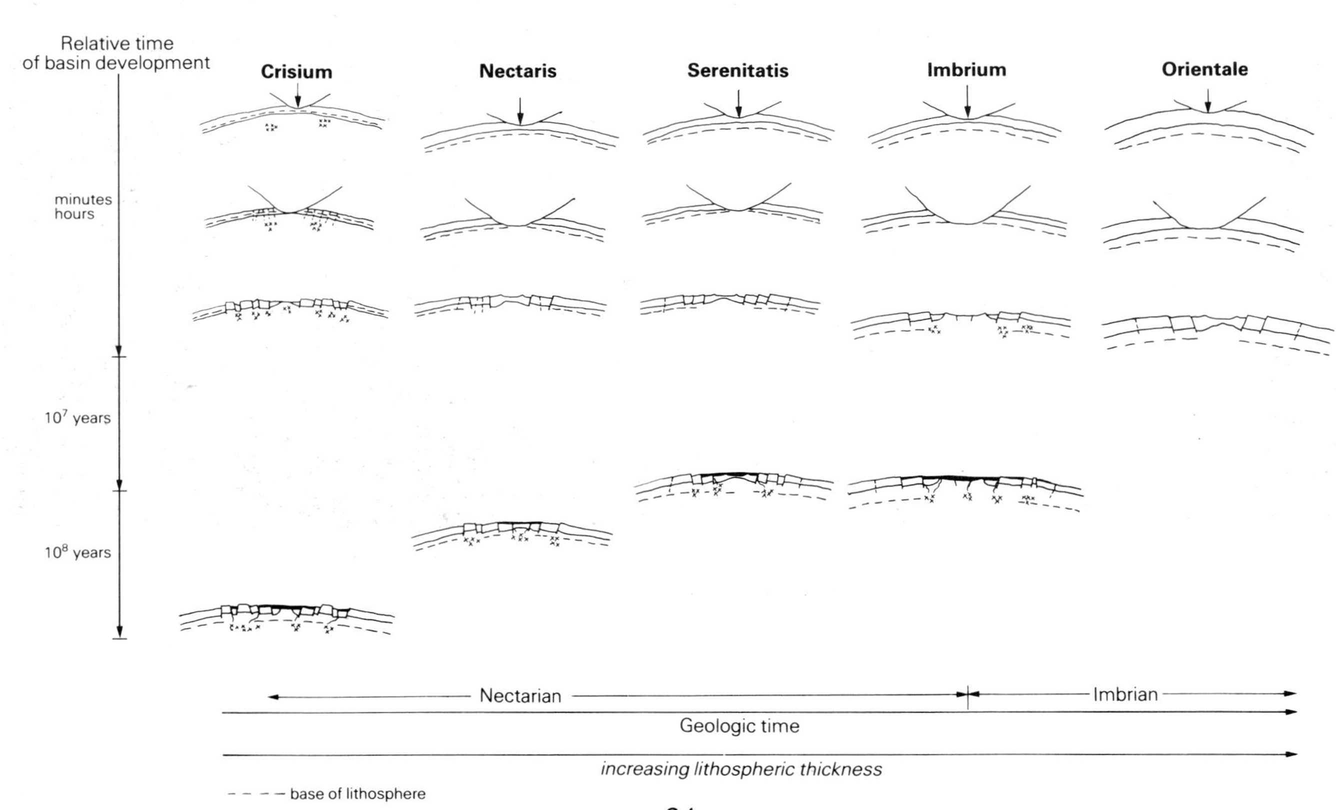

Figure 4.12 View of the Apollo 16 landing site showing the Cayley plains and the surrounding cratered terrain. The fresh crater (Dolland) in the lower left is about 10 km in diameter (AS16-M-439; courtesy C. Hodges).

Figure 4.13 Mosaic of Hasselblad photographs showing the southwest part of Mare Imbrium just north of the crater Euler showing multiple flow lobes and flow fronts. Some of the flows in this area can be traced for more than 1000 km and demonstrate that some mare units have been emplaced through repeated outpourings of lavas. Illumination from the right (AS17-155-23714 to 23716).

appear to have been emplaced at lower rates of effusion than the first phase, suggested by formation of various lava channels and lava tubes seen today in the form of lunar sinuous rilles (Fig. 3.7). The final stage involved lavas high in titanium, like the first phase, but erupted in lower volumes. Detailed mapping has shown that local sequences are much more complicated than this generalization.

The sources or vent areas for most lunar lavas are difficult to identify. In many cases, particularly the early-stage high-titanium lavas, the vents are presumed to have been fissures which the flood eruptions and later-stage lavas have buried. Mapping by Schaber (1973) of the well known Imbrium lavas (Fig. 4.13) shows that they originated near the zone defined by the outer basin ring and then flowed toward the center of the basin. Some of these flows exceed 1200 km in length and have flow fronts that are 10–65 m high. Thus, the flows formed rather thick units, despite their apparent fluidity.

Visible flow fronts and margins, however, are rare on the Moon. Either most lavas were too fluid to preserve flow fronts large enough to be visible, or they have been eroded by numerous small impacts. The margins of mare lavas in several areas, however, show distinctive benches which are interpreted to represent high stands of lava (Fig. 4.14).

Several areas on the Moon appear to be volcanic complexes which served as source areas for some of the mare lavas (Guest 1971, Guest & Murray 1976). The Aristarchus Plateau (Fig. 4.15), the Marius Hills (Fig. 4.16) and the Rümker Hills (Fig. 4.17) all appear to be areas that have experienced repeated volcanic activity.

4.2.5 Sinuous rilles

Lunar *sinuous rilles* occur in many areas on the Moon, although most are found around the outer margins of the mare-filled basins (Fig. 3.7). The origin of these features has been controversial, but most investigators now agree that they are former lava channels and/or collapsed lava tubes. As such, lunar sinuous rilles can be used to map the general source areas for some of the mare lavas and to give an indication of the flow directions that were involved in the emplacement of the lavas. Although lunar sinuous rilles are typically an order of magnitude longer and wider than their presumed counterparts on Earth, models based on estimated differences in the rate of eruption by Wilson and Head (1981) and considerations of differences in gravity show that some of the differences in size (compared to Earth) can be accounted for by the lunar environment.

The Apollo 15 astronauts visited the Hadley Rille, a prominent lunar sinuous rille that was one of the sources for the emplacement of lavas into the eastern

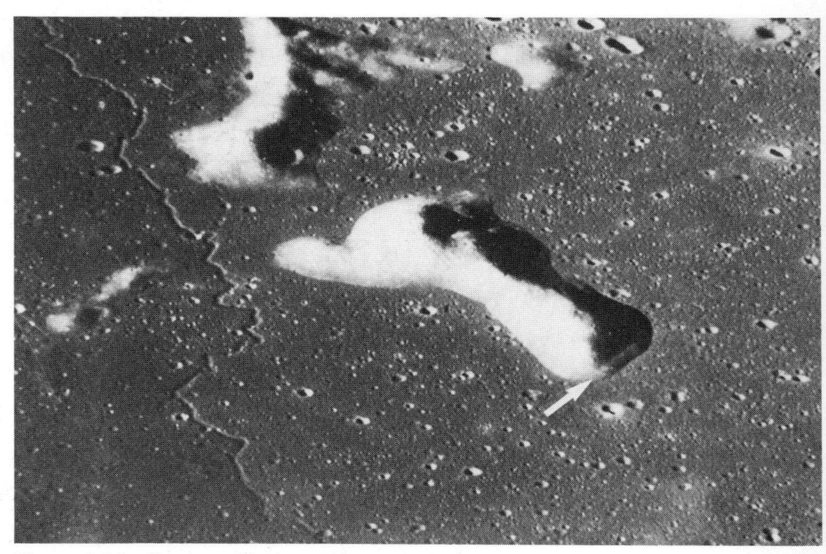

Figure 4.14 Topographic bench (arrow) along mare-highland contact interpreted as a high lava mark or terrace. Lavas may have been locally drained in this region by the sinuous rille, lowering the surface. Oblique view southward; near field width approximately 30 km, illumination from the left (AS16-19140; from Greeley & Spudis 1978a).

part of the Imbrium basin (Greeley 1971, Howard *et al.* 1972; Fig. 4.18). This rille stretches more than 130 km and originates in a cleft-shaped depression about 10 km long by 5 km wide at the base of the Apennine Mountains. It was a vent fed by magma erupted to the surface apparently in association with basin-related faults and fractures (Fig. 4.6).

In addition to lunar sinuous rilles, many other surface features are recognized on maria. Prominent among these are various so-called wrinkle or mare ridges. Mare ridges may extend tens of kilometers across the mare surfaces and usually are composed of a broad, basal arch, surmounted by a sharply crenulated, sinuous element (Fig. 4.19). Furthermore, Lucchitta (1976) has demonstrated that some mare ridges are directly related to highland scarps and are probably the result of vertical faulting, as shown in Figure 4.20.

The origin of mare ridges is enigmatic. On the one hand, there is evidence to show that they post-date the formation of the maria in which they occur – many of them intersect impact craters (Fig. 4.21), which would indicate that the mare crust had solidified sufficiently to preserve the crater. On the other hand, some segments show flow features that suggest lava was extruded from beneath the crust at the time the ridge was formed (Fig. 4.22), thus suggesting that the ridge formed relatively contemporaneously with the emplacement of the mare lavas. Figure 4.23 shows one possible explanation that accounts for both the preservation of craters and the extrusion of material from beneath the mare crust. Other ridges reflect underlying topography, such as buried crater rims (Fig. 4.24) and basin rings.

Figure 4.15 Oblique view taken by the Apollo 15 astronauts of the Aristarchus plateau and Schröter's Valley. The 40 km fresh impact crater in the upper left is Aristarchus. Schröter's Valley is more than 250 km long and exceeds 10 km in width. This rille originates in the "Cobra head" (arrow) and is considered to be a tectonic feature controlled by faulting combined with a volcanic flow channel which served as a vent for lavas emptying into Oceanus Procellarum to the upper right. The Aristarchus plateau is one of the major volcanic centres on the Moon. Illumination from the left (Apollo 15 AS15-M3-2611).

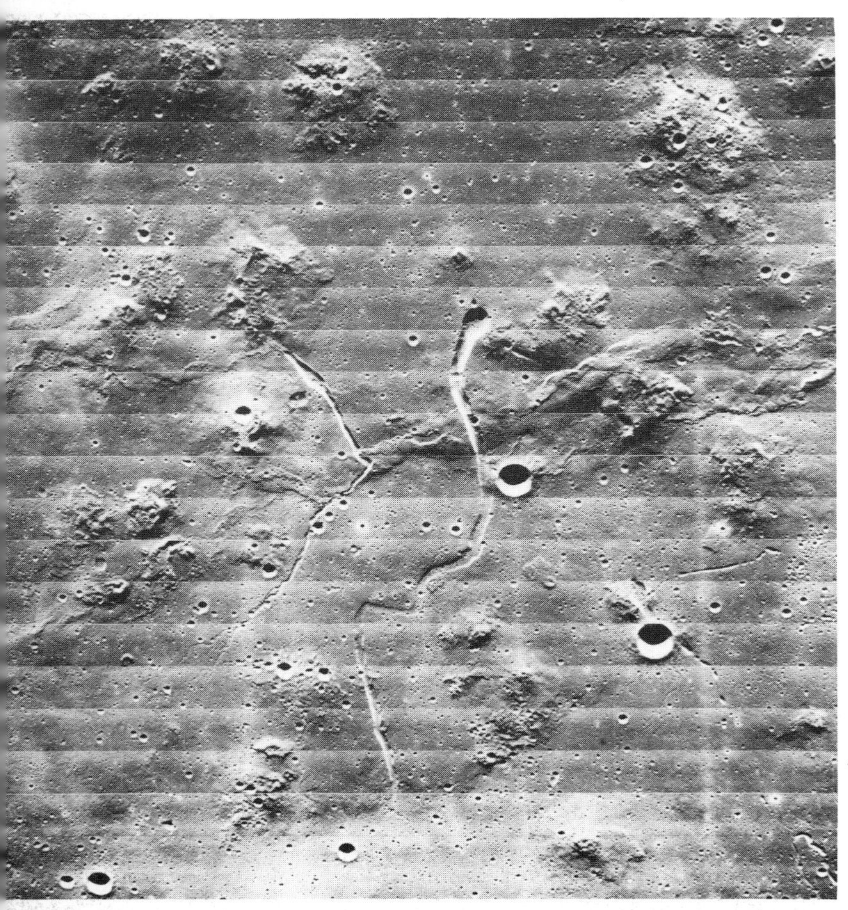

Figure 4.16 Lunar orbiter view of the Marius Hills, a volcanic center located in Oceanus Procellarum. Numerous domes, cones, and sinuous rilles occur in this area and are considered to be volcanic in origin; area shown is about 75 by 85 km (Lunar Orbiter V-M214).

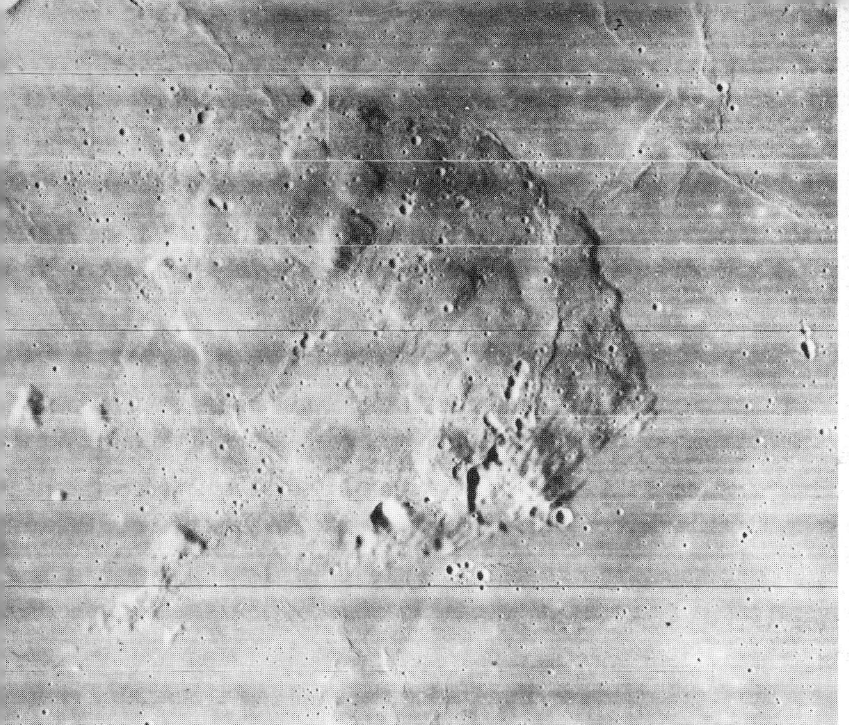

Figure 4.17 Lunar orbiter view of the Rümker Hills, a volcanic platform 70 km across in northwest Oceanus Procellarum; numerous domes, some of which exceed 10 km across, are considered to be volcanic (Lunar Orbiter IV-H170).

Figure 4.18 Apollo metric camera view of the Hadley Rille region, area of the Apollo 15 landing site (A). This site, situated in the Montes Apennines (mountainous areas on the right half of the photograph, H) was selected to sample rim material of the Imbrium basin and to examine the mare lavas associated with the rille (AS15-414). ▷

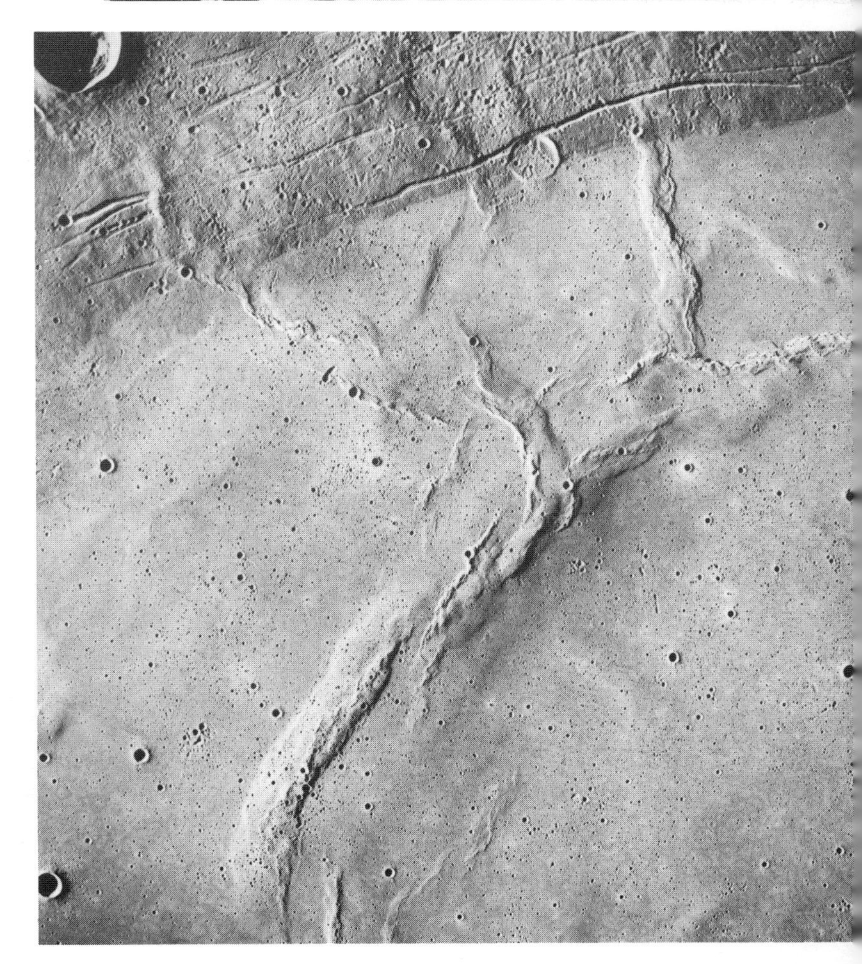

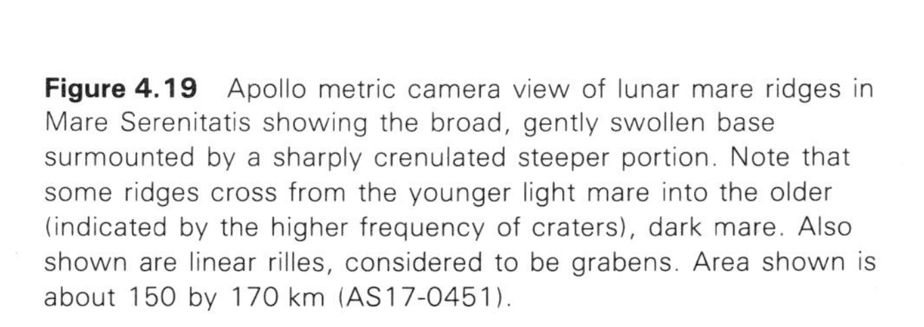

Figure 4.19 Apollo metric camera view of lunar mare ridges in Mare Serenitatis showing the broad, gently swollen base surmounted by a sharply crenulated steeper portion. Note that some ridges cross from the younger light mare into the older (indicated by the higher frequency of craters), dark mare. Also shown are linear rilles, considered to be grabens. Area shown is about 150 by 170 km (AS17-0451).

Figure 4.20 Part of panoramic camera photograph of the Apollo 17 landing site (arrow) and a bright deposit overlying the dark mare plain. This deposit covers more than 21 km^2 and originated from the massif to the south (bottom of picture), and may be either an avalanche, or secondary crater and bright ray material. Shown also is the Lee-Lincoln scarp which cuts across the mare valley and continues as a fault onto the highland block to the north. Area shown is 24 by 22 km, illumination from the right, north to the top (AS17 M-1220).

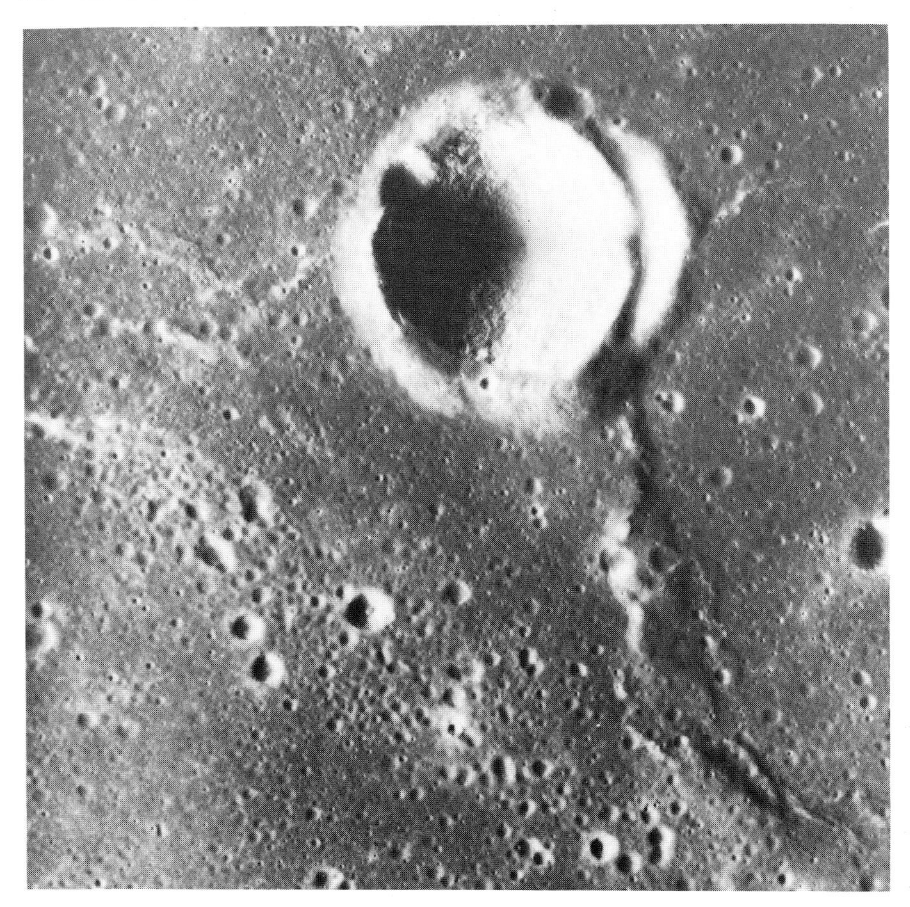

Figure 4.21 Impact crater (6 km in diameter) in Mare Cognitum that has been cut by a fault-ridge segment (from Lucchitta 1976; Apollo 16 panoramic frame AS16-5429).

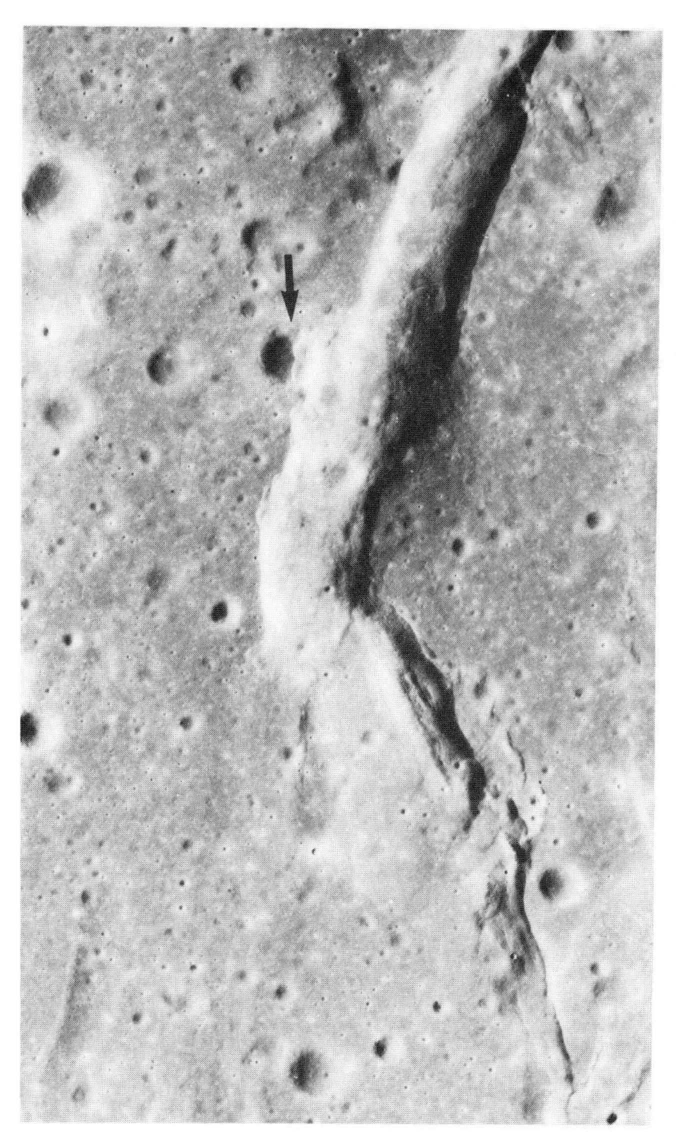

Figure 4.22 Apparent extrusion of lava from mare ridge superimposed on the crater wall (arrow) in eastern Mare Serenitatis, demonstrating the extrusive origin for parts of some mare ridges. Illumination from the left (from Hodges 1973; Apollo 15 panoramic frame AS15-9303).

Figure 4.24 "Ghost" crater (about 20 km in diameter) near the Herigonius rille complex in southeastern Mare Cognitum showing part of the rim buried by mare lavas and exposed as mare ridges (from Young *et al.* 1973; AS16-M2834).

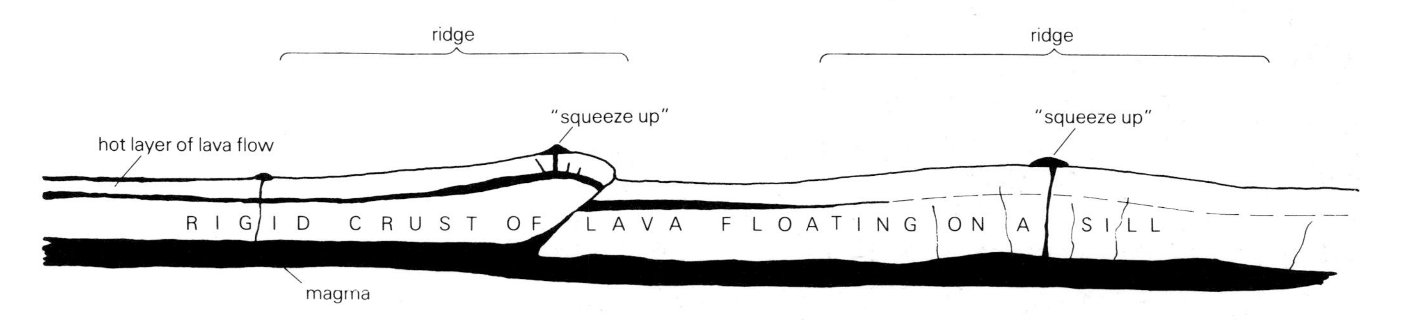

Figure 4.23 Diagram illustrating a hypothetical cross section through mare ridges of two different types. On the right is a ridge generated primarily by tectonic process; on the left is an asymmetrical ridge resulting primarily from igneous processes (from Guest & Greeley 1977).

Figure 4.25a Oblique view of the crater Alphonsus showing prominent dark-haloed craters (arrows) considered to be pyroclastic vents (some of which are associated with fractures) in the floor of this ~ 125 km impact crater. Device on the left side of photograph is the boom for the gamma-ray spectrometer carried on the spacecraft (AS16-M2478).

Figure 4.25b Lunar orbiter view of small structures in Oceanus Procellarum transecting the crater Hortensius D; these features are a few hundred meters across and are considered to be a row of spatter cones (Schultz 1976c; Lunar Orbiter IV-H133).

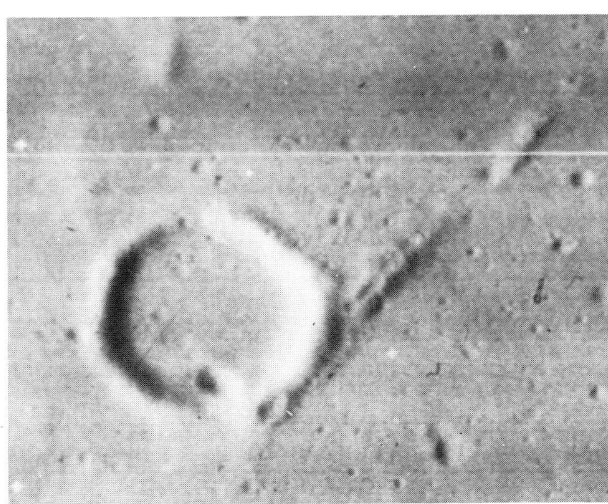

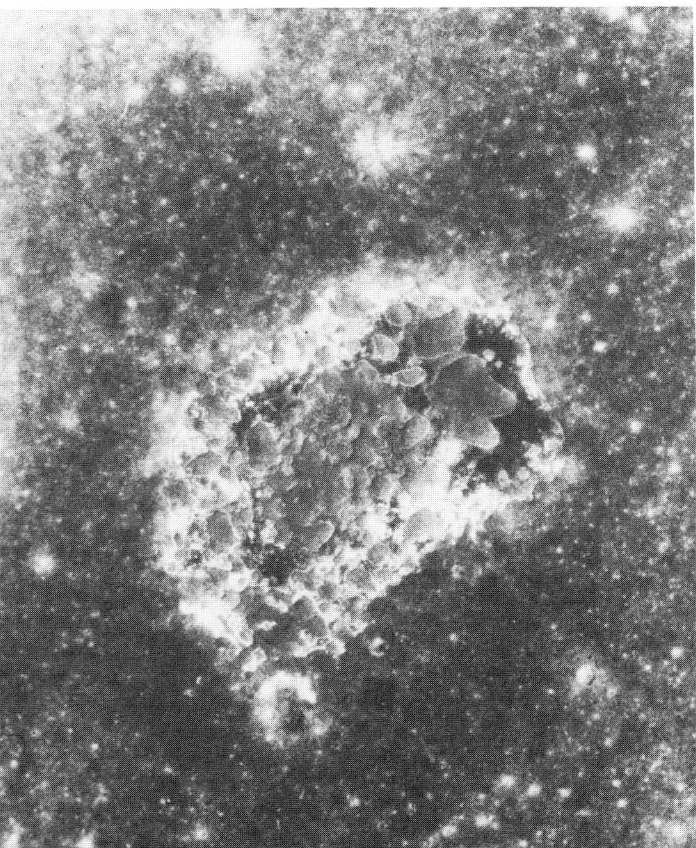

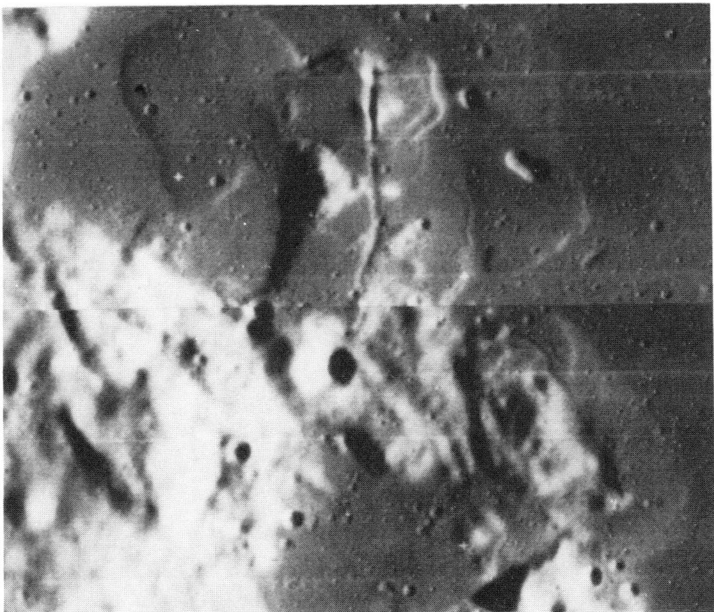

Figure 4.25d Small shield volcanoes about 10 km across in the Lacus Veris mare region of the Orientale basin. Illumination from the right (from Greeley 1976; Lunar Orbiter IV-H181).

Figure 4.25c Apollo 15 view of the so-called "Caldera D" structure (named Ina); this feature is unique on the lunar surface and its origin is enigmatic, although most investigators attribute its origin to volcanic processes (Whitaker 1972, Strain & El-Baz 1980; Apollo 15 panoramic frame AS15-10181).

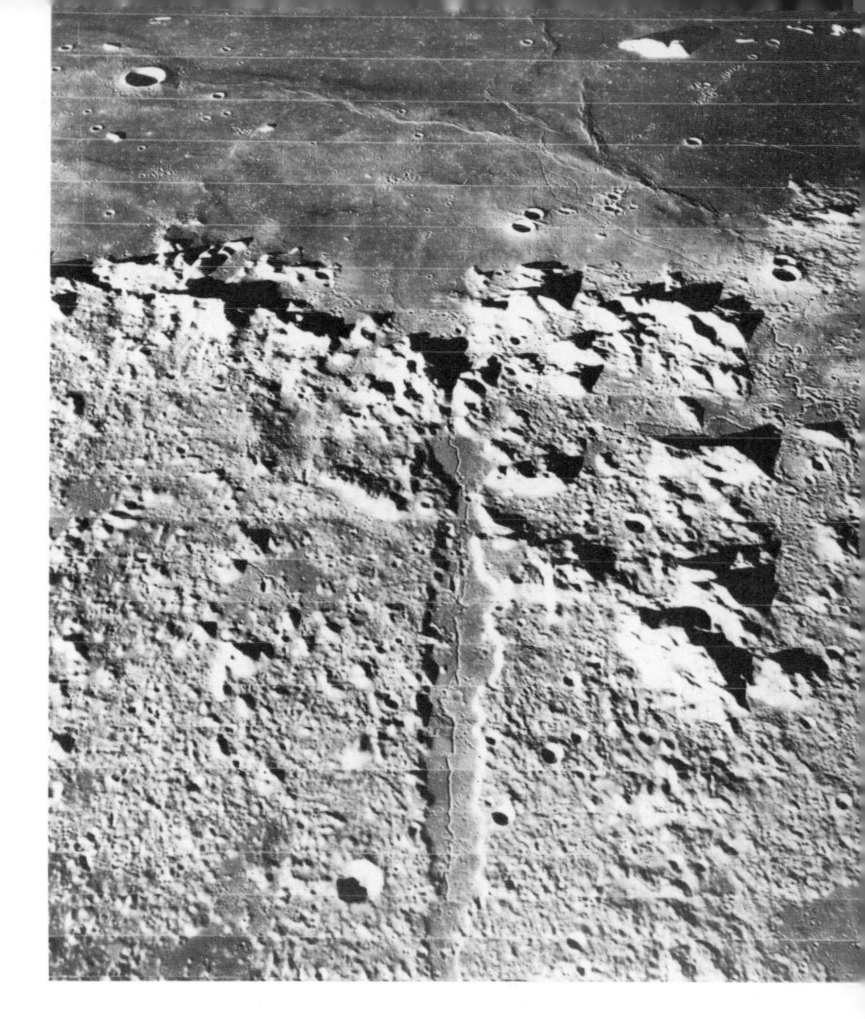

Figure 4.26 Oblique Lunar orbiter view of the Alpine Valley. This 150 km long by 10 km wide graben is radial to the Imbrium basin and is considered to be a tectonic feature associated with the formation of the basin. The floor of the graben has been flooded with mare lavas which have been emplaced through the rille seen in the center of the valley. View toward the south (Lunar Orbiter IV-M102).

4.2.6 Mare domes and other small features

Numerous domes (Fig. 4.17) have been identified on the Moon. They range in size from <1 to >20 km across. Head and Gifford (1980) have mapped more than 200 mare domes and recognize two types. The first type consists of low, relatively flat-topped circular structures which often have summit craters. They consider this type to be volcanic domes formed as a result of relatively low rates of eruption. Morphologic analyses by Schultz (1976c) show that many domes are furrowed and display features that could represent flow of viscous lavas. The second type of mare dome is considered to be high-standing "islands" of terrain remaining unflooded by younger mare deposits.

In addition to dome volcanics, other possible volcanic vents have been identified. As shown in Figure 4.25, these include dark halo craters which may be pyroclastic features, small shield volcanoes and possible spatter cones.

4.3 Tectonic features

Most tectonic features on the Moon result from crustal deformation associated with impact basins, or with adjustments inferred to be associated with the cooling and settling of mare lavas. The Alpine Valley is a graben 10 km wide which cuts across the Alpes mountains for 150 km (Fig. 4.26).

Numerous **linear rilles** criss-crossing the surface of the Moon attest to the tectonic deformation of the lunar crust. Unlike sinuous rilles which are generally controlled by topography, these rilles have the general form of grabens (Golombek 1979) and appear to be relatively deep-seated structures which cross both highlands and mare units (Fig. 4.27). Sequence and timing of the deformation can often be demonstrated from superposition and cross-cutting relations, as shown in Figure 4.28, where a rille is seen to cut part of the Montes Apennines (part of the rim structure for the Imbrium basin) and has in turn been flooded by mare units. Thus, it is clear in this region that the crustal faulting occurred after the formation of the basin but prior to the emplacement of the surface lava.

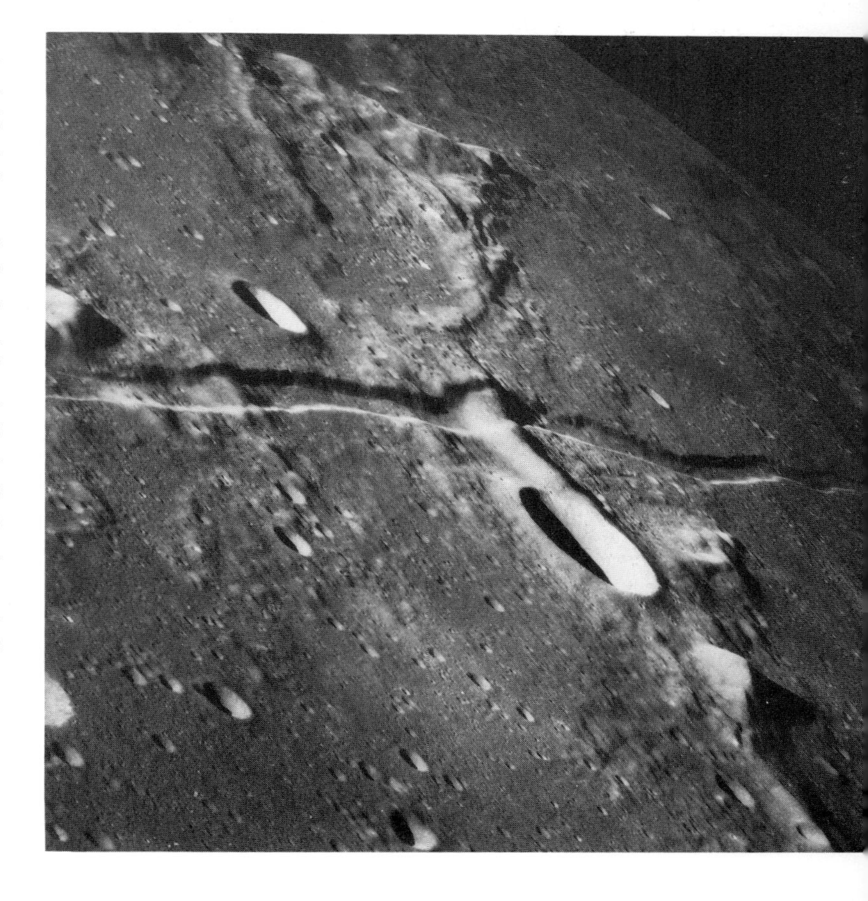

Figure 4.27 Apollo 10 oblique view of Rima Ariadaeus near the center of the lunar nearside. This linear rille is about 2 km wide and cuts across both mare and highland topography indicating that it is a deep-seated tectonic feature, probably a graben (Apollo 10 frame 10-4646).

Figure 4.28 Apollo 15 view of linear rilles (grabens) southeast of Hadley Rille (the sinuous rille in the upper right); illumination from the right (Apollo 15 frame M-1820).

Figure 4.29

Figure 4.29 Floor-fractured craters bordering western Oceanus Procellarum; largest crater is about 75 km across (Lunar Orbiter IV-H189).

Figure 4.30 Diagrams showing the development of floor fractured craters. The dashed line in the top figure shows the initial excavation cavity before adjustment of the floor and wall collapse immediately following impact (from Schultz 1976a).

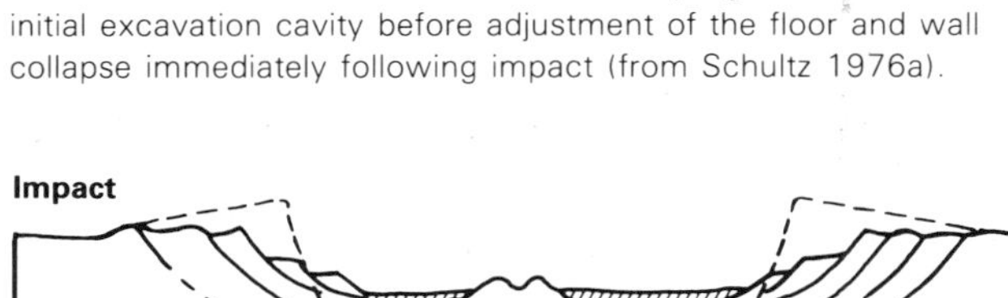

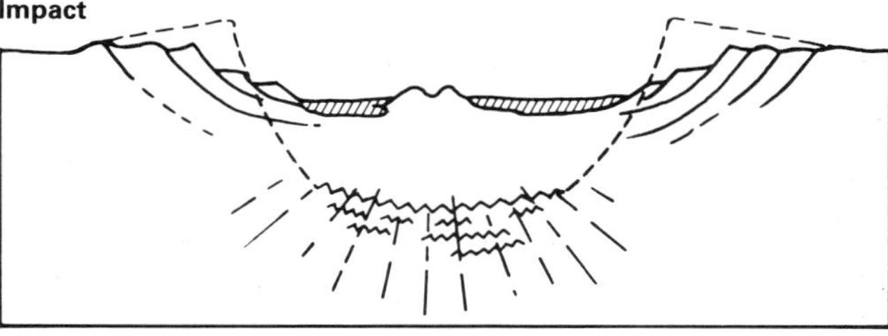

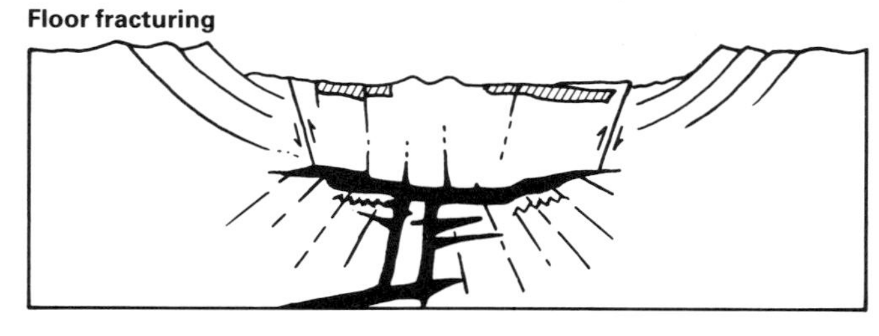

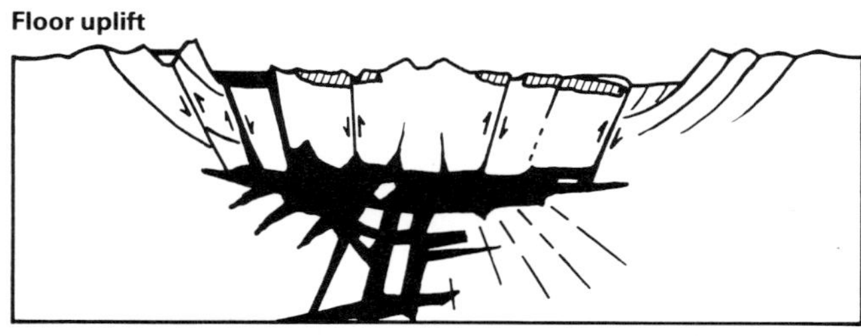

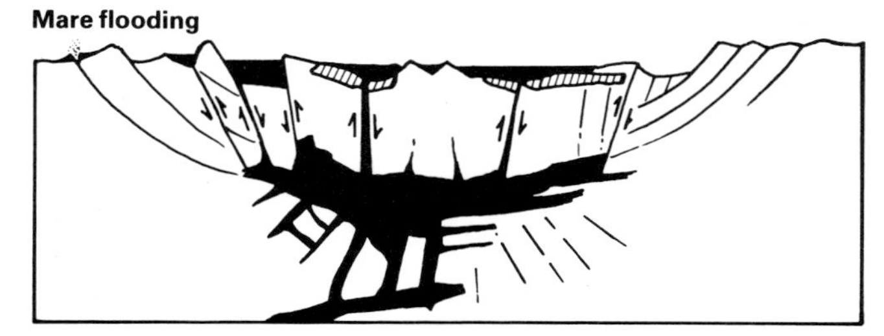

Figure 4.31 The crater Tycho (85 km in diameter) is an excellent example of a large, fresh, lunar crater. Typical of this crater class is a high central peak, slumped walls and hummocky rim deposits. The floor is extremely rough, with mounds and fissures. The rim displays prominent flow features and a strong concentric texture in many places (Lunar Orbiter V-M125).

Study of the global patterns of lunar grabens and their timing of formation as related to the emplacement of lavas (Lucchitta & Watkins 1978) shows: (a) most grabens formed after the emplacement of the early-stage mare lavas and considerably after basin formation; (b) the grabens formed on faults that were reactivated along older structures; (c) large graben systems ceased activity at about 3.6 ± 0.2 aeons; and (d) the grabens reflect a tensional stress field generated in the early history of the Moon.

The floors of many larger craters and small basins show evidence of tectonic deformation (Fig. 4.29). Termed by Schultz (1976a) **floor-fractured craters,** the deformation may have occurred as a consequence of magmatic intrusion and the emplacement of dikes, shown in Figure 4.30. Hall *et al.* (1981) argue on geophysical grounds that at least some floor-fractured craters could result from viscous relaxation of the crater topography.

4.4 Craters

Craters on the Moon can be classified as: (a) *impact basins* (Fig. 2.5), described above, (b) *large impact craters*, those larger than about 15 km across (Fig. 4.31), (c) *small impact craters* (Fig. 4.32), which range in size from micrometers to those 15 km in diameter, and (d) *non-impact craters*. The transition from large craters to basins on the Moon is generally considered to occur in the size range of 140–220 km.

Studies by Hale and Grieve (1982) show that the morphologic and volumetric characteristics of central peaks change in the range of crater diameters of 51–80 km and grade into peak rings. Peak ring impact structures grade, as their diameter increases, into ringed and multi-ringed basins – suggesting to some researchers that craters and basins are not fundamentally different and that basins should not be treated as a special category but simply as very large craters. This concept is consistent with various theories of the origins of central peaks, peak rings and interior multi-rings.

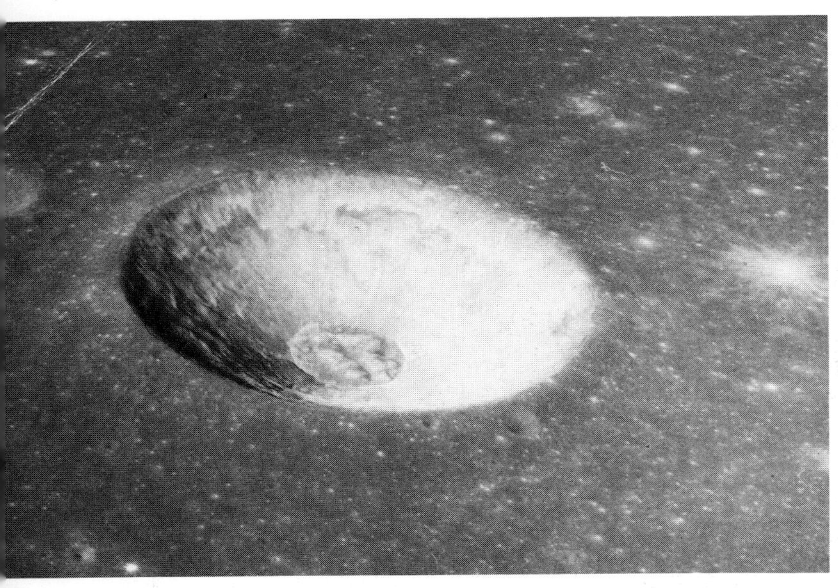

Figure 4.32 View of a typical small (8 km in diameter) bowl-shaped crater on the Moon (AS10-29-4253).

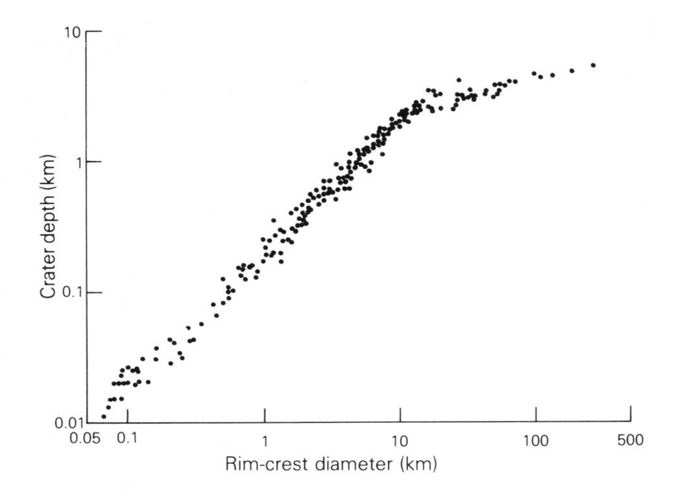

Figure 4.33 Depth to diameter ratios for 204 fresh lunar craters, as compiled from Apollo 15, 16 and 17 data. Distribution inflection at a diameter of about 15 km indicates the transition from small, simple craters to large, complex craters (from Pike 1974; *Geophys. Res. Let.*, copyright American Geophysical Union).

Figure 4.34 Oblique view of Necho crater on the lunar farside showing extensive wall slumping (AS10-28-4012).

The distinction between fresh, large craters and small craters can be seen in their cross-sectional geometry. For example, Pike (1980) and other workers have demonstrated a distinct break in the depth-to-diameter ratio, as shown in Figure 4.33. Generally, large craters are proportionally more shallow than small craters. The difference in the depth to diameter ratios between small and large craters (up to ~100 km) is explained by slumping and infill of large crater floors. In addition, large craters tend to have central peaks in a variety of arrangements (e.g. ridges and rings) and complex wall structures, including terraces formed by slumping (Fig. 4.34). In many cases the floors of large craters appear to have been modified by various post-impact processes which may include volcanic eruptions that could form domes and smooth floor deposits, as seen on the floor of Copernicus (Fig. 4.35). Crater floors are frequently covered with blocks, fissures and mounds. Long fissures commonly parallel the floor–wall contact, while shorter, more irregular ones have variable directions. Some mounds are surrounded by symmetrical rings or shells which most researchers think may be due to the flowing of shock-fluidized material off the surface of the mounds.

The zones outside the rims of large craters often show extensive modifications, including scouring and erosion by ejecta and ejecta-related melt deposits, some of which flowed down the flanks of the crater to form smooth "pools" where it ponded (Figs 4.36 & 4.37). An apronlike expanse of striated material beyond the northwest rim of the prominent mare-filled farside crater Tsiolkovsky is interpreted as a giant landslide (Fig. 4.38), similar to the Sherman landslide in Alaska (Fig. 3.31d) which was triggered by an earthquake. In contrast, small craters tend to be simple, bowl-shaped depressions, although modifications of the wall by slumping are also visible in some cases.

Figure 4.35 Lunar orbiter view of the floor of Copernicus showing numerous domical structures (some of which have summit craters), fissures, and smooth floor deposits, all possibly indicative of magmatic activity; the largest domical feature is ~ 1 km across (Lunar Orbiter V-H154).

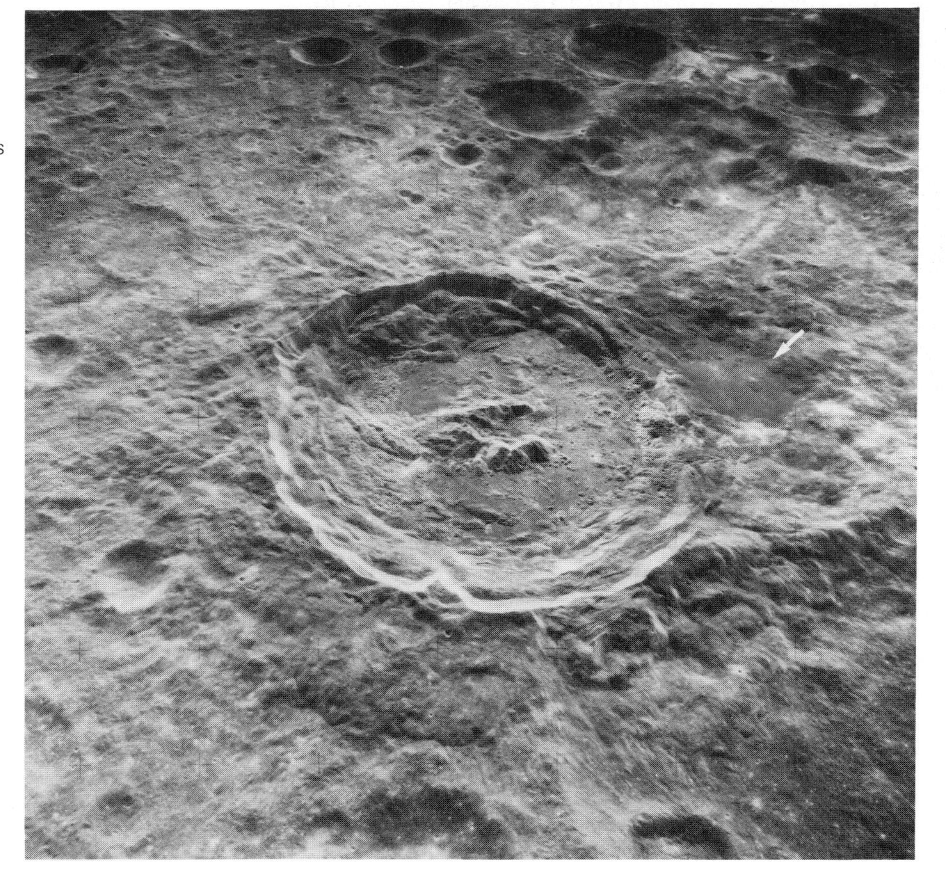

Figure 4.36 Oblique view of King crater (~ 75 km) on the lunar farside showing smooth deposits (arrow) considered to be melted ejecta associated with the impact that formed the crater. Similar flow deposits are observed around many craters of this size (AS16-M0488).

Figure 4.37 A closeup of the "pools" of material observed beyond the north rim of crater King in Figure 4.36. Cracks and channels in the pools and numerous small ridges suggest that the material was molten (AS16-5000).

Figure 4.38 A large lobate sheet of striated material on the northwest flank of Tsiolkovsky crater (visible on the right) interpreted as a giant landslide. The distance between the head scarp, just beyond the rim of the crater and the outer edge of the flow is ~80 km (AS 17 M-2608).

As discussed in Chapter 3, impact cratering involves the point-source transfer of energy from the bolide to the target surface. This typically results in craters and crater ejecta deposits that are radially symmetrical. However, for angles of impact lower than about 15° from the horizontal, both craters and ejecta deposits tend to become asymmetric with rays that frequently form a "wing" pattern, as seen by several examples on the Moon (Fig. 4.39).

Ejecta deposits for all fresh lunar craters larger than a kilometer can be divided into three relatively well-defined regions which are of increasing distance and usually (with the above-noted exceptions) concentric about the rim crest. The inner zone is characterized by blocks and concentric fractures. Larger craters may exhibit "pools" of smooth-surfaced units and dune-like features. The middle zone is a region of continuous ejecta which exhibits a hummocky surface with radial and concentric patterns, some of which appear dune-like. The outermost zone is of discontinuous ejecta which is composed of filamentary rays and numerous secondary craters which frequently display a characteristic herringbone pattern. Large impact craters, such as Copernicus (Fig. 4.40), possess both concentric and radial patterns of secondary crater chains.

Some volcanic craters are morphologically similar to impact craters. Thus, while the vast majority of lunar craters are considered to be of impact origin, it is possible that volcanic craters are also present. Some linear arrays of craters almost certainly are of internal origin. Chains of craters of possible non-impact origin are classified (e.g. Eppler & Heiken 1975, Pike 1976) into: (a) collapsed lava tubes, (b) cinder-cone craters and (c) drainage pits. This latter category includes the Hyginus Rille (Fig. 4.41) which exhibits a series of rimless craters several kilometers across, all contained within a graben. These are generally considered to be collapse craters, possibly formed in association with magmatic withdrawal.

4.5 Degradational features

Because the Moon lacks an atmosphere and has never experienced liquid water on the surface, lunar degradational processes are restricted primarily to mass wasting and various impact-related events. Mass wasting has occurred in several forms, including possible landslides

Figure 4.39 View of Messier (1) and Messier A (2) in Mare Fecunditatis; the shape of these craters and the asymmetric ejecta pattern are considered to be the result of a low-angle impact in which the bolide either broke into two components, or "skipped" across the surface to form two craters. Messier is about 8 by 15 km (AS15-M3-2674).

Figure 4.40 An oblique view across southern Mare Imbrium looking toward Copernicus, the large (93 km in diameter) crater near the horizon. The distance from the lower edge of the picture to the center of Copernicus is 400 km. The mountains at the edge of the mare are the Montes Carpatus, a major ring of the Imbrium basin. Copernicus is one of the youngest of the Moon's large craters. The many chains, loops, and clusters of small irregular craters and associated bright streaks extending across the mare are caused by the secondary impact of ejecta from Copernicus (AS17-M2444).

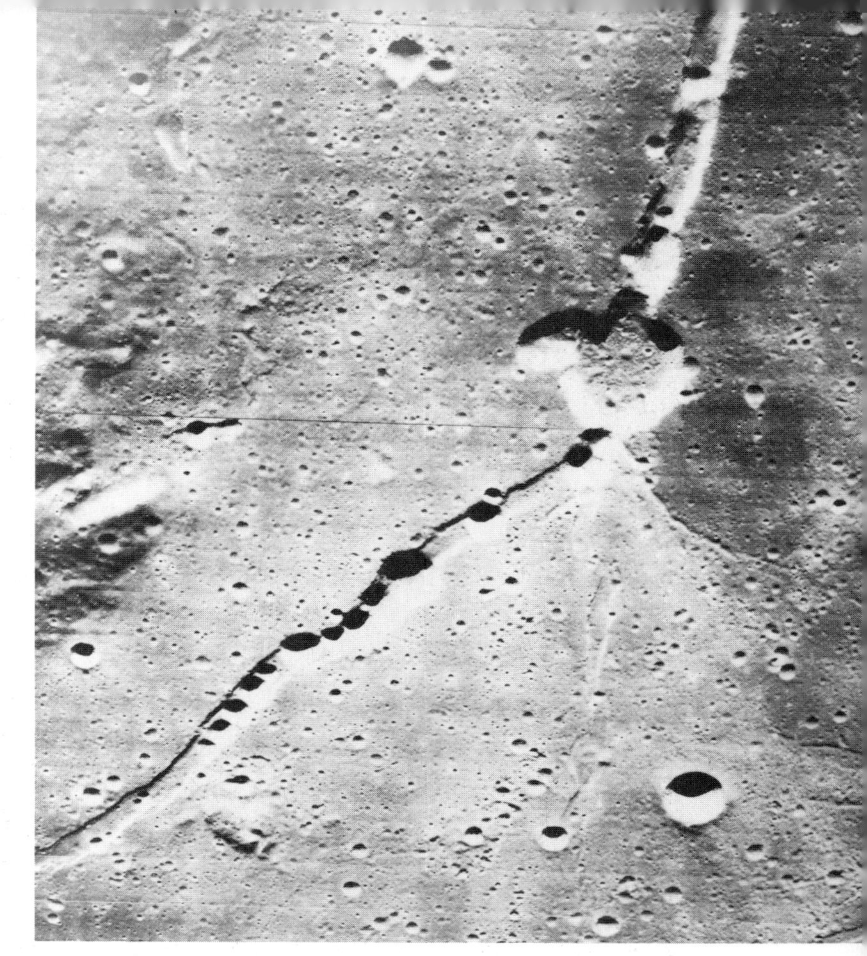

Figure 4.41 Lunar orbiter view of the Hyginus Rille. This rille extends for more than 200 km and contains a series of rimless craters (some as large as 12 km in diameter) considered to be similar to volcanic pit craters and ascribed to endogenic processes (Lunar Orbiter V-M97).

(Howard 1973; Fig. 4.20), rock-falls and slides (Fig. 4.42), and surface creep leading to **treebark textures**. Treebark texture, shown in Figure 4.43, is considered to result from small-scale impact cratering. Slumping of crater walls is particularly important in the degradation of craters (Fig. 4.36). Impact cratering is also responsible for the generation of lunar regolith deposits. Lunar surfaces are constantly being fragmented by impact cratering at all scales. This constant pulverization has led to the formation of regolith deposits, the thickness of which is directly proportional to the age of the surfaces. Thus, young mare lavas are blanketed with a fragmental surface layer estimated to be only a few meters thick, while older mare surfaces may have regolith deposits exceeding 10 m in thickness (Oberbeck & Quaide 1968). Figure 4.44 shows the lunar surface viewed by the Apollo astronauts and gives a general indication of the formation of regolith through impact processes.

4.6 History of the Moon

Photogeologic mapping and remote sensing, analysis of various surface features and incorporation of results from lunar sample studies have yielded a general model for the evolution of the lunar surface. Figure 4.45 illustrates the major stages in the history of the Moon, keyed to the lunar stratigraphic timescale (Wilhelms 1980) in which systems recognized from oldest to youngest are: (1) pre-Nectarian, (2) Nectarian, (3) Imbrian, (4) Eratosthenian, (5) Copernican. The pre-Nectarian system represents all of the older cratered terrains which were formed prior to the impact of the Nectaris basin. This interval of time records the final stages of the accretionary process in the formation of the Moon. Impact-generated heat is considered responsible for the melting of the outer few hundred kilometers and the separation of an anorthositic crust. By about 3.9 aeons, the outer crust was rigid and volcanism was deep-seated.

The Nectarian system begins with the formation of the Nectaris basin and the emplacement of its ejecta blanket, the Janssen Formation. This interval of time

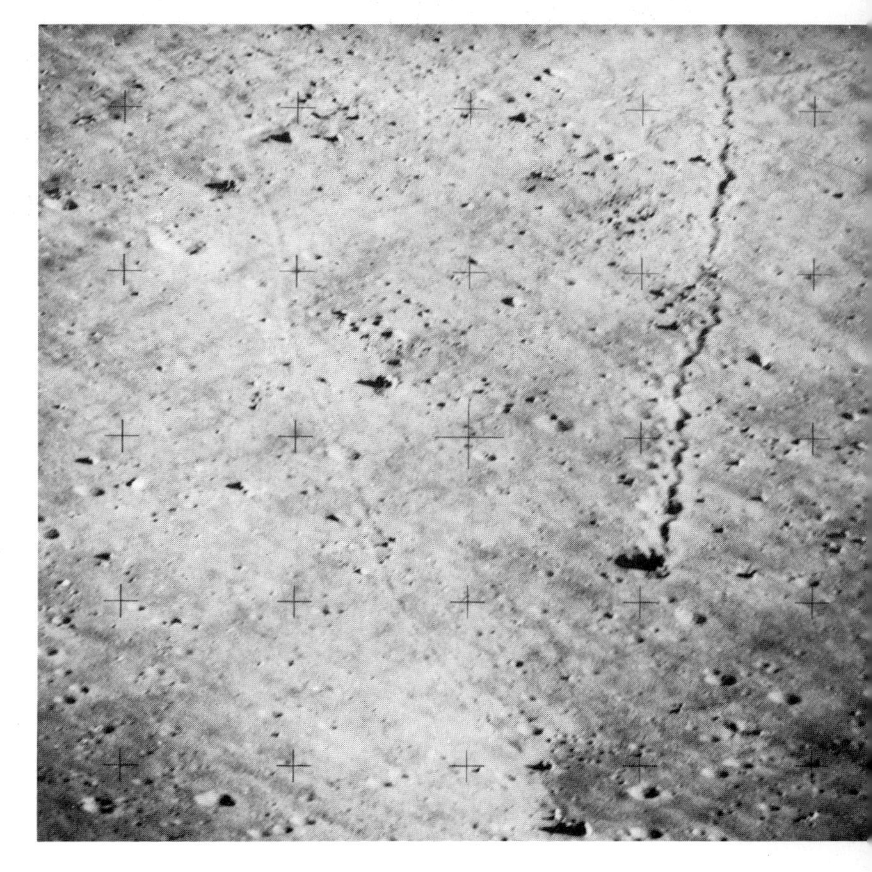

Figure 4.42 View obtained by the Apollo astronauts using a telephoto lens to photograph boulders and boulder tracks on the slopes of North Massif at the Apollo 17 landing site. The largest boulders are about 5 m across; illumination is from the right (AS17-144-21991).

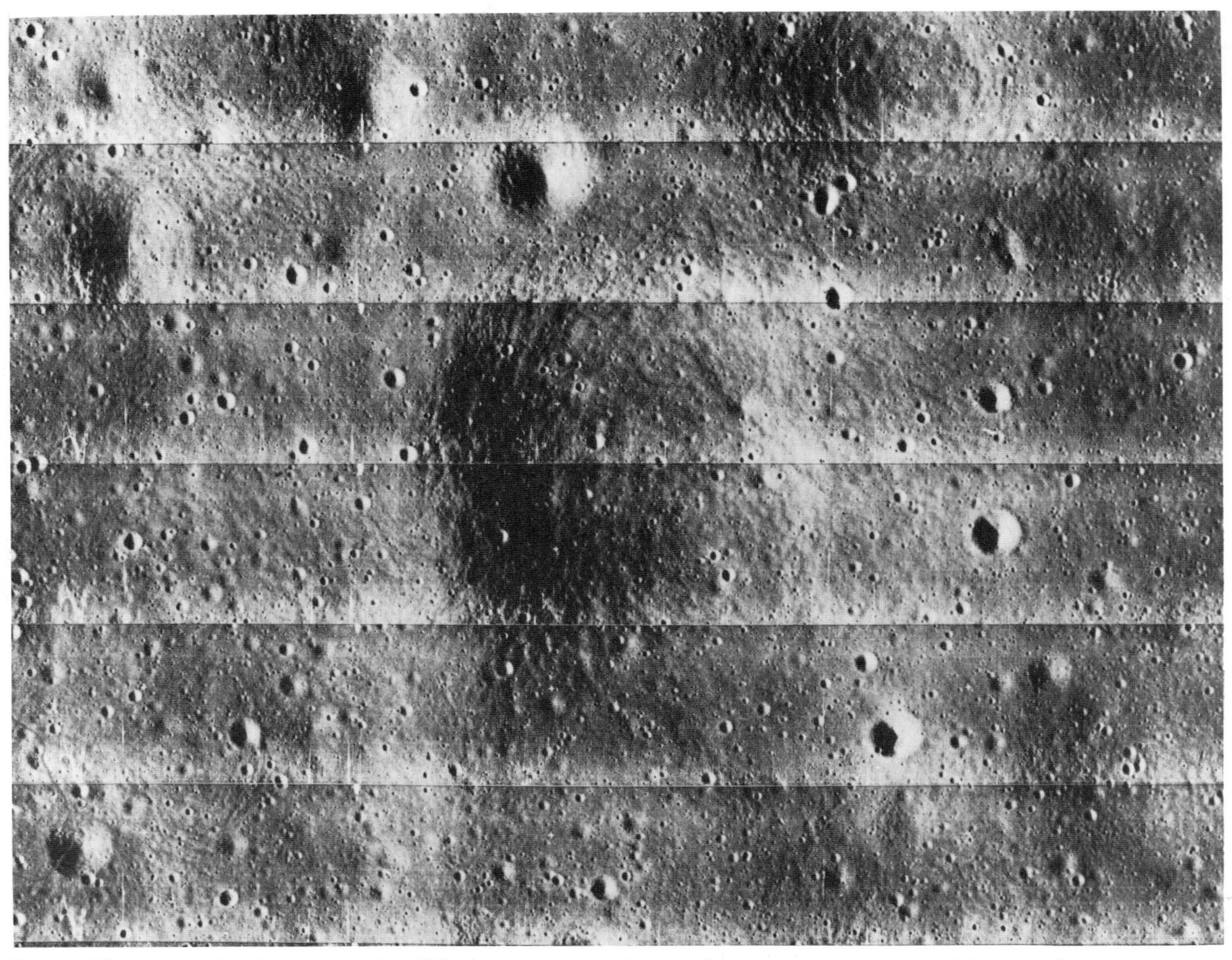

Figure 4.43 Lunar orbiter view of a small (~600 m) crater showing "treebark" texture on the crater wall; this pattern is considered to result from impact-generated creep (Lunar Orbiter II-H56).

includes the formation of other older basins – including Humorum and Crisium – and eruptions of high-aluminum mare lavas and lavas rich in potassium (K), rare earth elements (REE), and phosphorus (P) collectively termed KREEP and KREEPy lavas. Studies by Hawke and Head (1978) show that KREEP volcanism began prior to 4.1 aeons and continued into early to mid-Imbrian time.

The impact that formed the Imbrium basin (Fig. 4.45a) occurred about 3.85 aeons and marked the beginning of the Imbrian system. This interval of time includes the formation of the Orientale basin and the eruption of about two-thirds the volume of the visible maria (Fig. 4.45b).

The Eratosthenian system is exemplified by impact craters similar to Eratosthenes which have well-preserved secondary craters but which lack bright-ray deposits. During this interval of time about one-third the remaining portion of the visible maria were emplaced. The generation of late-stage high-titanium mare lavas, plus some low-titanium, intermediate-stage lavas also occurred during this period.

The last episode of lunar history is identified as the Copernican system, which includes all fresh, large craters that retain bright rays. Photogeologic mapping of some presumed mare lava flows superimposed on Copernican-age craters showed that some lunar volcanism may have extended to relatively recent times (post ~2.0 aeons) on the lunar surface.

Figure 4.44 View near the Hadley Rille in the Apollo 15 landing site area, showing typical block-littered surface in which the rocks are being physically weathered by impact cratering (AS15-82-11147).

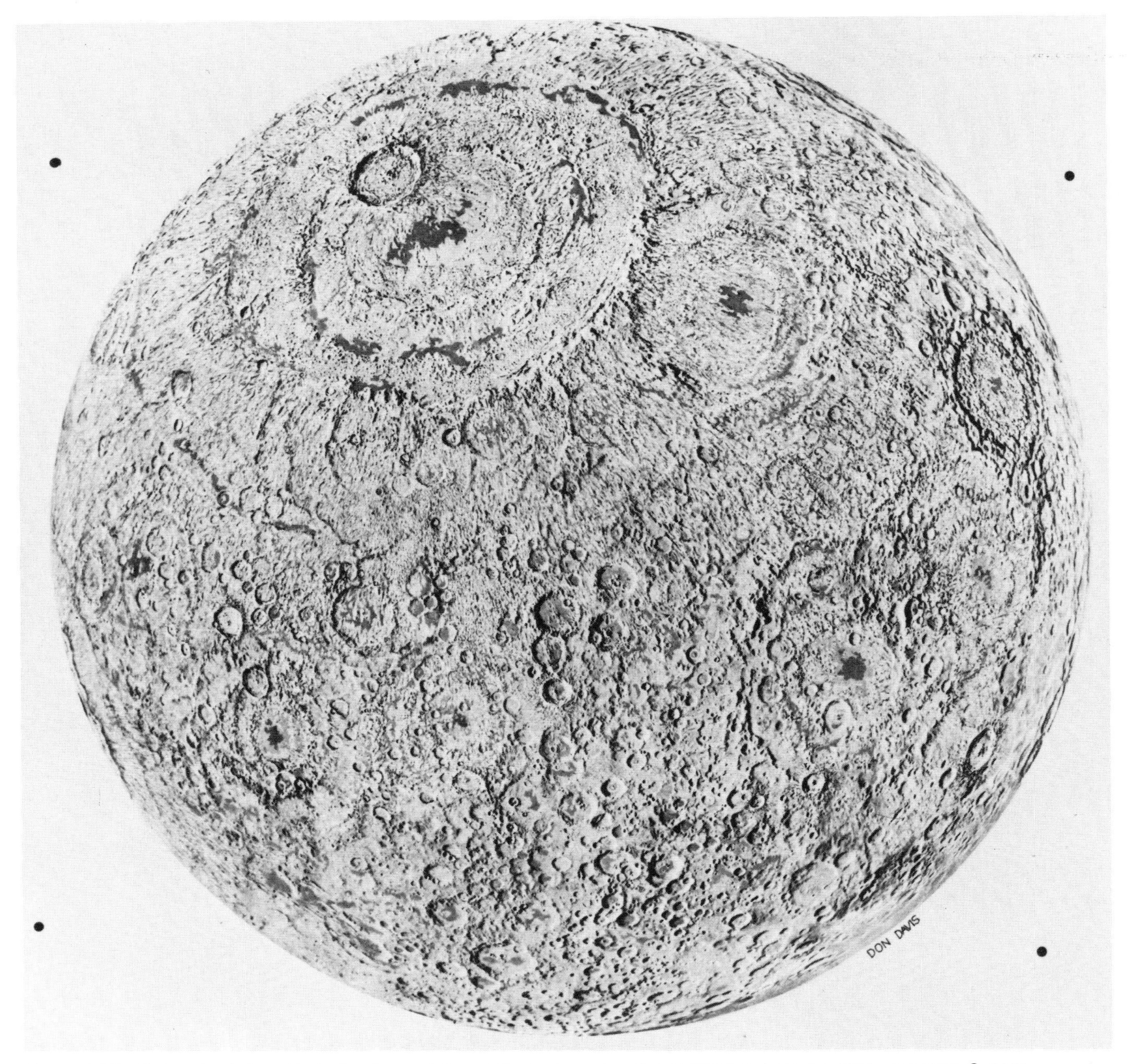

Figure 4.45 Series of paintings illustrating the evolution of the lunar surface (from Wilhelms & Davies 1971, copyright © 1971 Academic Press). **(a)** The Moon during the middle of the Imbrium period, after the formation of most of the large impact basins, but before extensive flooding by mare lavas; the Imbrium basin is in the upper left.

Figure 4.45b The Moon after extensive eruptions of mare lavas at about 3.3 aeons ago, but before the formation of craters such as Eratosthenes, Tycho and Copernicus.

Figure 4.45c The present Moon.

5 Mercury

5.1 Introduction

Mercury, innermost planet of the Solar System, is easily visible to the naked eye and has figured prominently in folklore and mythology through the ages. Mercury derives its romanized name from the Greek god, Hermes, the Olympian patron of trade, travel and thieves. The first recorded observation of Mercury is by Timocharis, who noted its position in the heavens in 265 BC. As viewed from Earth, Mercury is never more than 28° from the Sun and, consequently, it can be seen only in the twilight hours. Thus, even though it is an easy object to see, telescopic scientific observations are difficult. The Italian astronomer, Zupus, first recorded the phases of Mercury in 1639, while surface markings were not noted until 1800 when the astronomers Schröter and Harding reported albedo patterns.

Prior to acquisition of Mariner 10 results in 1974–5, relatively little was known of Mercury geologically. Estimates of its bulk density suggested that Mercury was more like the Earth than the Moon (Table 2.1), which led to speculation that there could be twice as much iron

Figure 5.1a

Figure 5.1b

Figure 5.1c

Figure 5.1 Mosaics of Mariner 10 images of Mercury obtained on the three flybys of Mariner 10 (courtesy NASA-Jet Propulsion Laboratory).

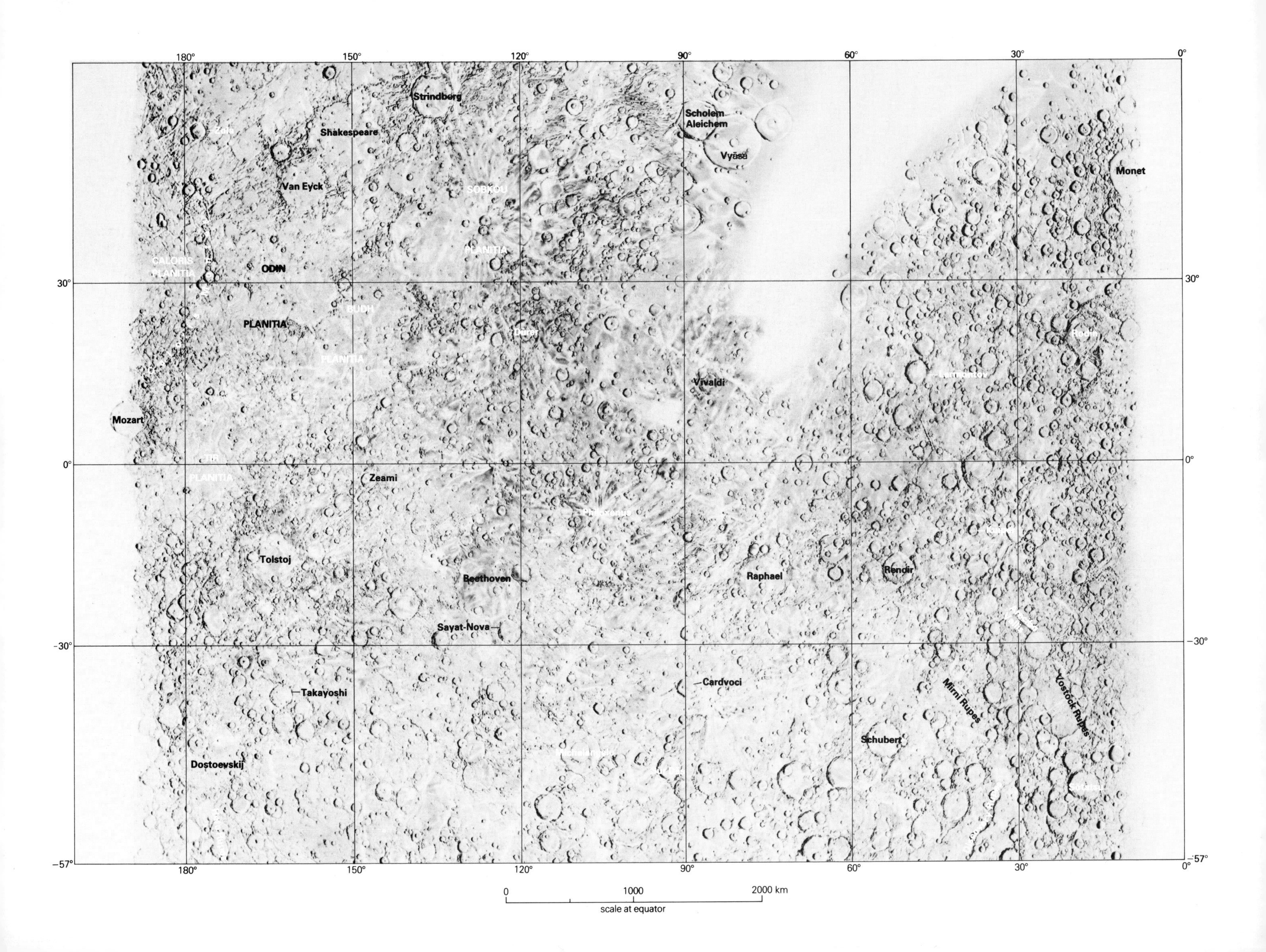
180°
150°
120°
90°
60°
30°
0°
30°
0°
−30°
−57°
Strindberg
Scholem Aleichem
Vyāsa
Shakespeare
Monet
Van Eyck
SOBKOU
PLANITIA
CALORIS PLANITIA
ODIN
PLANITIA
BUDH
PLANITIA
Vivaldi
Mozart
TIR PLANITIA
Zeami
Tolstoj
Beethoven
Raphael
Renoir
Sayat-Nova
Cardvoci
Takayoshi
Mirni Rupes
Vostock Rupes
Schubert
Dostoevskij
0
1000
2000 km
scale at equator

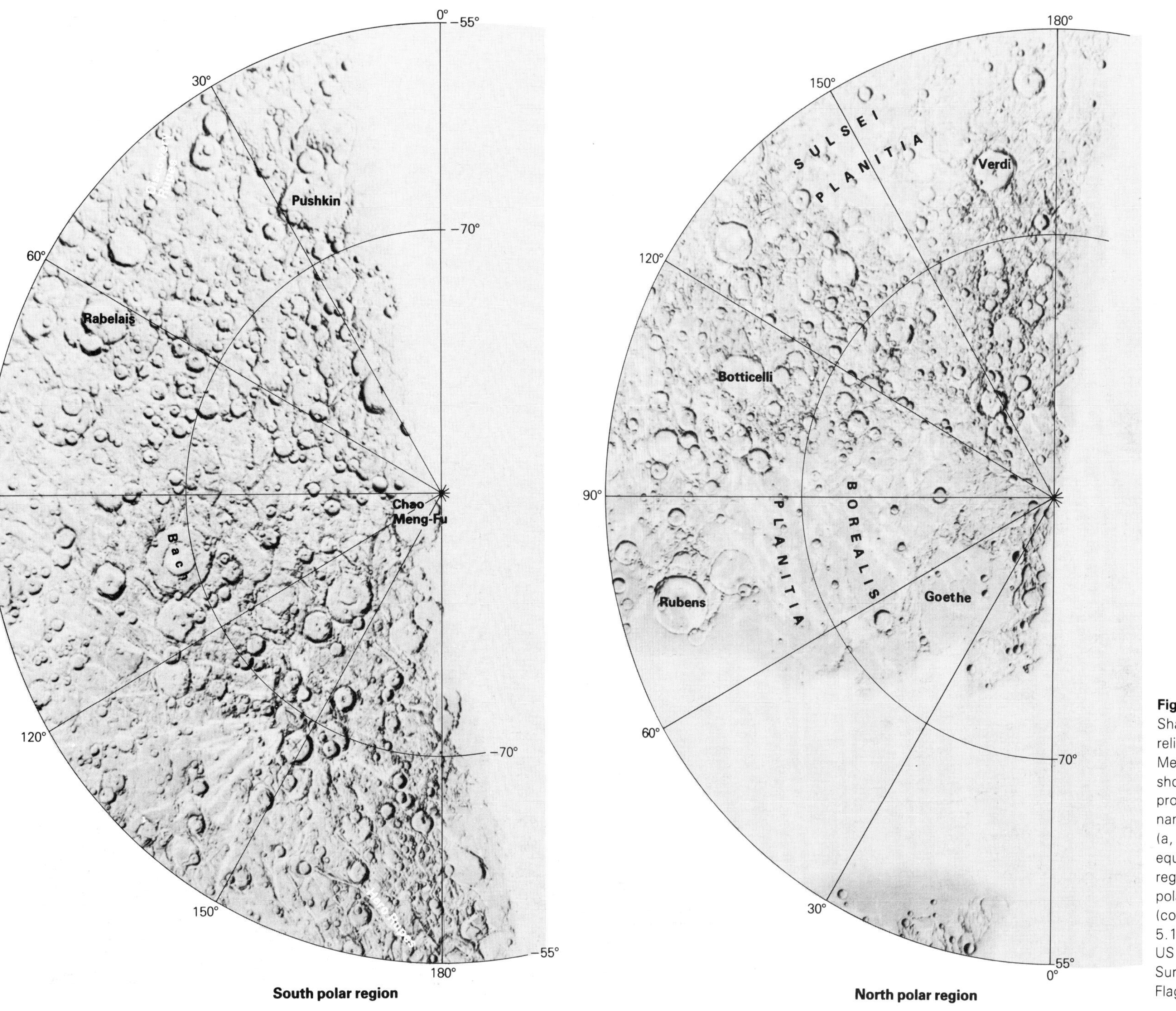

Figure 5.2 Shaded airbrush relief map of Mercury, showing prominent named features: (a, top) equatorial region, (b, left) polar areas (compare to Fig. 5.1). (Courtesy US Geological Survey, Flagstaff.)

in the core of Mercury in proportion to its size than any other planet in the Solar System. Earth-based observations of the reflective and thermal properties of the mercurian surface yielded results that were similar to those from the Moon's surface. This led to the suggestion that its surface is covered with a layer of dark, porous, fine-grained material which, as on the Moon, could have been generated by impact cratering processes.

It was not until radar observations were made in the early 1960s that the rotation period for Mercury was determined accurately. Mercury was found to have a two-thirds resonance with respect to its orbital period; that is, Mercury rotates on its axis precisely three times for every two orbits around the Sun. Various observations prior to the Mariner 10 mission also showed that Mercury has no significant atmosphere; thus, surface weathering or erosional processes involving running water or wind to modify its surface were not expected. Of course, it was not known if such processes might have occurred in the geologic past.

Mariner 10 opened Mercury to geologic inspection by providing the first (and as yet the only) detailed information for its surface (Murray *et al.* 1975, Strom *et al.* 1975a). Figure 5.1 shows a series of mosaics obtained on the three flybys of Mariner 10, the closest of which occurred at a distance of only 700 km. In all, approximately half of Mercury was covered and more than 2700 useful images were obtained (Murray 1975). Although these images have 5000 times better resolution than Earth-based telescopic views, it is important to keep in mind that most Mariner 10 images of Mercury are comparable to Earth-based telescopic views of the Moon (Table 2.3). On first inspection Mercury appears to be a Moon-like object (Fig. 5.2). In recognition of its similarity to the Moon, a major conference comparing the Moon and Mercury was held in 1976, with the proceedings published in a special issue of the journal, *Physics of Earth and Planetary Interiors* (1977; Table 1.2).

Like the Moon, the mercurian surface is dominated by impact craters. In addition, there is an asymetry in the distribution of terrains: smooth plains dominate one part of the planet (Fig. 5.1a), whereas heavily cratered terrains (Fig. 5.1b) occur elsewhere, similar to the concentration of mare on the lunar nearside (at least for the half of Mercury for which data are available). Closer examination of the surface, however, reveals significant differences between Mercury and the Moon. First, there is no correlation between albedo and topography. The albedo of both the smooth plains and the cratered terrain is remarkably similar on Mercury, in contrast to the Moon with its dark-toned mare lowlands and light-toned highlands. An extensive terrain, termed **intercrater plains**, some of which may represent primordial crust, has been little modified throughout the evolution of the mercurian surface and has no direct lunar counterpart. Some aspects of the morphology of impact craters on Mercury are different in comparison to lunar craters. Finally, numerous scarps cut across plains, craters and cratered terrain on Mercury and suggest compressional crustal deformation.

In addition to the return of images, Mariner 10 made other measurements important to the interpretation of the geologic history of Mercury. It was found that the planet has a magnetic field, although it is very weak in comparison to Earth: 600 gammas versus 50 000 gammas for Earth. The origin of this magnetic field remains an enigma. It could be a remnant from some event in the past, a weak active field generated by magnetohydrodynamic processes, or a combination remnant/active field. In any event, the presence of the field, when combined with models of the interior that include a high proportion of iron and the suggestion of volcanism inferred from Mariner 10 images, suggests that Mercury underwent chemical differentiation.

In addition to the references given in Table 1.2, several important reviews of Mercury have been published, including Guest and O'Donnell (1977) and Gault *et al.* (1977). Robert Strom's review (1979) is very extensive and is especially recommended. Davies *et al.* (1978) provide a synopsis of the history of studies of Mercury, a brief outline of its geology and cartography, and a superb atlas of photographs and charts keyed to the US Geological Survey system of quadrangles.

5.2 Physiography

Mariner 10 obtained images for only about half of the planet and thus the analysis of Mercury cannot be applied globally. Early in the analysis of the images, a geologic terrain map was derived and the major physiographic features were defined (Trask & Guest 1975). Following this preliminary mapping, the US Geological Survey embarked on a geologic mapping program based on 1:5 000 000 charts. While this mapping has allowed a refinement of the units and has filled in detail, the major units and the basic ideas regarding their origin have not changed significantly. A synthesis of all mapping is currently underway by the US Geological Survey which will lead to publication of a planet-wide geologic map in the late 1980s.

Figure 5.3 is taken from Trask and Guest (1975) and shows the major terrain types and features on Mercury. Four major terrains are recognized: smooth plains, intercrater plains, heavily cratered terrain, and hilly and lineated terrain.

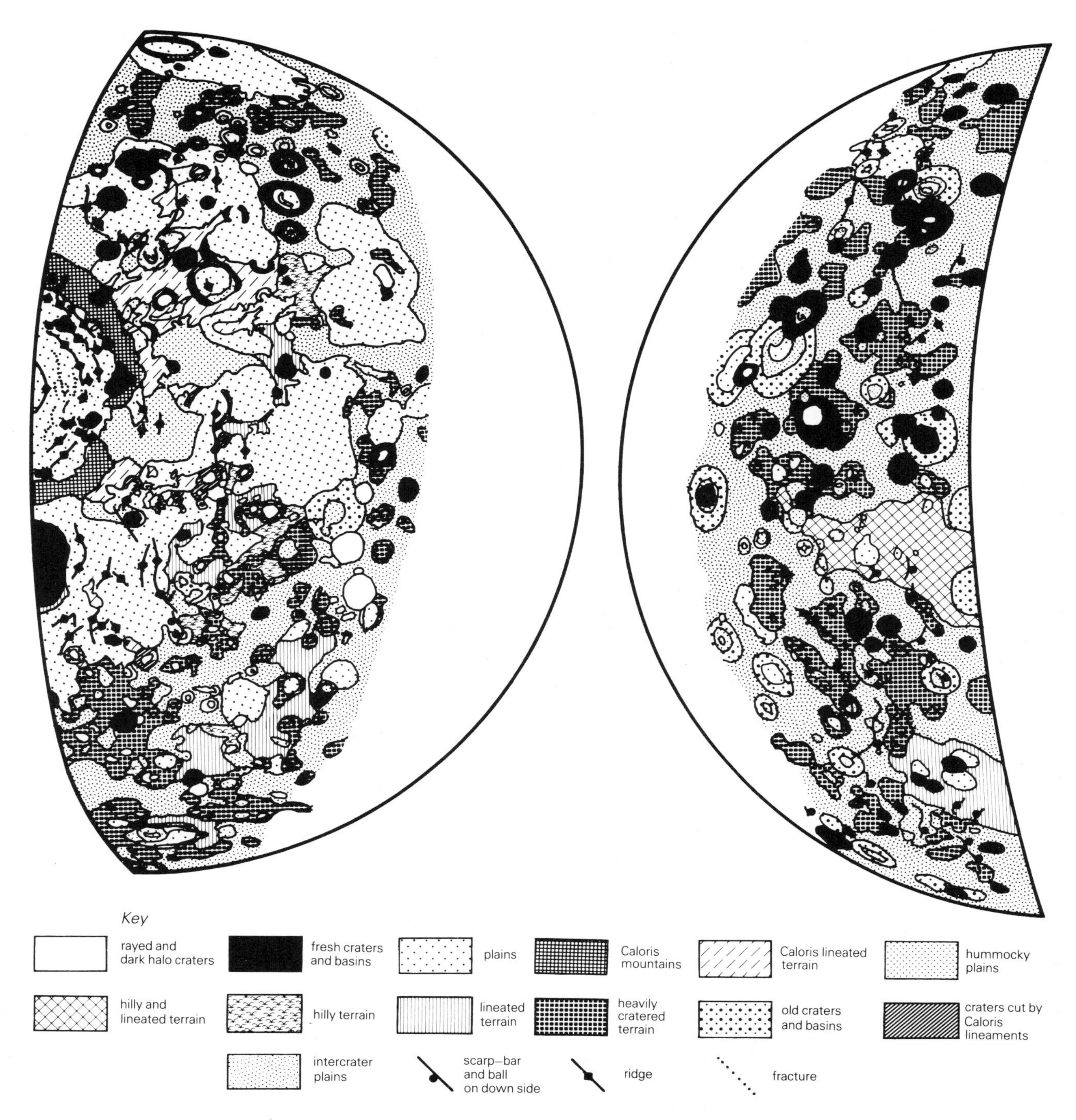

Figure 5.3 Map showing the principal terrain units on Mercury, based on the mosaics shown in Figures 5.1a & b (from Trask & Guest 1975, *J. Geophys. Res.*, copyright American Geophysical Union).

Figure 5.4 Smooth plains southeast of the Caloris basin showing a mare-like ridge; area shown is about 100 km across (Mariner 10 FDS 70).

5.2.1 Smooth plains

Smooth plains form relatively flat, sparsely cratered surfaces which resemble lunar maria and display bright-halo, sharp-rimmed craters generally smaller than ~10 km. Many smooth plains have sinuous, lobate ridges which are similar to lunar mare ridges but lack the upper crenulated ridge (Fig. 5.4). The plains often fill large craters and embay older terrains (Fig. 5.5) with well-defined contacts. From these contact relations and their lack of large, superposed craters, the smooth plains are considered to be among the youngest stratigraphic units on Mercury.

Smooth plains are found predominantly in the northern hemisphere between 120° to 190° longitude, where they form an approximate annulus around the Caloris basin (Fig. 2.18). However, smooth plains also occur as isolated patches elsewhere on Mercury, including the floors of large craters and basins. Regardless of location, smooth plains have about the same frequency of superposed impact craters, which suggests that the plains were formed at about the same time. Their spectral properties (color) are approximately the same at all locations, which suggests a homogeneous composition, unlike lunar maria or the Cayley plains. Moreover, smooth plains have about the same albedo as the heavily cratered terrain and intercrater plains in which they occur.

The origin of smooth plains is a matter of debate. Most of the controversy concerns the smooth plains found in the vicinity of Caloris. Based on the characteristics listed above, Wilhelms (1976) and Oberbeck *et al.* (1977) noted that the smooth plains east of Caloris are similar to the Cayley Formation and other lunar highlands and suggested that the mercurian smooth plains are also impact related. The difficulties with this interpretation are the apparent volume of material involved and the timing. From crater frequency distributions, there appears to be a substantial interval between the formation of the Caloris basin and the emplacement of the plains. Thus, it would seem unlikely that the plains could be either impact melt or ejecta-related deposits of Caloris.

The same difficulty applies to study of smooth plains in other areas. Figure 5.5 shows Borealis Planitia, a circular depression (basin?) some 1000 km across, within which occurs Goethe, a 340 km impact basin. The plains flooding the depression and Goethe have a crater density significantly lower than the ejecta units from these large impact structures, indicating that the smooth plains post-date both the circular depression and Goethe and are unlikely to be associated with either impact event. However, it has been argued that these plains are related not to the Borealis basin or to Goethe, but to some unknown basin or basins hidden from view in the part of Mercury not imaged by Mariner 10.

Most investigators consider the smooth plains to be volcanic. Although there is no direct evidence, the following reasons are rallied for a volcanic origin.

(a) Smooth plains are widespread over Mercury and are unlikely to have been the result of a single impact event; moreover, the smooth plains near Caloris appear to be of too large a volume to be solely Caloris ejecta.
(b) The morphology and inferred physical properties of craters in smooth plains suggest similarities to lunar mare (Cintala *et al.* 1977). The presence of lunar-like mare ridges also suggests a similarity; because lunar mare are volcanic, the analogy is drawn for Mercury.
(c) Stratigraphic relations indicate that the smooth plains are much younger than the impact craters and basins with which they are associated.

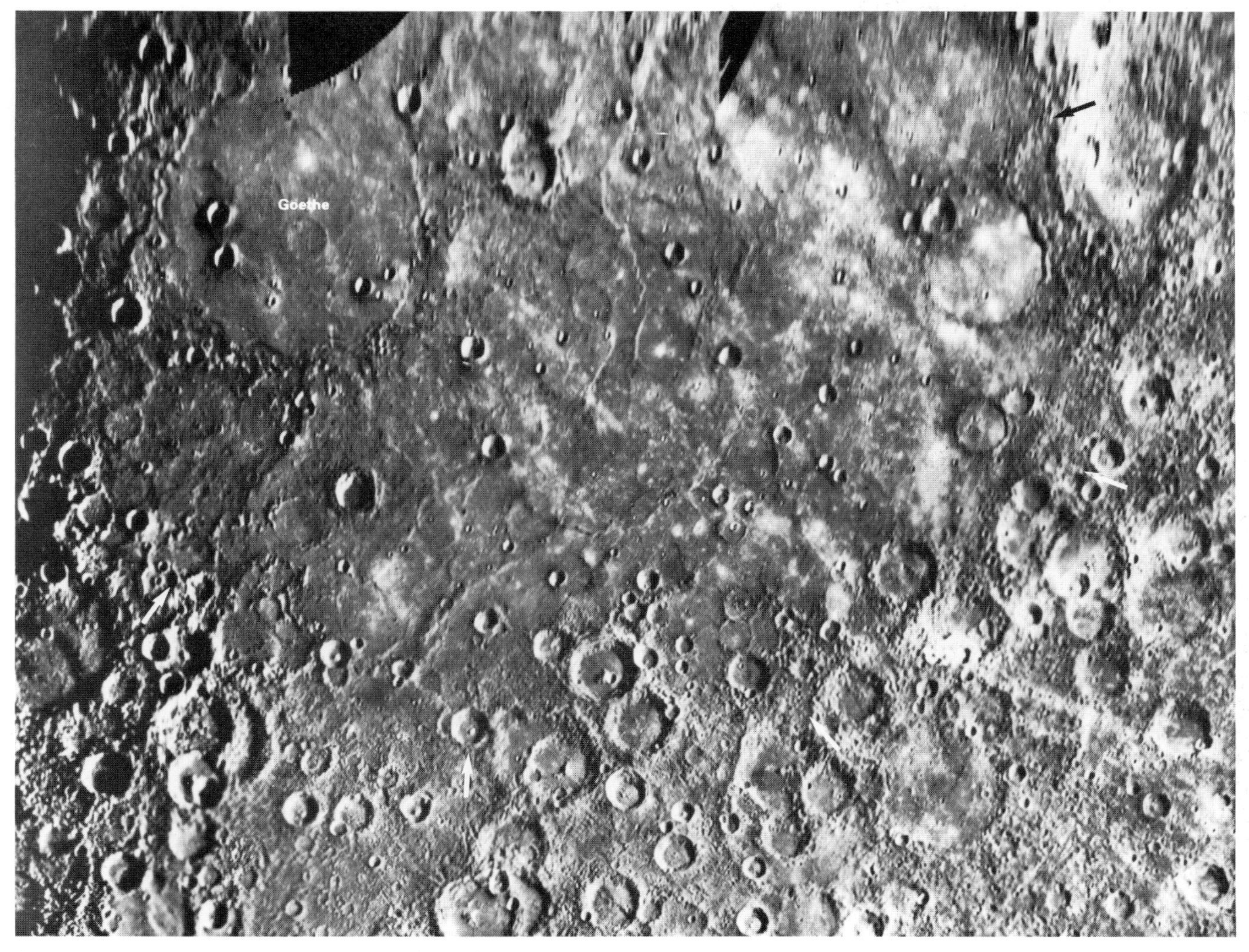

Figure 5.5 Photomosaic of Mariner 10 images in the north polar region of Borealis Planitia showing the Goethe basin (340 km diameter) and a ~1000 km unnamed basin (arrows mark rim), both of which have been flooded by smooth plains. Frequency distributions of superposed craters show that the plains are much younger than either Goethe or the large basin and hence are not related to the impacts which formed them (from Strom 1979).

5.2.2 Intercrater plains

The intercrater plains, shown in Figure 5.6, are the most widespread unit mapped and constitute about one-third of the area imaged by Mariner 10. This terrain consists of level to gently rolling plains peppered with craters smaller than 5–10 km, many of which appear to be secondary craters. The density of superposed craters is much higher than that of smooth plains. Preliminary mapping suggested that most intercrater plains were overlain by the heavily cratered terrain units and large, old craters. Consequently, intercrater plains were interpreted to be the oldest unit on Mercury, pre-dating the period of heavy impact bombardment and perhaps representing relatively undisturbed primordial crust. Malin (1976) and Strom (1979), however, have discovered that some heavily cratered terrain is embayed (i.e. superposed) by intercrater plains (Fig. 5.7). Furthermore, Spudis and Strobell (1984) find a large population of ancient multi-ring basins which underlie the intercrater plains on a planet-wide basis, suggesting that the plains are not primordial crust.

It is important to note that the intercrater plains should not be treated as a stratigraphic unit (Guest & O'Donnell 1977); this terrain unit probably represents different events separated not only geographically but by time. Leake (1981) analyzed the intercrater plains and relative ages of cratered terrain and individual large, old craters and concluded that at least some intercrater plains were being formed concurrently with heavy impact cratering. Thomas (1980) and Thomas *et al.* (1982) studied the relationships between the plains and nearby impact craters as a function of their stage of degradation and concluded that intercrater plains formation has occurred throughout the history of Mercury.

Figure 5.6 High oblique view showing large areas of intercrater plains and heavily cratered terrain. The apparent "tear" toward the top of the picture was caused by a loss of image data and is an artifact. This image was taken at a distance of 77 800 km from Mercury (Mariner 10 FDS 27328).

Figure 5.7 High-resolution Mariner 10 image showing typical intercrater plains and the superposition of numerous craters; arrow points to a 90 km diameter crater that has been embayed by intercrater plains and, thus, pre-dates the emplacement of the plains. The lobate scarp, Santa Maria Rupes, post-dates the intercrater plains. Image is 400 km wide (from Strom 1979; Mariner 10 FDS 27448).

Strom (1979) carried this possibility one step further and concluded that the intercrater plains and the very young smooth plains may represent a continuous process of plains formation, beginning with the earliest observable history of Mercury and extending through to the latest events.

The origin of the intercrater plains, like that of smooth plains, remains an enigma. Stratigraphic relationships led to the conclusion that intercrater plains are very old, perhaps representing primordial crust. However, recognition that intercrater plains span a wide range of ages led to the idea that the plains are not a single unit and may not have a single mode of origin.

Like the mercurian smooth plains, the resemblance of intercrater plains to lunar maria suggests a volcanic origin. Some of the intermediate-age plains fill large degraded basins and, based upon the appearance of ghost craters in these plains, may be relatively thin units (DeHon 1978), such as thin flood lavas.

The intercrater plains have been likened to the pre-Imbrian pitted plains as mapped by Wilhelms and McCauley (1971) on the Moon and, like the lunar plains, alternative origins to volcanism have been proposed for the mercurian plains. Wilhelms (1976) and Oberbeck *et al.* (1977) suggest that the plains are generated by impact cratering processes and could be ejecta deposits. Although this origin could explain the interleaving of intercrater plains with the formation of large craters, the difficulty with this interpretation is the lack of source basins for the ejecta.

Figure 5.8 Image of the south polar area centered at 69°S, 104°W, showing Bach, a ~200 km diameter ringed basin filled with smooth plains (Mariner 10 mosaic).

5.2.3 Heavily cratered terrain

Heavily cratered terrain consists of closely packed and overlapping large craters (Fig. 5.6), along with individual craters ranging in size from 30 km to several hundred kilometers across. In most of the heavily cratered terrains, clearly defined ejecta deposits and fields of secondary craters are not identified, although some of the craters on the intercrater plains may be secondaries from craters in this unit (Trask & Guest 1975). Many of the large craters in this terrain are filled with smooth, much younger, sparsely cratered plains (Fig. 5.8). The similarity between the heavily cratered terrain and the lunar highlands leads most investigators to consider this unit to be the consequence of the terminal phase of heavy bombardment which evidently occurred throughout the inner Solar System (Gault *et al.* 1977).

5.2.4 Hilly and lineated terrain

Hilly and lineated terrain consists of an unusual landscape covering more than 250 000 km^2, centered at about 20°S, 20°W. Informally termed the "weird terrain" by the Mariner 10 imaging team, this landscape consists of hills 5–10 km across and 0.1–1.8 km high. Most of the crater rims in this area have been broken into numerous massifs, hills and depressions as shown in Figures 5.9 and 5.10.

The hilly and lineated terrain is antipodal to the Caloris basin, a location which suggested to Schultz and Gault (1975) that it may be related in some way to the Caloris impact. They analyzed the partitioning of seismic energy that would be related to an impact of the magnitude required to form the Caloris basin and predicted that there would be a tremendous focusing of energy at the antipode (Fig. 5.11), resulting primarily from the presence of a large planetary core. Schultz and Gault estimate that the mercurian surface would have been lifted tens of meters by seismic waves at the antipode which could account for the broken and "jostled" appearance of the hilly and lineated terrain.

Figure 5.9 Hilly and lineated ("weird") terrain centered at about 27°S, 24°W antipodal to the Caloris basin. Crater rims and intercrater areas have been broken into hills and valleys, whereas most of the smooth plains filling Petrarch (the ~150 km crater on the right) have not been modified. This suggests a sequence beginning with cratering, followed by terrain modification, and ending with the emplacement of smooth plains. The episode of terrain modification correlates with the formation of the Caloris basin. Arrow points to crater shown in Figure 5.10 (from Strom 1979; Mariner 10 FDS 27370).

5.3 Craters and basins

Impact craters, basins and related structures are the dominant landforms on Mercury. In this section the morphology of impact craters and basins is discussed and compared to lunar craters and the results from crater counts are assessed to determine the possible sequence of events in the evolution of Mercury.

5.3.1 Morphology

Initial examination of mercurian impact craters suggests a similarity to craters on the Moon. This is not surprising since both objects are airless bodies. Closer scrutiny, however, shows significant differences between mercurian and lunar craters, especially in regard to rays, ejecta and secondary craters. Bright rays are more prominent (Fig. 5.1a), the continuous ejecta facies seems closer to the crater rim and secondary craters appear to be deeper, have sharper rims and occur closer to the parent crater rim than on the Moon (Fig. 5.12). Gault *et al.* (1975) attribute some of these differences to the higher gravitational acceleration on Mercury by arguing that ejecta would travel a shorter distance than in comparison to the Moon (Fig. 3.6). They show that ejecta should travel only about 0.65 as far on Mercury as on the Moon (Fig. 5.13). Thus, the gravitational effects would cause the ejecta to form a much thicker deposit near the rim of mercurian craters than on the Moon.

Figure 5.10 High-resolution image of hilly and lineated terrain; prominent crater on the left is about 40 km across and is identified in Figure 5.9 (Mariner 10 FDS 27463).

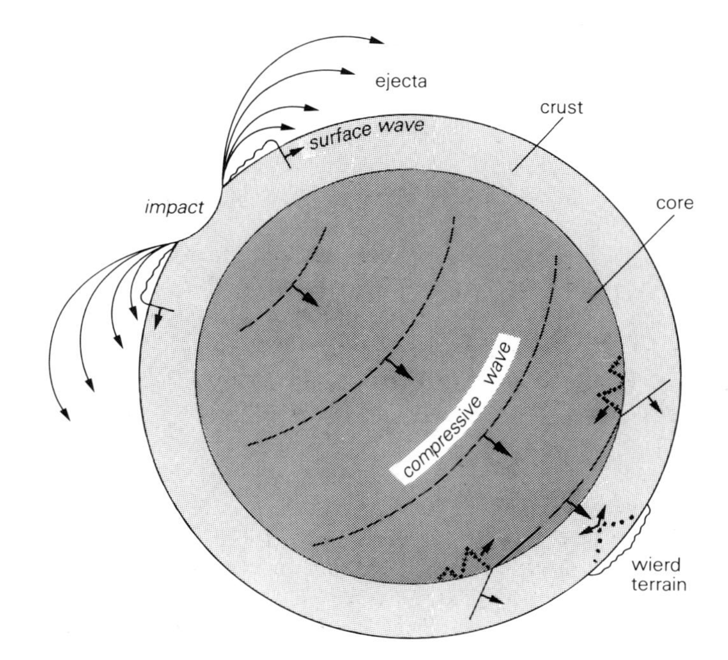

Figure 5.11 Diagram showing partitioning of energy resulting from major impact on a differentiated object such as Mercury; surface waves and compressive waves focus on the antipode and can result in severe seismic deformation (from data of Schultz & Gault 1975; courtesy of P. Schultz, Lunar and Planetary Institute).

Figure 5.12 Mariner 10 view of a 140 km diameter crater in the Shakespeare region showing the narrow hummocky rim, radial grooves and ridges, and field of secondary craters typical of large mercurian craters (Mariner 10 FDS 166).

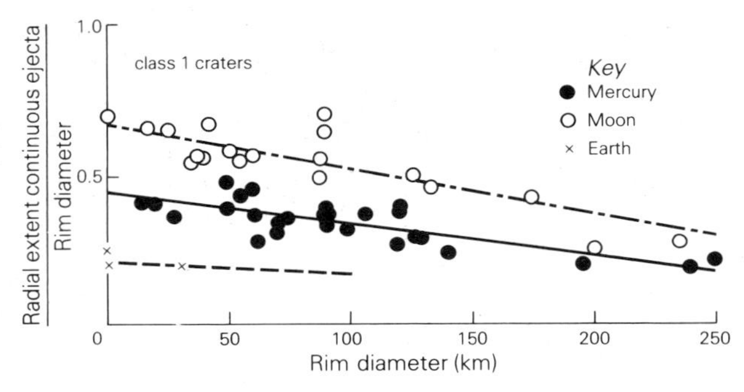

Figure 5.13 Average radial extents of continuous ejecta deposits around fresh craters on Mercury and the Moon (after Gault *et al.* 1975; *J. Geophys. Res.*, copyright American Geophysical Union).

Scott (1977) also considered secondary impact cratering velocities and found that due to the higher gravity on Mercury, velocities would be about 1.5 times greater than on the Moon. He attributed the sharper rims and apparently deeper secondary craters on Mercury to these more energetic impacts.

Mercurian craters display size-dependent morphologies that begin with small, bowl-shaped craters and progress to larger craters that have flat floors, central peaks, wall terraces (Fig. 5.14) and more complicated ejecta deposits. At approximately 140 km, most mercurian craters begin to display concentric rings (Fig. 5.8) and the transition to basins occurs. With a diameter of 1300 km, the Caloris basin (Fig. 2.18) is the largest known basin on Mercury.

The depth-to-diameter ratio for mercurian craters is similar to that for lunar craters except that the transition to more shallow forms for large craters occurs at a diameter of 7–8 km rather than ~14 km, as on the Moon (Fig. 4.33). Although Gault *et al.* (1975) attribute this to the higher gravity on Mercury, Malin and Dzurisin (1977 & 1978) find that the difference in the depth-to-diameter ratio between mercurian and lunar craters is small and consider gravitational effects to be minimal.

Several investigators have examined the morphology of crater interior features, such as central peaks, of the mercurian craters and attempted to correlate the results with terrain types for comparisons with the Moon. Malin and Dzurisin (1977 & 1978) concluded that the onset of central peaks occurs at about the same diameter on Mercury as on the Moon and that their formation appears to be independent of the terrain type. Cintala and co-workers (1977) conducted a more detailed analysis and arrived at the same conclusion. In addition, they found that the morphologies of craters formed in mercurian smooth plains and heavily cratered terrain is essentially the same as the morphology of craters in lunar maria. However, they found a greater difference in crater morphology for the lunar highlands in comparison to mercurian cratered terrain, leading them to conclude: (a) that differences in gravity have not dominated the formation of crater morphologic features, such as central peaks; (b) that the lunar mare/mercurian smooth plains similarities suggest that the smooth plains are volcanic, or at least have similar physical properties to the lunar maria; and (c) that morphologic differences in craters found in the heavily cratered terrain on

Figure 5.14 Oblique view of a 98 km diameter crater in the Shakespeare region, showing well-developed wall terraces and central peak (from Gault *et al.* 1975; Mariner 10 FDS 80).

Mercury and lunar highlands imply that these two terrains are dissimilar in their physical properties.

5.3.2 Crater states of preservation

At least four processes degrade craters on Mercury: (a) erosion from impact processes, including ballistic sedimentation; (b) burial or modification by plains units, such as lava flows; (c) isostatic adjustment (this process is particularly appropriate for the larger craters and basins); and (d) seismic events which may induce wall failures and crater rim degradation.

Several schemes to classify craters by their state of preservation have been devised. Most of these schemes are similar to those proposed for the Moon and are based on factors such as presence or absence of bright rays, preservation of secondary craters and sharpness of the rim. Classification involving three or more classes has been proposed but, unfortunately, the class numbering schemes have not been consistent. In some schemes, Class 1 refers to the freshest (i.e. youngest) crater, while in other schemes, Class 1 refers to the most degraded craters. The US Geological Survey mapping program utilizes a five-fold scheme with Class 5 representing the freshest craters.

Schultz (1977) examined mercurian craters and noted the similarity of some modified craters to the floor-fractured craters on the Moon. As with the lunar forms, he considers these to be impact craters that have been modified by volcanic or magmatic processes. The evidence for internally modified craters includes: (a) presence of very shallow floors with floor fractures which may indicate uplift and suggest intrusion; (b) smooth, sparsely cratered plains on the crater floors which indicate emplacement long after the impact event and suggest volcanic extrusion (Fig. 5.15); and (c) the

Figure 5.15 Mariner 10 image of the 120 km diameter crater Zeami (148°S, 2°S), showing smooth dark plains and bright patches on the floor. The shallow floor, the possible rimmed moat and the floor deposits are similar to volcanically modified lunar craters; arrows indicate outer boundary of dark halo (from Schultz 1977).

Figure 5.16 Mariner 10 image showing well-preserved secondary crater field surrounding an old, degraded 190 km crater, Ma Chih-Yuan, the Discovery Quadrangle; arrows mark crater rim (from Schaber *et al.* 1977).

presence of dark-halo craters and dark deposits associated with the floor units of craters, similar to dark-halo craters observed on the Moon (Fig. 4.25a) which are considered to be volcanic. Schultz also noticed that there are spectral contrasts between the floor units and the surrounding terrain, which suggested to him differences in composition which could not be attributed to impact events but rather suggest volcanic processes.

5.3.3 Crater frequency distributions

Guest and Gault (1976) recognized three classes of craters based on stages of degradation and analyzed their frequency distribution. They found a general deficiency of craters ~50 km in diameter which they attributed to: (a) a change with time in the distribution of bolides which could have caused a lack of crater formation in certain size ranges; (b) during one stage of crustal evolution, the mercurian crust may have been semi-molten and incapable of preserving craters above a certain diameter range; and/or (c) an erosional process occurred which preferentially obliterated some craters of certain size ranges. Their analysis also showed that fresh craters smaller than 20–30 km appear to be lacking in the cratered terrain surrounding the Caloris basin. They speculated that the Caloris event (or impacts which formed other "hidden" basins) might have caused widespread seismic obliteration of the population of craters in question. Schaber and his colleagues (1977) noted that Ma Chih-Yuan is one of many basins/large craters on Mercury which appear to be in an advanced state of isostatic adjustment but have well preserved secondary craters (Fig. 5.16). They cite these as evidence for impacts into a thermally active crust of relatively low viscosity.

Initial analysis of the Mariner 10 data suggested that there were few basins 300–1000 km in diameter on Mercury (Schaber *et al.* 1977, Frey & Lowry 1979). Subsequent studies by DeHon (1978) and Spudis and Strobell (1984), however, have revealed a number of basins in this size range that have been partly buried or modified by various plains units and hence are difficult to see. There may not, therefore, be a significant lack of basins on Mercury.

5.4 Scarps and ridges

One of the most important discoveries on Mercury is the presence of scarps and ridges apparently formed by compressional tectonism (Fig. 5.17). Classifying the ridges and scarps and mapping their distribution and orientation is important in understanding the evolution of Mercury because they provide clues to styles and timing of tectonic processes. The most extensive study of these features is provided by Dzurisin (1978) who classified their general morphology (scarps, ridges or troughs) and planimetric form (arcuate, lobate, irregular or linear).

5.4.1 Arcuate scarps

Arcuate scarps form curved cliffs that may be locally sinuous or irregular in outline. They range in length from 100 to 600 km and may be as high as 1.1 km. Arcuate scarps cut across both intercrater plains and large craters, and post-date most other tectonic features. Hero Rupes (Fig. 5.18), a typical arcuate scarp, has a radius of curvature of about 370 km, is more than 500 km long and stands 0.7–1.0 km high. From cross-cutting relations with large craters, its formation appears to be contemporaneous with the period of heavy impact cratering.

Vostock is one of the most important arcuate scarps on Mercury. As shown in Figure 5.19, this scarp cross-cuts a 75 km diameter crater, shortening the crater circumference and suggesting 5–7 km of thrust-fault

Figure 5.17 (top right) Photomosaic of Discovery Rupes centered at 38°, 52°N. The scarp is more than 500 km long and is about 2 km high, making it one of the largest scarps on Mercury. Two craters (a) 55 km diameter Rameau and (b) 35 km unnamed crater are cut by the scarp, suggesting thrust-faulting; arrow points to dome shown in Figure 5.30 (from Strom 1979).

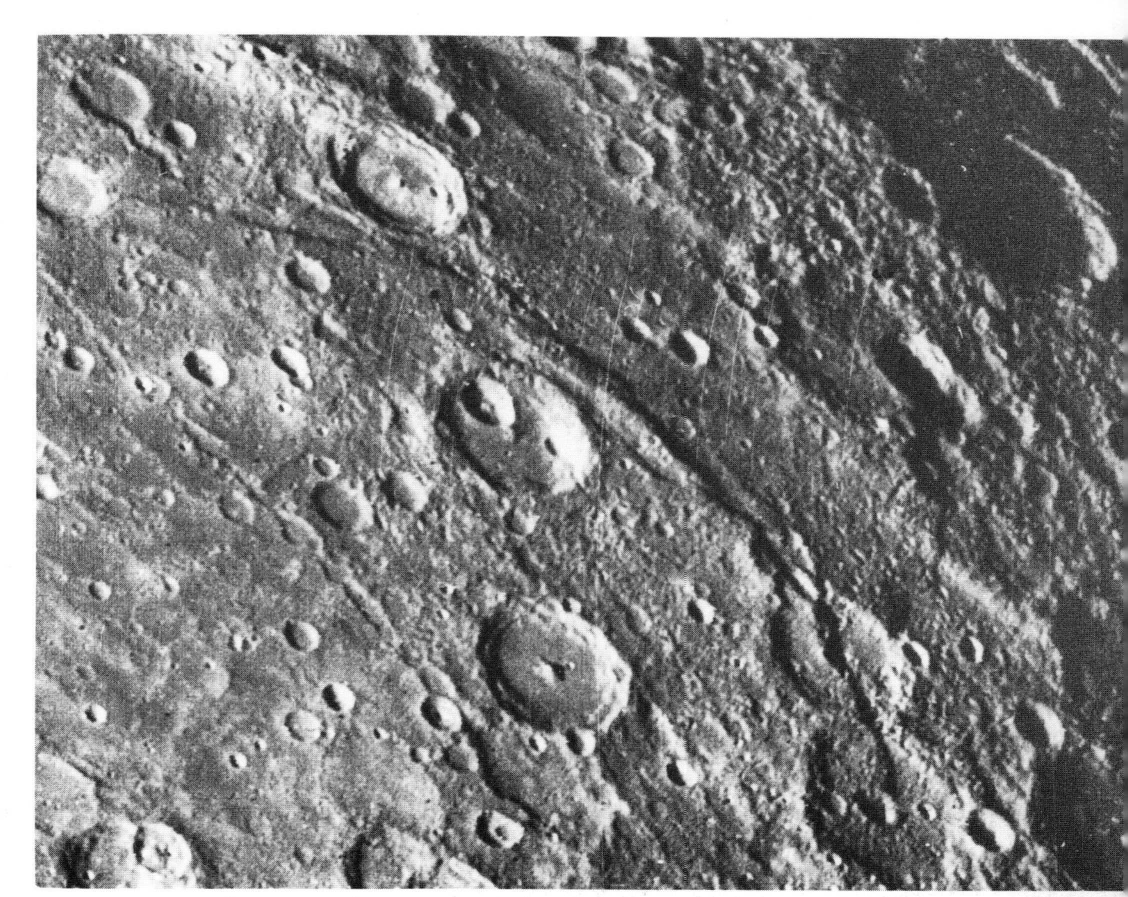

Figure 5.18 Oblique view of Hero Rupes; this typical arcuate scarp may outline an old basin and crosses smooth plains and transects several large craters. Area shown is nearly 700 km wide (from Dzurisin 1978; Mariner 10 FDS 166842).

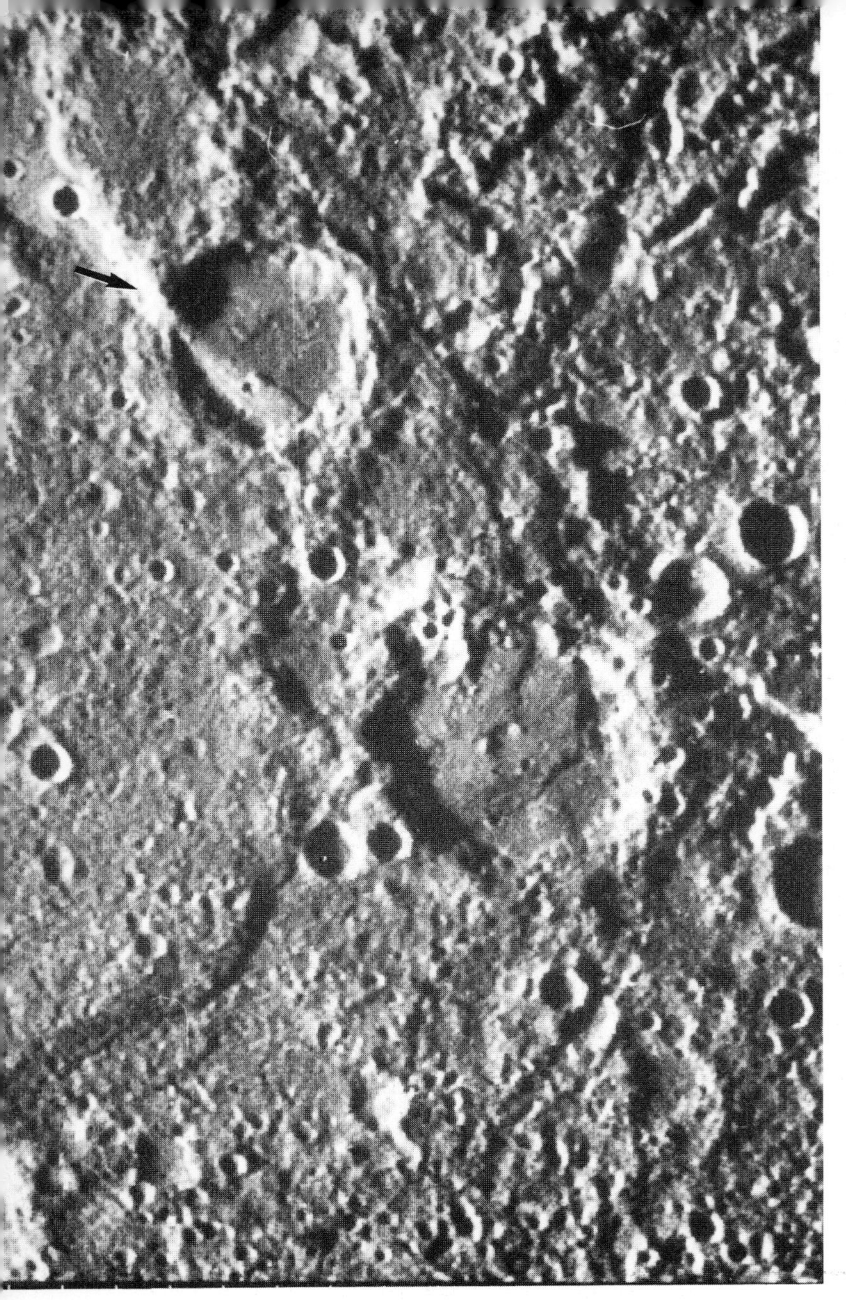

Figure 5.19 View of Vostock Rupes cutting ~75 km diameter Guide D'Arezzo crater; foreshortening of the rim (arrow) suggests 5–7 km of thrust-fault displacement and evidence of compression of the mercurian crust (Mariner 10 FDS 27380).

movement; there is no evidence of strike-slip faulting along this scarp. Analysis of the geometry implies a fault-plane dip of about 11°.

Study of all mercurian arcuate scarps suggests that they are tectonic features which resulted from compressional deformation. From the magnitude and timing of deformation, most investigators infer widespread compression of the crust early in the history of Mercury. Strom *et al.* (1975b) suggest that this compression was the result of cooling and shrinkage of Mercury's large iron core and calculate that a 1–2 km decrease in planetary radius could account for the observed crustal foreshortening. An independent analysis performed by Solomon (1976), based on models of lithospheric cooling, leads to a similar amount of crustal foreshortening. In both models the pattern of crustal deformation should be spherically symmetrical, which would generate randomly oriented compressional features. The random orientation assumes that there were no preexisting fractures.

Melosh (1977) considered the consequences of tidal despinning for Mercury which might have been generated by solar tidal torques. He speculated that compressional stresses would have been oriented E–W, with NW–NE linear shear and N–S thrusting in the crust. Unfortunately, analysis of scarp orientations is not conclusive in testing these models and is further hampered by the lack of global image coverage for Mercury. Most investigators conclude that the arcuate scarps are tectonic features which resulted from global compressive stresses generated by a combination of tidal despinning and planetary cooling.

5.4.2 Lobate scarps

Lobate scarps form the margins of elevated lobes or "tongues" of terrain and have fronts that are 500–700 km long and stand 0.5–1.0 km high. Like arcuate scarps, lobate scarps transect intercrater plains and large craters (Fig. 5.20), which led Dzurisin (1978) to suggest that they result from tectonic uplift. He noted that linear segments of some lobes are parallel to nearby lineaments and inferred that their orientation might have been controlled by some pre-existing structural "fabric" such as fractures from large basins or from stresses generated by planetary cooling or tidal spindown.

Other investigators have suggested that lobate scarps are widespread lava flows. However, the lobes are so large and have such high fronts (implying flow thicknesses of up to 1 km), it is difficult to envision how such a viscous flow could be emplaced.

5.4.3 Irregular intracrater scarps

These scarps occur on smooth, flat crater floors and may be 100 km long and 400 m high. They exhibit no consistent planimetric form, some being linear, others being sinuous. As shown in Figure 5.21, some intracrater scarps cut crater walls, whereas others appear to be diverted by topography such as crater central peaks. In some cases, the scarps separate differences in floor textures and they appear to represent a contact of some sort; Strom and colleagues (1975b) suggest that these scarps may represent viscous lava flows. Alternatively, intracrater scarps may have resulted from local tectonic processes associated with isostatic adjustments of the crater floor, or from magmatic intrusion.

Figure 5.20 Lobate scarp (arrows) which cuts intercrater plains for ~700 km; Dzurisin considers this and other lobate scarps to be tectonically uplifted blocks of crust; other investigators suggest that lobate scarps mark the flow fronts of viscous lava flows (from Dzurisin 1978; Mariner 10 FDS 166591).

5.4.4 Linear ridges

Linear ridges are straight, positive relief features 50–300 km long and 0.1–1 km high. Those found in the intercrater plains tend to have rounded profiles and may transform into linear scarps. Linear ridges in the Caloris area which are radial to the basin typically are broad and have flat crests. Figure 5.22 shows Mirni Rupes, which has been analyzed by Strom (1979) who suggests that it is actually a ridge (in contrast to "rupes" which designates scarps) which pre-dates the plains in which it occurs. His proposition is based on crater counts, and he suggests that the ridge may have formed early in the history of Mercury and then was later flooded by plains-forming materials.

5.4.5 Linear troughs

Linear troughs occur in the hilly and lineated terrain. They are 40–150 km long, 5–15 km wide, 100–500 m deep and are generally open-ended with the valley floor merging onto the surrounding terrain. Figure 5.23 shows Arecibo Valles and the sculpturing of the walls which may represent slumping and mass wasting. Strom *et al.* (1975b) suggested that linear troughs result from tensional forces and, therefore, are grabens.

5.5 The Caloris basin

Caloris (Fig. 2.18) is the largest structure observed on Mercury. Named because it lies near the subsolar point during Mercury's closest approach to the Sun, the Caloris basin is more than 1300 km across and bears many similarities to the lunar Imbrium basin. Ejecta and ejecta-related deposits dominate much of the terrain observed during the flyby of Mariner 10 (Fig. 5.1b). The

Figure 5.21 Slightly oblique Mariner 10 image showing three intracrater scarps; scarp at (A) is directed around central peak of a ~80 km diameter crater; scarp (B) cuts smooth plains in a 90 km diameter crater; scarp (C) is atypical of this class in that it extends across the crater rim into the intercrater plains to the north; also, the features consist of two almost parallel scarps and resemble an uplifted block or horst (modified from Dzurisin 1978; Mariner 10 FDS 27428).

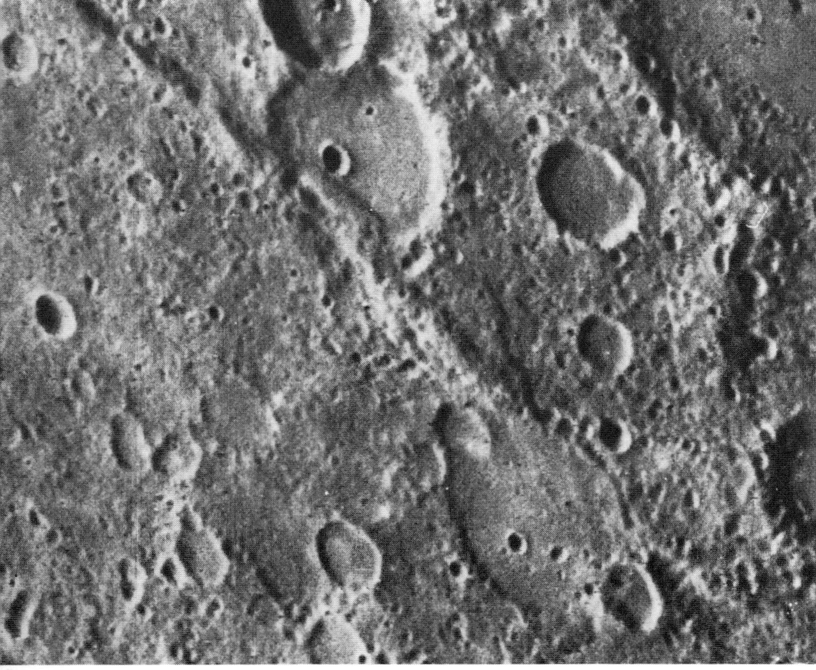

Figure 5.22 Oblique view of Mirni Rupes, a ridge in the Discovery Quadrangle at 37°S, 45°W, showing the broad, rounded cross section typical of linear ridges. Area shown is about 300 km across (Mariner 10 FDS 27420).

formation of the Caloris basin marks an important time-stratigraphic event in the history of Mercury and its ejecta blanket is a critical datum plane, similar to the Fra Mauro Formation on the Moon.

Ejecta deposits extend outward from the Caloris basin as far as one basin diameter (Fig. 5.24). On the basis of geologic mapping of the Shakespeare and Tolstoj quadrangles, McCauley and his colleagues (1981) derived a formal rock-stratigraphic sequence and proposed the name Caloris Group for the mappable units associated with the basin.

The rim of the basin is defined by the Caloris Mountains, composed of the Caloris Montes Formation which consists of rectilinear massifs as much as 30–50 km across (Fig. 5.25) standing as high as several kilometers above the surrounding terrain. The mountains are probably uplifted blocks of pre-basin crust which are veneered by late-stage ejecta from Caloris. A weakly defined outer scarp lies 100–160 km beyond the Caloris Mountains northeast of the basin and forms the outer margin of the structure.

The Nervo Formation consists of patches of hummocky to rolling plains that occur in depressions within Caloris Montes and in the zone between the mountains and the outlying scarp. In some places it appears to be draped over older terrain. This unit is interpreted as fallback ejecta which may include some impact melt.

The Odin Formation is an extensive unit which lies beyond the Caloris Mountains and was mapped by Trask and Guest (1975) as consisting of hummocky plains and low, closely spaced to scattered hills

Figure 5.23 High-resolution image of Arecibo Valles in the hilly and lineated terrain, centered at about 26°S, 28°W. The walls of the ~13 km wide valley are scalloped and may have been modified by slumping and other mass-wasting processes (Mariner 10 FDS 27470).

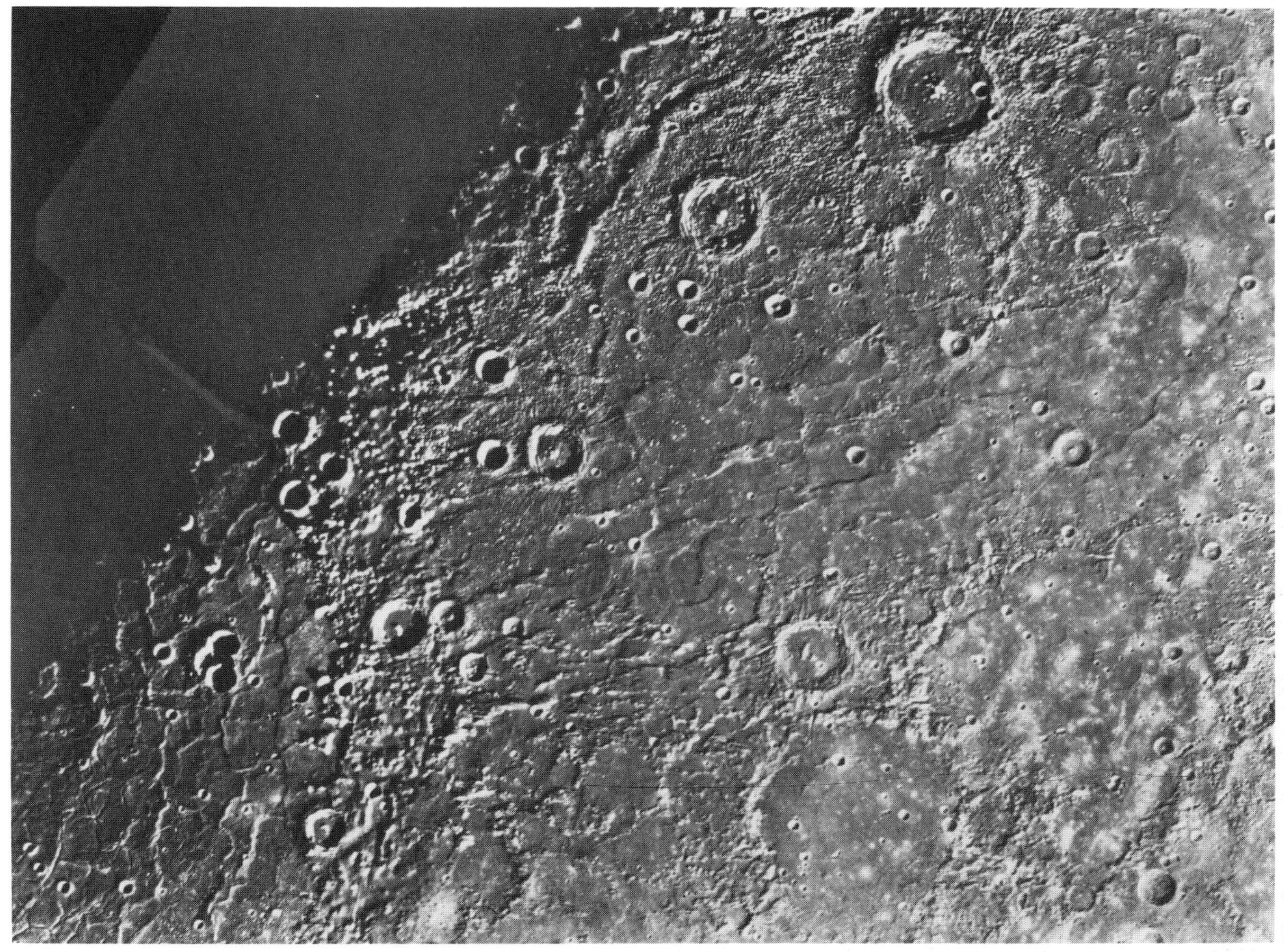

Figure 5.24 Mosaic of Mariner 10 images showing the northeast part of the Caloris basin and the Shakespeare basin.

0.3–1 km across. In some places, the hills are aligned concentrically with the basin rim and give a corrugated appearance. This unit can be traced as far as 1100 km from the basin rim and constitutes one of the major ejecta deposits of Caloris.

The Van Eyck Formation is similar to the lunar Imbrium basin Fra Mauro Formation and has a lineated facies and a secondary crater facies. The lineated facies extends as far as 1000 km from the Caloris Mountains and consists of long, hilly ridges and grooves radial to the basin (Fig. 5.26). The Van Eyck Formation is considered to represent the outermost ejecta facies associated with the Caloris basin.

The Caloris basin is floored with smooth plains that morphologically resemble those outside the basin, except that they have been severely modified by tectonic processes that have generated sets of ridges and grooves (Fig. 5.27). The plains are not considered to be impact ejecta. Because of their similarity to smooth plains exhibited elsewhere on Mercury and their resemblance to lunar mare plains, the Caloris interior smooth plains are considered to be lava flows and the interpretation is supported by the presence of possible lava channels (Fig. 5.28).

The ridges within the Caloris basin have rounded crests and are 50–300 km long, 1–12 km wide, and 100–500 m high. They form roughly polygonal patterns not seen in other mercurian basins or on the Moon. The ridges tend to be concentric with the basin rim, radial with respect to the basin center, and more numerous and larger toward the basin center. The troughs are 1–10 km wide, 100–500 m deep, and generally have flat floors and steep walls and form polygonal patterns. The troughs tend to be concentric with the rim and slightly radial to the basin center. Although some troughs transect the ridges, ridges are never seen to transect troughs (Fig. 5.29).

Based on geometry and transection relations, Strom *et al.* (1975b) suggest that the ridges formed as a result of compression as the floor of the Caloris basin subsided. Dzurisin attributes this subsidence to magma withdrawal beneath Caloris as lavas were extruded out-

Figure 5.25 Southeast side of Caloris basin showing the type locality for the Caloris Montes Formation, the uplifted crustal blocks which form the main ring of the Caloris basin. Image width is about 730 km (Mariner 10 FDS-229).

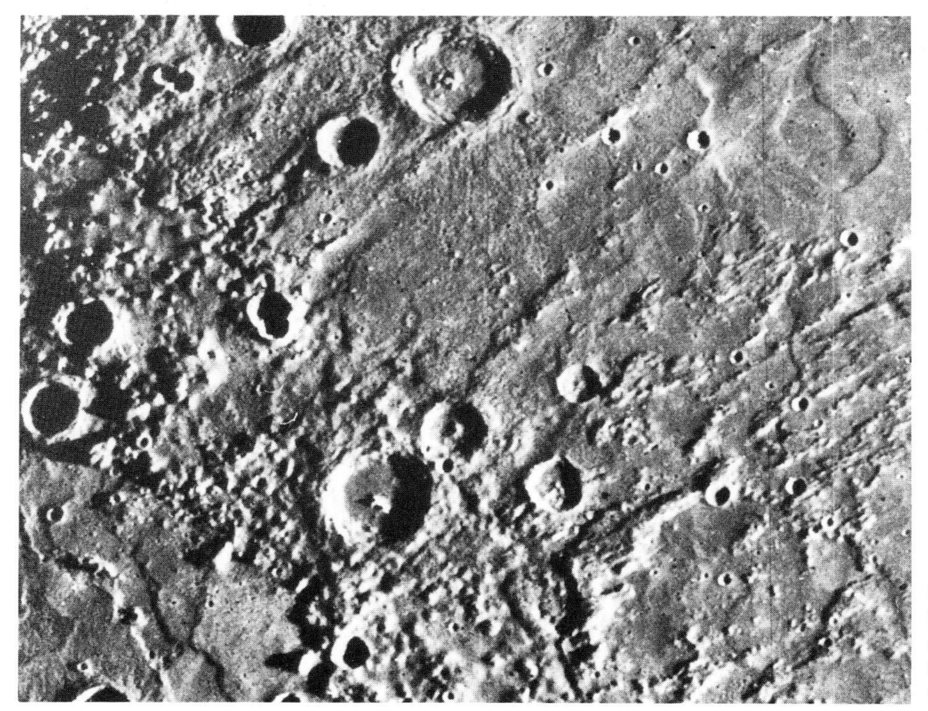

Figure 5.26 Northeast side of Caloris basin showing the radial lineated facies of the Van Eyck Formation beyond the Caloris Montes. Image width is about 590 km (Mariner 10 FDS 193).

Figure 5.27 Southeast side of Caloris basin showing the ridged and grooved smooth plains of the interior of the basin. Image width is about 275 km (Mariner 10 FDS 110).

side the basin to form the extensive smooth plains. Floor subsidence was followed by uplift of the basin center – perhaps in response to isostatic adjustment of the basin cavity – leading to tensional forces and the formation of the troughs.

5.6 Volcanism

Volcanism is repeatedly called upon to explain various features observed on Mercury and is inferred to have occurred throughout its history. Yet very few features, such as domes, cones and shield volcanoes, can be unequivocally identified as volcanic. Schultz (1977) and Malin (1978) have both considered volcanic features as might be seen on the Mariner 10 images and concluded that it would be very difficult to make positive identifications. For example, Malin examined images of volcanic features on the Moon obtained at resolutions comparable to Mariner 10 and found that their identification was far from definitive.

Malin did, however, suggest that some positive relief features could be volcanic domes (Fig. 5.30), although these features could be basin massifs. In addition, Dzurisin has suggested that segments of some ridges may be volcanic features and inferred that the fractures associated with ridge and scarp formation could have served as avenues for magma extrusion.

However, until better images are available, the evidence for volcanism on Mercury remains inconclusive.

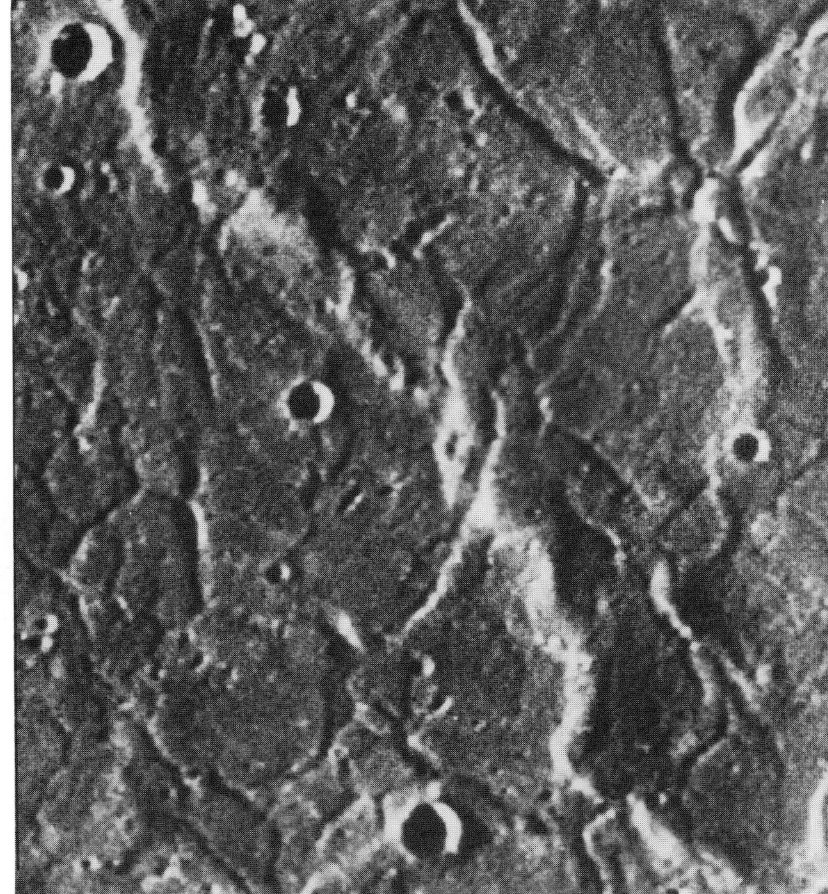

Figure 5.29 Smooth plains of the Caloris basin showing troughs transecting ridges. Image width is about 50 km; north is toward bottom (Mariner 10 FDS 106).

Figure 5.28 Interior of the Caloris basin showing possible lava channel in the smooth plains (P. D. Spudis, personal communication, 1983). Image is about 30 by 100 km (Mariner 10 FDS 528996).

Figure 5.30 High-resolution image of Discovery Dome (arrow), a 15 km diameter feature which may be a volcanic dome (Malin 1978), or remnants of a basin rim.

5.7 Geologic history

Analyses of the geologic and terrain mapping and consideration of the tectonic and thermal history exhibited by various landforms, such as arcuate scarps, lead to a synthesis of the evolution of the mercurian surface (from earliest to latest; modified from Dzurisin 1978).

(a) The formation of a solid lithosphere occurred prior to the end of heavy impact bombardment. The crust of Mercury may have been semi-molten and may not have recorded all impact events that were taking place; this may account for some of the intercrater plains and the possible lack of craters in some size ranges. Tidal spindown and/or crustal contraction (resulting from either lithospheric cooling or shrinkage of an iron-rich core) generated early-stage global tectonic fractures.

(b) Continued crustal contraction resulted in compression to form arcuate scarps, some of which postdate large craters (as evidenced by cross-cutting relations), and the rejuvenation of early-formed tectonic patterns to generate lobate scarps. This period was accompanied by extrusion of volcanic materials to form additional intercrater plains. Some ridges and scarps may have served as avenues for magmatic extrusion.

(c) The final stage of heavy impact cratering was marked by the formation of the Caloris basin. Ejecta from the basin produced radial grooves and ridges extending outward from the basin rim. Various smooth plains were emplaced both inside and outside the basin following the impact. Irregular ridges formed on the interior smooth plains as a result of compression, possibly as a consequence of basin floor subsidence, which occurred in response to magma withdrawal and extrusion of lavas to form the smooth plains outside the basin. Isostatic adjustment of the basin interior resulted in central uplift and the formation of extensional features to form troughs.

Concurrent with the impact of the Caloris basin was the formation of the hilly and lineated terrain in the region antipodal to Caloris. This terrain is considered to have resulted from focusing of seismic energy resulting from the impact.

(d) With the decline in the Solar System flux of bolides, Mercury experienced a gradually decreasing rate of impact cratering, similar to that observed on the Moon.

(e) The youngest recognizable event in the evolution of the mercurian surface (exclusive of the continued formation of small impact craters) was the emplacement of smooth plains deposits. Ridges in some of the plains suggest an origin similar to lunar mare plains, that is, related to basaltic flood volcanism.

6 Venus

6.1 Introduction

Venus has been called both the "evening star", since it is prominent in the evening sky just after sunset, and the "morning star", since it shines brightly during the dawn of the day. After the Sun and the Moon, Venus is the brightest object in the sky. However, even in ancient times, Venus was set apart from stars because of its "wandering", or planet-like, motion.

In 1610, Galileo recognized, with the aid of his telescope, that Venus has lunar-like phases. This discovery lent support to the Sun-centered concept of the Solar System proposed by Copernicus. With the improvements in telescopes over the years, bright and dark markings were seen in the atmosphere of Venus, and by the mid-1800s observers noted that Venus has a gray halo when silhouetted against the Sun. Lomonosov, a Russian astronomer, used this information to infer the presence of an atmosphere on Venus. At about the same time, estimates of the diameter of Venus were made, which led to the idea of Venus and Earth being "twin" planets. This, in turn, aroused speculation that Venus was hospitable to life.

Visual and photographic spectroscopy were applied to Venus in the 1930s and resulted in the tentative identification of carbon dioxide as the principal component of the atmosphere. Kuiper (1962) extended these observations into the infra-red portion of the spectrum and found nearly 40 CO_2 absorption bands. By this time the presence of an extensive atmosphere, including clouds, was well established. Although Michael Belton and coworkers (1968), using infra-red spectroscopy, ruled out the possibility that the clouds were water vapor or ice crystals, microwave observations made in the 1960s and some early Soviet Venera results (Table 6.1) showed the presence of some water vapor, but not in sufficient quantities to account for the clouds. Results from microwave observations also yielded brightness temperatures of about 600 K at wavelengths greater than 3 cm, and it was inferred that these could result from either high surface temperatures or ionospheric effects. Measurements from Mariner 2 resolved this problem by showing that the surface is hot. At about this same time, Sagan (1961) and Pollack and Sagan (1965) formulated the "greenhouse model" for Venus in which solar energy penetrates the venusian atmosphere, is reflected from the surface and re-radiated from clouds back to the surface, entrapping infra-red radiation. This could explain Venus's high surface temperature.

Until the mid-1970s most studies of Venus concentrated on the evolution of the atmosphere and various astronomical considerations such as the period of rotation. With the landings of Venera spacecraft in the 1970s, estimates of the composition of surface materials were made and images were taken of the surface. In addition, Earth-based radar observations shed light on global surface properties which, in combination with the Venera results, led to the initiation of geologic studies. Geologic and geophysical analyses were further enhanced with data obtained from the Pioneer Venus mission. Although designed primarily to obtain atmospheric data, the Pioneer Venus orbiter carried a radar altimeter on board to measure radar "roughness" and to determine global topography, and this has allowed some interpretations of the geophysical characteristics of Venus.

The combination of Earth-based radar, Pioneer Venus altimetry and results from the Venera missions has generated speculation on the general geology of Venus and has placed constraints on various geophysical models of its interior. Because Venus is so like Earth in its size, density and position in the Solar System, this "twin" of Earth remains of high interest in comparative planetology, as reviewed by George McGill (1983). Hunten and co-editors (1983) provide an excellent collection of papers on all aspects of Venus known through the time of the Venera 13 and 14 landers in 1982.

6.1.1 Radar data

Information on the geomorphology of Venus is derived from radar observations. As discussed in Chapter 2, radar images differ markedly from optical images and require knowledge of radar geometry and remote sensing for proper interpretation. Radio waves (including radar) emitted or reflected from planetary surfaces yield a variety of information, including estimates of: (a) surface roughness (e.g. rocks, lava textures), shown by radar-bright images which typically indicate "rough" (or high radar backscattering) surfaces and radar-dark images which typically indicate smooth surfaces; (b) average surface slopes, indicated by root-mean-square (r.m.s.) values at scales ~10 times the radar wave-

Table 6.1 Missions to Venus (adapted from Colin 1980).

Mission	Arrival	Type	Encounter characteristics
Mariner 2	14 Dec. 1962	flyby	Closest approach: 34 833 km.
Venera 4	18 Oct. 1967	bus entry probe	Burn-up. Hard lander, nightside.
Mariner 5	19 Oct. 1967	flyby	Closest approach: 4100 km.
Venera 5	16 May 1969	bus entry probe	Burn-up. Hard lander, nightside.
Venera 6	17 May 1969	bus entry probe	Burn-up. Hard lander, nightside.
Venera 7	15 Dec. 1970	bus entry probe	Burn-up. "Soft" lander, nightside.
Venera 8	22 July 1972	bus entry probe	Burn-up. "Soft" lander, dayside.
Mariner 10	5 Feb. 1974	flyby	Closest approach: 5700 km.
Venera 9	22 Oct. 1975	orbiter entry probe	Periapsis: 1560 km; apoapsis: 112 200 km; period: 48 hours, 18 m; inclination: 34°10′. "Soft" lander, dayside.
Venera 10	25 Oct. 1974	orbiter entry probe	Periapsis: 1620 km; apoapsis: 113 900 km; period: 49 hours, 23 m; inclination: 29°30′. "Soft" lander, dayside.
Pioneer Venus 1	4 Dec. 1978	orbiter	Periapsis: <200 km; apoapsis: 66 000 km; period: 24 hours; inclination: 105°.
Pioneer Venus 2	9 Dec. 1978	bus entry probes	Burn-up, dayside. 4 hard landers, dayside and nightside.
Venera 11	25 Dec. 1978	flyby entry probe	Closest approach: 25 000 km. "Soft" lander, dayside.
Venera 12	21 Dec. 1978	flyby entry probe	Closest approach: 25 000 km. "Soft" lander, dayside.
Venera 13	March 1982	flyby entry probe	"Soft" lander with imaging, dayside.
Venera 14	5 Mar. 1982	flyby entry probe	"Soft" lander with imaging, dayside.
Venera 15	10 Oct. 1983	orbiter	Radar imaging.
Venera 16	14 Oct. 1983	orbiter	Radar imaging.

length; and (c) other physical and chemical characteristics such as the dielectric constant (which is a function of many complex, interrelated parameters, including surface composition).

Earth-based radar observations of Venus which began in the 1960s have steadily improved in quality. Most of these observations have been obtained from four facilities: (a) the Arecibo radar system in Puerto Rico which has an antenna 300 m in diameter and operates at wavelengths of 12.6 and 70 cm; (b) the deep-space tracking antenna at Goldstone, California, which has a diameter of 64 m and operates at a wavelength of 12.6 cm; (c) the Haystack radar system in Massachusetts which has an antenna diameter of 43 m and operates at 3.8 cm; and (d) the Soviet Academy of Science which operates a 39 cm wavelength system. In general, the larger the antenna, the higher the spatial resolution. Thus, the Arecibo system can obtain data for as small as 1.3 km radar pixel on Venus. Under optimum conditions, about 30 percent of Venus might be imaged by radar at a spatial resolution of ~2 km, but because of the orbital geometries of Earth and Venus, the area imaged is concentrated on either side of the equatorial region on the hemisphere that faces Earth during closest approach (Fig. 2.21).

The Pioneer Venus spacecraft was placed into orbit in 1978 and data from the Pioneer Venus radar altimeter (Pettengill *et al.* 1980) have enabled the gross topography of Venus to be derived (Fig. 6.1a). These data cover the region between 74°N to 63°S and include about 93 percent of the surface with an elevation accuracy of about 200 m, integrated over an area about 100 by 100 km. In addition, coarse radar images (Figs 6.1b & 6.2) were generated from the same radar system for the region between 45°N to 15°S and were used with other radar data to portray the general relief of Venus.

The Soviet Veneras 15 and 16 were placed in orbit around Venus in late 1983 and began obtaining radar images for the northern hemisphere. With a spatial resolution of ~1–2 km, plans call for mapping about 35 percent of the surface with the 8 cm radar system.

6.2 Physiography

Although relatively flat over most of its surface, Venus has substantial topographic relief in some areas. Two features, named Alpha and Beta, were identified on the basis of their high radar backscatter properties in Earth-based radar images (Fig. 6.3). One smaller radar feature, named Eve, is southwest of Alpha and is a radar-bright ring encircling an intense radar-bright central spot (Fig. 6.4). This feature is used as the planet-fixed reference for longitude; because Venus is in

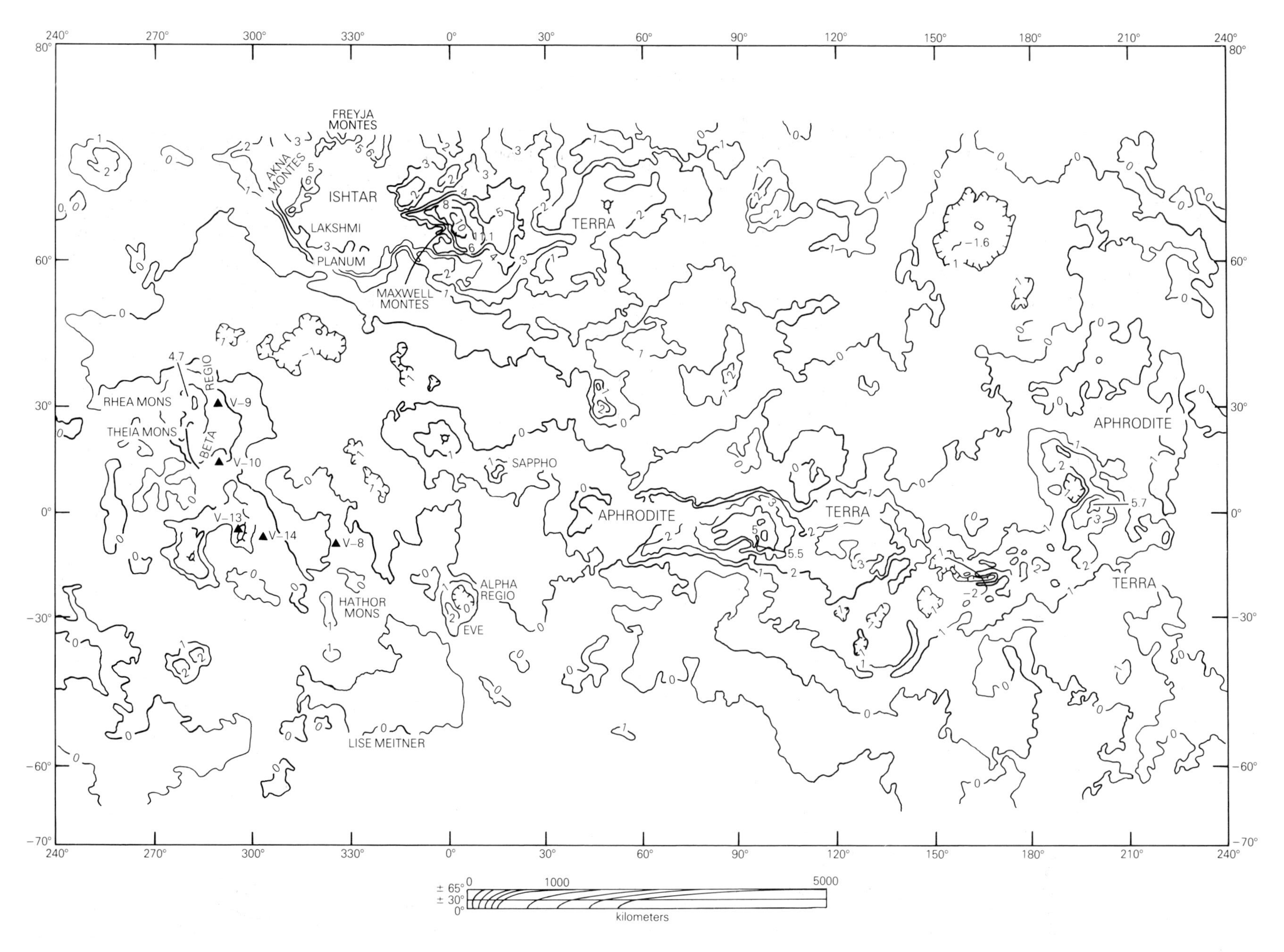

Figure 6.1a Generalized topographic map of Venus showing elevations of surface features; contour interval 1 km (courtesy US Geological Survey).

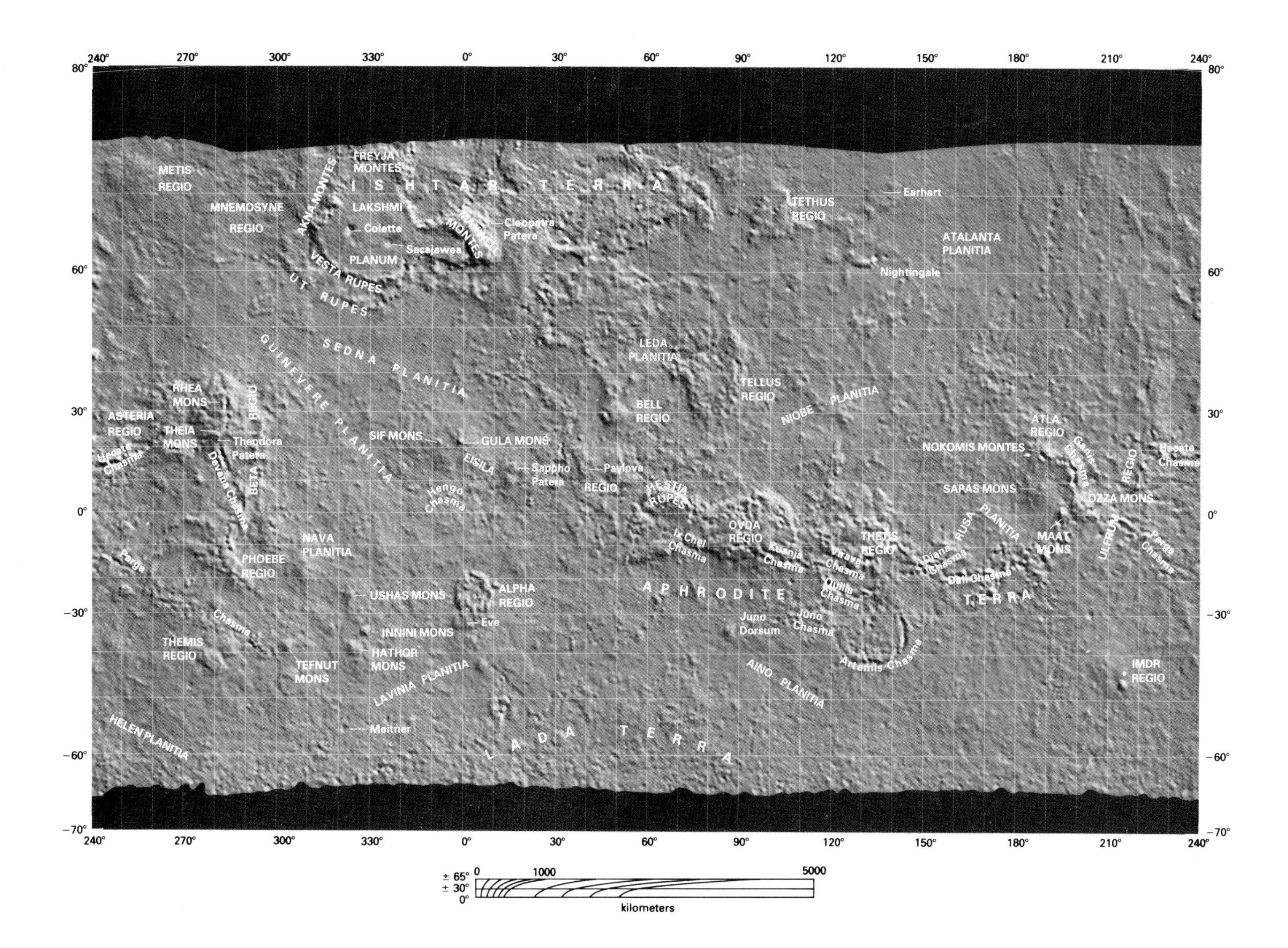

Figure 6.1b Shaded relief map of Venus based on altimetric data from the Pioneer Venus orbiting spacecraft (courtesy US Geological Survey).

Figure 6.2 Surface roughness at a scale of 1 : 10 m; areas with the greatest roughness are shown as white ("radar-bright") and decreasing roughness is indicated by progressively darker tones (courtesy US Geological Survey).

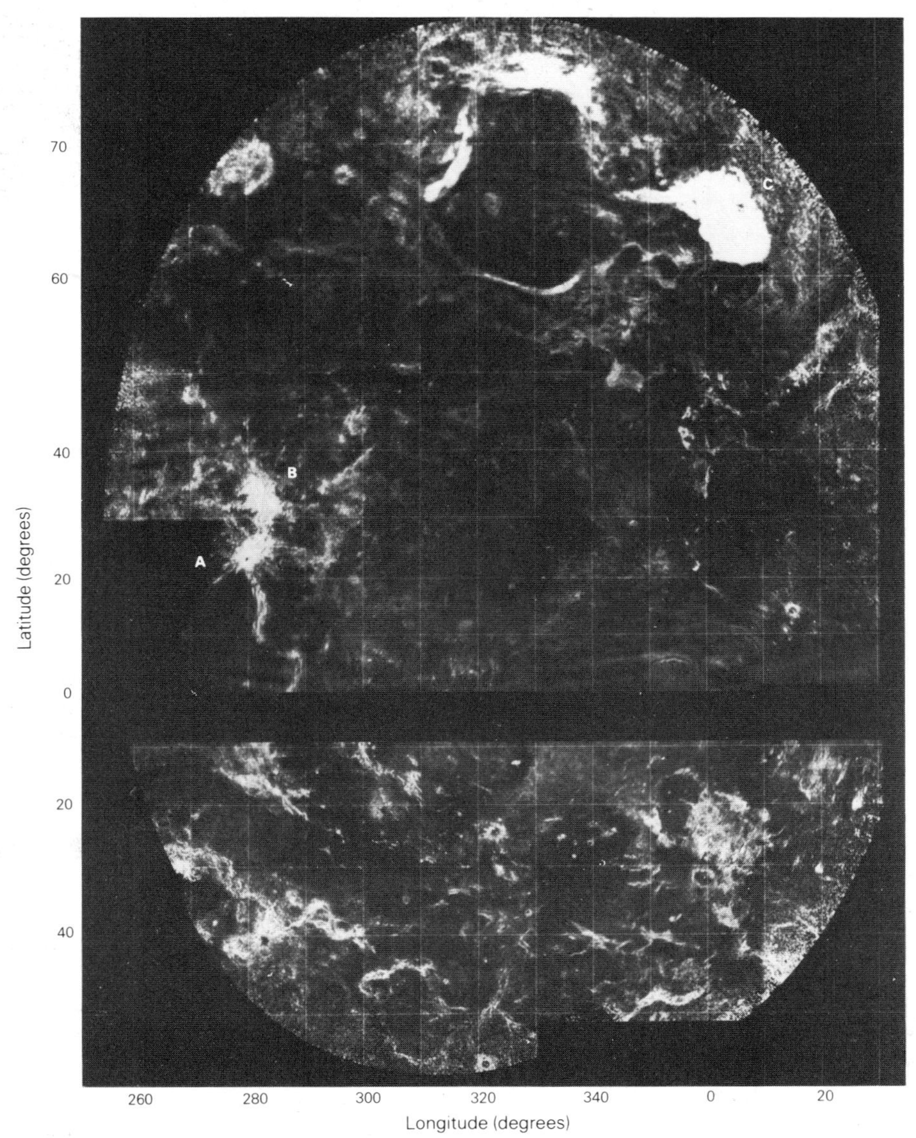

Figure 6.3 Mosaic of radar images from the Arecibo facility, showing some of the prominent named features on Venus. Note the bright patterns, suggested to be rough lava flows, radiating from Theia Mons (A) and Rhea Mons (B) in Beta Regio. Maxwell Montes (C) is in upper right. Resolution of the component images ranges from 10 to 20 km (from Campbell & Burns 1980, *J. Geophys. Res.*, copyright American Geophysical Union).

retrograde rotation, longitude increases toward the east by International Astronomical Union convention. Except for Alpha, Beta and a few other features, most geographic features on Venus are given women's names such as those of goddesses from mythology (Fig. 6.1).

The highest point on Venus occurs within Maxwell Montes and rises at least 11.5 km above the datum (datum is based on the mean planetary radius). For comparison, Mt Everest rises about 12 km above the Earth's median elevation (which is about 3 km below sea level). The lowest area on Venus is in Diana Chasma, a trench centered at 14°S, 156°E. The floor of the trench is at least 2 km below datum, equivalent to the Dead Sea Rift on Earth; however, it is less than one-fifth the depth of the Mariana Trench which at − 8 km below the Earth's median elevation (− 11 km below sea level) is the lowest place on Earth. Thus, the total topographic relief on Venus is about 13 km, considerably less than Earth's maximum relief of 20 km. However, the smoothing effect of the "footprint" (which ranges in size from ~100 by 100 km for high latitudes to ~25 by 10 km for equatorial regions) for the Pioneer Venus altimetry data undoubtedly masks areas that are both higher and lower than these values so the total relief on Venus probably exceeds 13 km.

Hypsometric curves show the percentage of global topography lying at different elevations and are useful for planetary comparisons. As shown in Figure 6.5, Venus is strongly unimodal in its distribution of topography, Earth is strongly bimodal and Mars is trimodal. The bimodal distribution of topography on Earth results from two distinct zones: the continental platforms (including the continental shelf) and the deep-sea basins. Early workers, including G. K. Gilbert and Alfred Wegener, speculated that if Earth's crust consisted of homogeneously distributed materials, then the hypsographic curve should be a simple Gaussian distribution. However, recognition of two distinct topographic levels led them to consider the compositional and density differences between continental crustal materials (sial) and oceanic crustal materials (sima) and the consequences of isostatic adjustments. Except for the maximum relief noted above, Venus is extremely flat, with 60 percent of the surface falling within 500 m of the modal planetary radius (~6051.5 km), leading some investigators to speculate that the venusian crust may not be differentiated in a manner similar to Earth's crust.

Masursky and colleagues (1980) have subdivided Venus into three provinces based on elevation (Fig. 6.6): (a) *upland rolling plains* which lie between 0 and + 2 km elevation and constitute 65 percent of the surface; (b) *highland provinces* which are higher than + 2 km elevation and constitute 8 percent of the surface; and (c) *lowland areas* which are lower than the 0 km datum and

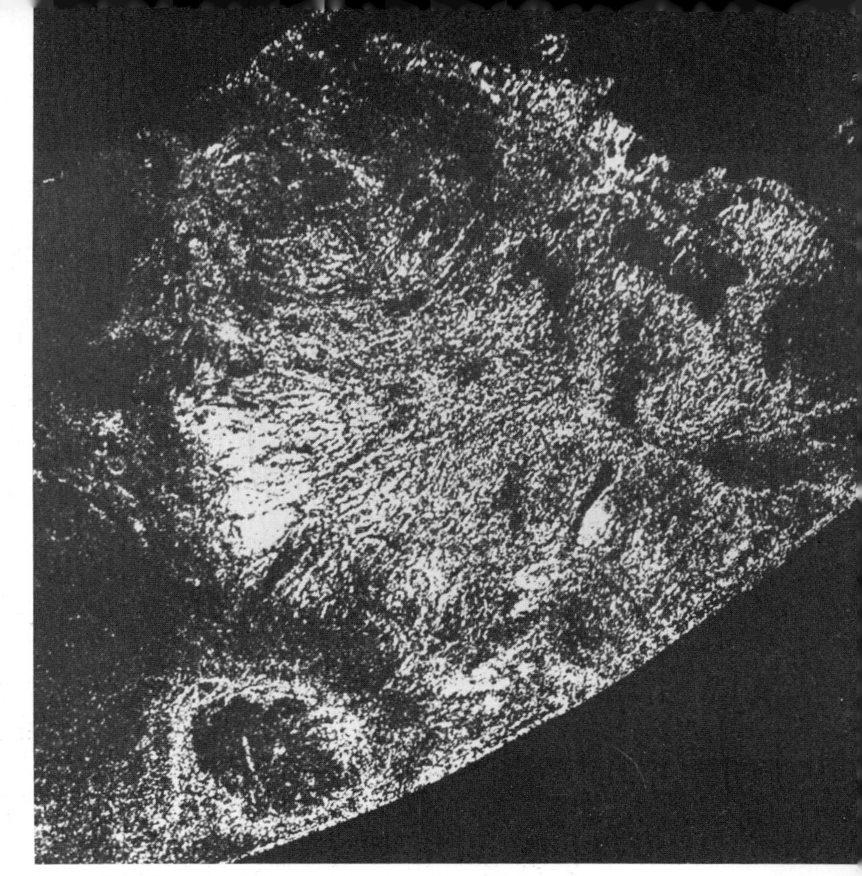

Figure 6.4 High-resolution radar image of the Alpha region showing Eve, the bright spot in the ring-shaped feature; the ring is about 200 km across and may represent an impact crater (courtesy D. B. Campbell, Arecibo Observatory).

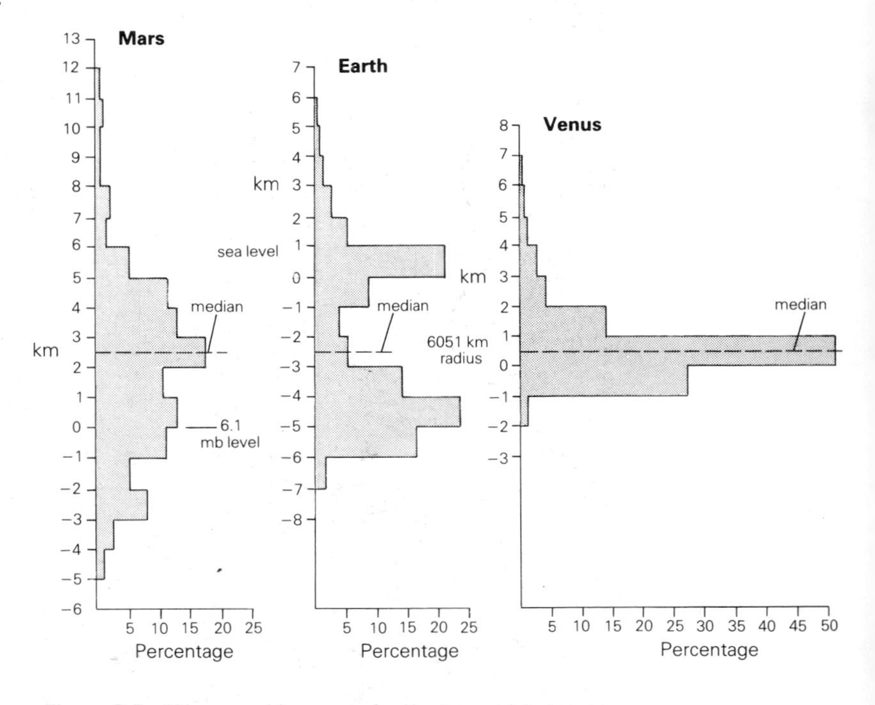

Figure 6.5 Hypsographic curves for the terrestrial planets, showing Venus (unimodal distribution), Earth (bimodal distribution reflecting continent and ocean basins) and Mars (trimodal distribution for the northern lowlands, the intermediate-level cratered terrain and the high volcanoes).

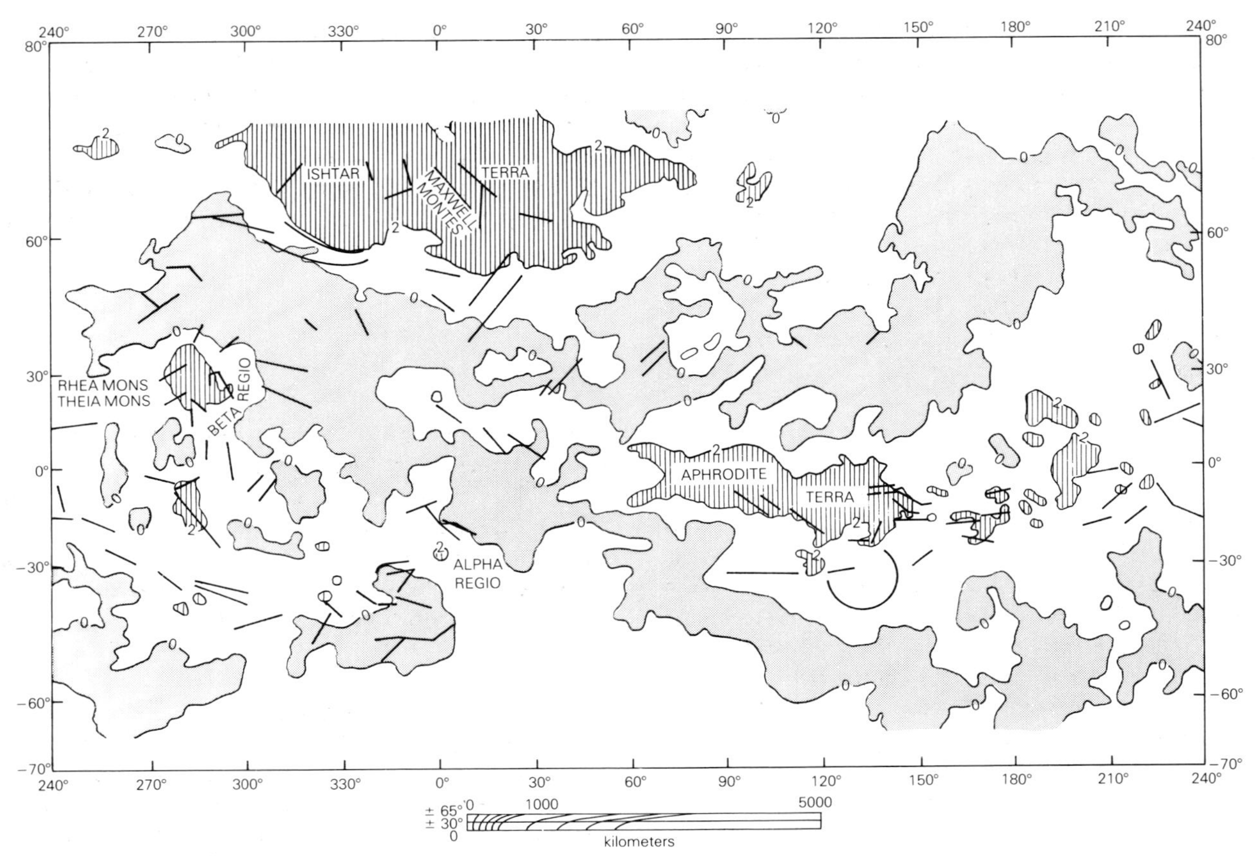

Figure 6.6 Map showing the distribution of topographic provinces of Venus: rolling plains (no pattern), 0–2 km above datum; highlands (ruled) higher than 2 km; and lowlands (shaded) lower than 0 km, and tectonic lineaments thought to be scarps, trenches or ridges visible on topographic, reflectivity, or roughness maps (modified after Masursky *et al.* 1980).

constitute 27 percent of the surface. While these provinces are based solely on elevation, there is also a general correlation between elevation and radar-brightness (i.e. surface roughness; Fig. 6.2): the upland rolling plains are of intermediate brightness, the highland provinces are radar-bright and the lowlands are radar-dark. This correlation may be related to surface processes in which high-standing areas are eroded and fine-grained materials are removed, leaving blocky surfaces. These fine-grained sediments are then deposited in the lowlands to produce smooth, radar-dark surfaces.

Owing to the topographic relief of the surface, atmospheric temperature and pressure vary over the various topographic provinces. In the lowlands, temperatures are approximately 465 °C and atmospheric pressure is 96 bar; at the elevation of the highlands, the temperature and pressure are lower: 374 °C and 41 bar, respectively (Warner 1983). This wide range of conditions means that impact, volcanic, weathering and mass transport processes may operate differently in the highlands and the lowlands.

6.2.1 Rolling plains

The rolling plains constitute most of the venusian surface. Radar data suggest slopes ranging between 2.2° and 3.5° on a local scale of 1 : 10 m. Numerous circular features have been discovered on the rolling plains. Most of these features are shallow and have flat floors. Some display bright central spots. The circular features could be impact craters, volcanic features or tectonic structures. Other circular radar-bright features, such as Hathor Mons and Sappho, are 200–300 km in diameter and stand more than 1 km above the surrounding plains. Some of these features have radar-dark centers and are postulated to be volcanoes (Masursky *et al.* 1980).

Alpha Regio is a radar plateau having a range of radar brightnesses centered within the rolling plains at 25°S, 5°E. It consists of a circular central region having a mean elevation of 0.5 km, surrounded by a discontinuous rim rising 2.5 km above datum. Alpha Regio displays high r.m.s. slopes (4.5° to 10°) and numerous parallel, anastomosing lineaments which run northeast to southwest for more than 1300 km. Masursky and colleagues (1980) note that Alpha Regio resembles the highly fractured Tharsis plains of Mars and may be a tectonically modified plateau, an uplifted block of ancient crust or block-faulted terrain similar to the North American Basin and Range Province on Earth.

Various arcuate features of possible tectonic origin also occur within the rolling plains. Artemis Chasma is a huge curvilinear feature (Fig. 6.1b), consisting of two asymmetric ridges separated by a trough, similar to some terrestrial rift valleys. Another possible depression, Surali Chasma, centered at 2°N, 3.5°E, consists of two radar-bright curved features separated by a dark zone, similar to Artemis Chasma.

6.2.2 The highlands

The venusian highlands form prominent radar-bright areas which occur mostly in three regions (Fig. 6.6): Ishtar Terra and Aphrodite Terra, both of which are continent-size masses, and Beta Regio. Ishtar Terra, centered at 65°N, 350°E, is an Australia-size mass composed of three elements: (a) Lakshmi Planum which forms a vast western plateau and contains two depressions, Colette and Sacajawea; (b) Maxwell Montes which forms the central part of Ishtar Terra and includes the highest-standing mountains on Venus; and (c) an area east of Maxwell that may be fractured, as indicated by a complex array of ridges and troughs. Preliminary results from the Venera 15 and 16 missions show several craters 10–40 km in Lakshmi Planum. Sacajawea was described by the Soviets as resembling a caldera during early discussions of mission results.

Aphrodite Terra, centered at 5°S, 120°E, is an Africa-size mass also composed of three sections: (a) a western mountainous area which rises to 5.5 km; (b) a central mountain complex standing 4 km above datum; and (c) an eastern section (Atla Regio) which forms a curved mountainous belt, parts of which rise to 5.7 km. The central and eastern parts are separated by a low, broad saddle which contains complex ridges and troughs, such as Diana and Dali Chasmae.

Beta Regio is highly variable in both radar brightness and r.m.s. slopes. This high-standing area is dominated by two shield-shaped masses, Theia Mons and Rhea Mons, both of which rise above 4.5 km. Saunders and Malin (1977) proposed that Theia Mons could be a volcanic complex, a speculation supported by data from Veneras 9 and 10 which landed on the flanks of Beta Regio and yielded results suggesting basaltic compositions for the surface.

6.2.3 The lowlands

The lowlands, defined as areas lower than the 0 km datum, occur in several areas (Fig. 6.6), notably in the Guinevere–Sedna Planitia area between Ishtar Terra and Beta Regio, and in Atalanta Planitia, a region east of Ishtar Terra that is the size of the Gulf of Mexico. In general, the lowlands are radar-dark to variable and have low r.m.s. slopes, suggesting smooth surfaces. Venera 15 and 16 images showing marelike ridges and the scarcity of abundant impact craters suggest that they may be young, lowland plains analogous to lunar mare or ocean basins on Earth.

6.3 Craters

Impact craters have been identified on all the other terrestrial planets and their presence is to be expected on Venus. Radar images reveal numerous circular features that could be impact craters (Cutts *et al.* 1981; Fig. 6.7). In this section, the radar observations are presented and the potential effects of the unique venusian environment on impact cratering are discussed.

6.3.1 Radar observations

Campbell and Burns (1980) analyzed Arecibo radar data and recognized two classes of circular features. Class I features are quasi-circular, diffuse, radar-dark patterns ranging from 200 to 1300 km across. Twelve such features were identified, most of which have small, radar-bright spots in the interior, suggestive of central peaks. Some of these features were described by Malin and Saunders (1977) as having low, but distinctive, raised rims, based on analysis of Goldstone radar images. Although topographic data are lacking for most of these features, Masursky *et al.* (1980) note that some have low depth-to-diameter ratios and give depths of 200 m for one 400 km feature and 700 m for a 600 km feature. Furthermore, they found that class I features are characterized by smooth surfaces and low r.m.s. slopes, suggesting that these features could be impact basins filled with lava flows, similar to some lunar basins. Radar images obtained by the Venera 15 spacecraft early in the mission reveal numerous features that were interpreted by the Soviets and Boyce *et al.* (1984) to be multi-ringed impact basins.

Class II features include 33 radar-bright rings ranging from 20 to <300 km across. They are crisp, well defined, circular features of low radar backscatter surrounded by a high backscatter zone, suggesting rough terrain. The lower size limit is a function of resolution; because at least four resolution cells are required for detection, 20 km is about the smallest possible feature that could be defined. Some of these features have central radar-bright spots similar to central peaks which are typical for impact craters in this size range on other planets.

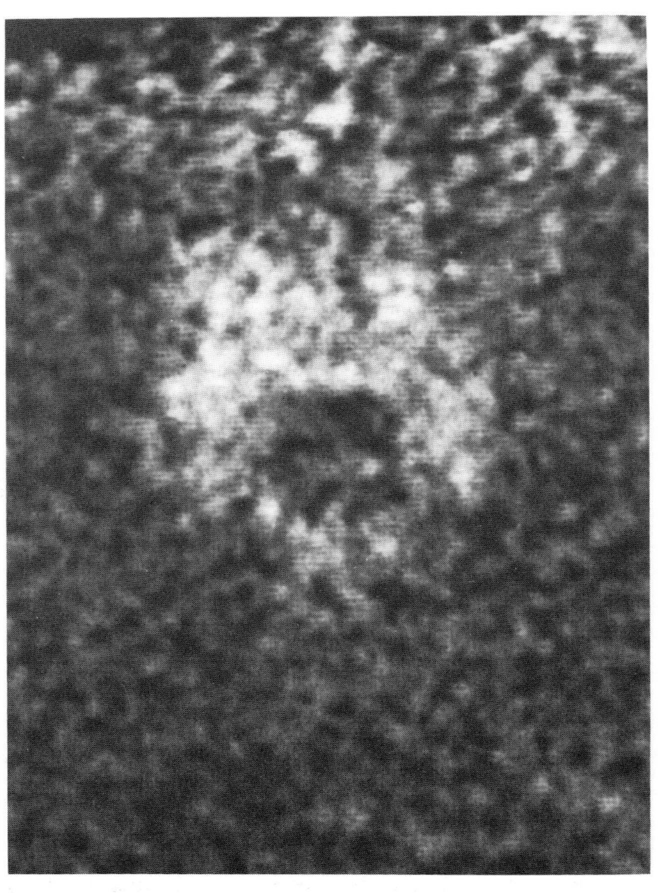

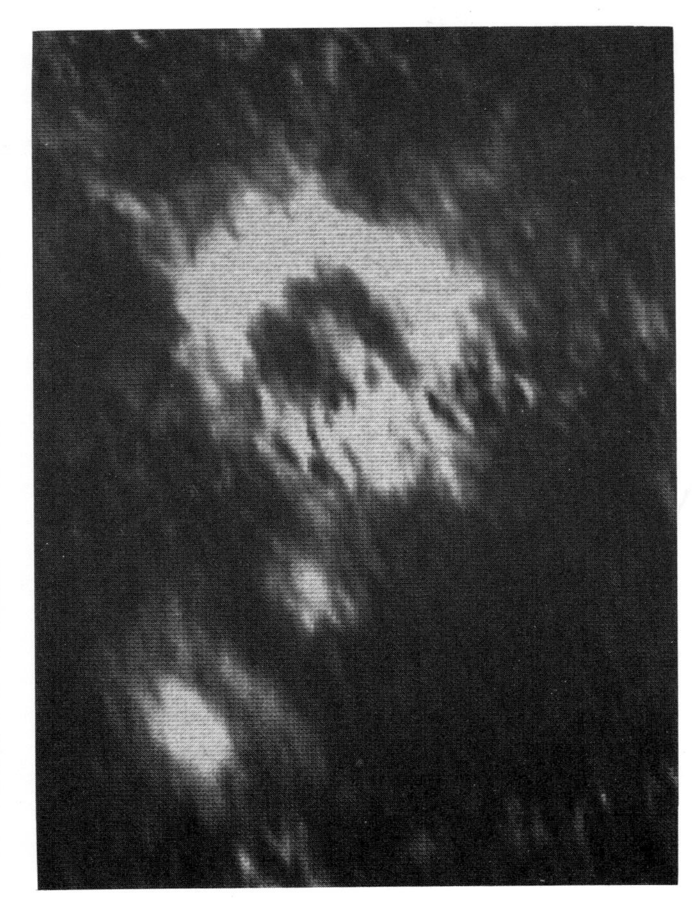

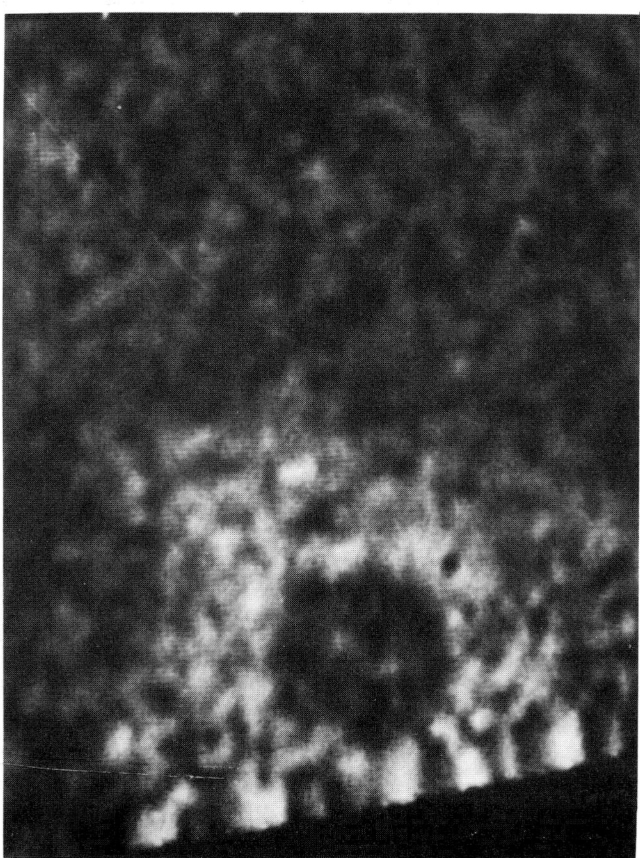

Figure 6.7 Examples of circular radar patterns on Venus: (a, top left) dark zone encircled by radar-bright ring at 18°N, 327°E; (b, above) 120 km diameter feature at 14°N, 16°E; (c, left) two central bright features in dark area surrounded by circular bright zone at 56°S, 321°E (from Campbell & Burns 1980, *J. Geophys. Res.*, copyright American Geophysical Union).

6.3.2 *Venus's environmental effects on impact processes*

The high temperature and dense atmosphere on Venus (Table 2.1) may influence impact cratering processes in several ways: (a) by slowing the incoming bolide; (b) by impeding the emplacement of some ejecta; and (c) by enhancing post-impact modifications of the crater. Tauber and Kirk (1976) assessed the effects of the venusian atmosphere on impacting bolides and noted that small objects will break apart before impacting the surface. Fracturing in flight would be caused by frictional heating and/or air pressure acting directly on bolides at high speeds. They calculated that the smallest stony meteoroid that would survive is ~80 m in diameter, whereas the smallest iron meteoroid would be 30 m. From scaling relationships based on impact kinetic energies, Tauber and Kirk estimated that the smallest craters would be 300–400 m for stony bolides and 150–200 m for iron bolides, thus placing lower limits on the sizes of impact craters expected on Venus.

The atmosphere also affects the emplacement of impact ejecta. Based on calculations of atmospheric drag, Tauber and co-workers (1978) at NASA-Ames Research Center note that the range of some ejecta fragments would be substantially less on Venus than on planets with thin or no atmospheres. Their calculations, however, do not take into account the possibility of turbulent flow within the ejecta cloud, and Schultz and Gault (1979) and Schultz (1981) have speculated that ejecta fragments <10 m could be carried along the surface as entrained flow. They also suggest that ejecta from large (>100 km) craters might not be slowed appreciably upon ejection, but during re-entry fragments as large as 10 m would be decelerated so that fragmentation upon impact would be minimal. Thus, one might expect an annulus of coarse blocks to surround larger impact craters on Venus following their formation.

The high temperature on Venus is expected to influence the amount of impact-generated melt and vaporization (Grieve & Head 1982). They estimate that as much as 50 percent more melt and vapor would be produced on Venus than a comparable impact on Earth. The high ambient temperature also may have an important influence on large craters in the post-impact modification stage. Solomon and co-workers (1982) have calculated that viscous relaxation of large topographic features may be enhanced in the high temperatures on Venus in comparison to other terrestrial planets. They conclude that impact basins that are larger than several hundred kilometers and older than 3 aeons would have "relaxed" to the point where they would have little topographic relief at present, as shown in Figure 6.8.

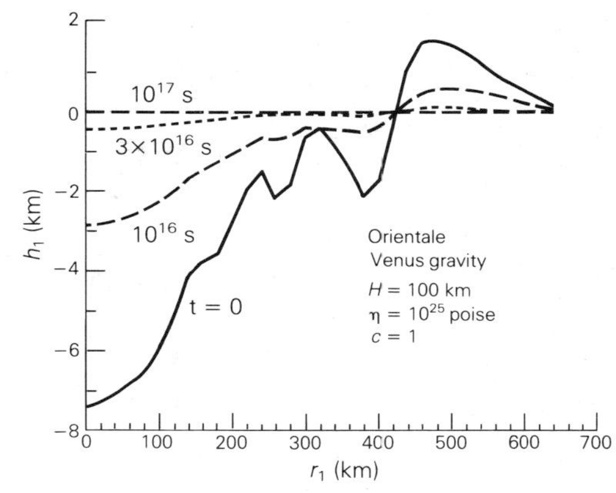

Figure 6.8 Viscous relaxation of topography for an impact basin on Venus with an initial profile similar to the Orientale basin on the Moon (Fig. 2.5); times shown scale linearly with the assumed viscosity (from Solomon *et al.* 1982, *J. Geophys. Res.*, copyright American Geophysical Union).

6.3.3 *Discussion*

Despite numerous analyses of available radar images and attempts to scale impact crater processes to Venus, it is still difficult to make positive identification of features as impact craters. Nonetheless, comparison of venusian features with radar images of lunar craters (Fig. 6.9) shows some close similarities, and it is likely that as better-quality images become available for Venus, impact craters will be found. Identification of basin-size impact features on Venus is even more problematical. Schaber and Boyce (1977) identified 12 circular features >600 km as possible impact basins, some of which are included in the class I category. These features lack appreciable relief, as would be expected according to the viscous relaxation model. One class II feature, Atalanta Planitia, however, exhibits greater topographic relief than would be expected for an ancient impact basin. Thus, either Atalanta is exceedingly young (not consistent with known impact fluxes), or it resulted from processes other than impact. Solomon *et al.* (1982) suggest that Atalanta may be similar to crustal platform basins produced on Earth by lithospheric extension and thermal subsidence.

Despite the uncertainties in the origin of the circular features observed on Venus, it is useful to consider the implications for surface history based on their size-frequency distributions. Campbell and Burns (1980) concluded that the surface of Venus may be relatively young. Based on the size-frequency distribution of class II features, they suggest that the surface is younger than 600×10^6 years (Fig. 6.10), assuming that all of the features are of impact origin. During presentation of preliminary results from Veneras 15 and 16, Basilevsky estimated a surface age of about 1 aeon for the area north of 30° latitude based on impact crater frequencies. If some of the features are non-impact, then an even younger age is implied.

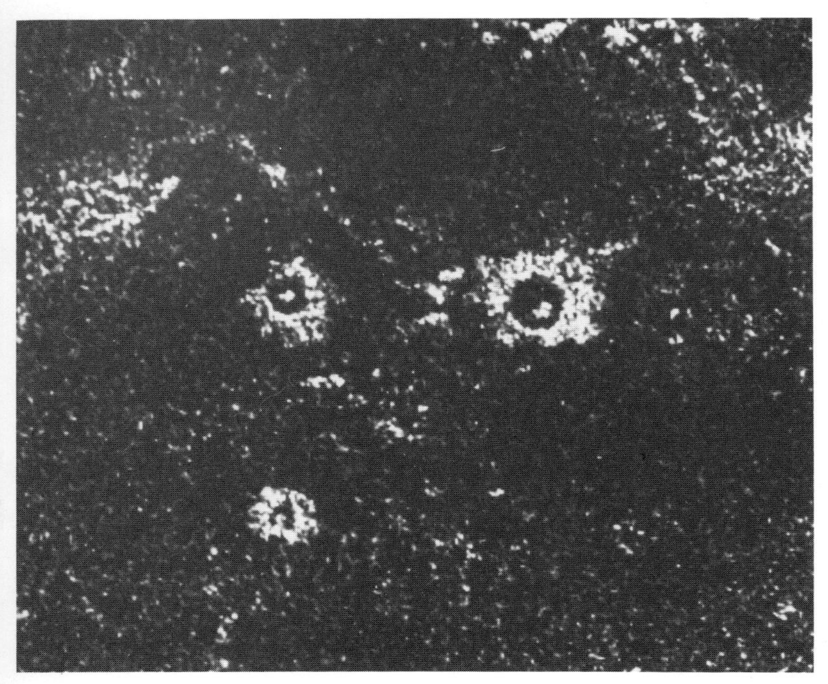

Figure 6.9 Radar image of circular features on Venus. The image has a resolution of 5 km and covers an area 500 km wide, centered at 27°S, 340°E (from Campbell & Burns 1980, *J. Geophys. Res.*, copyright American Geophysical Union).

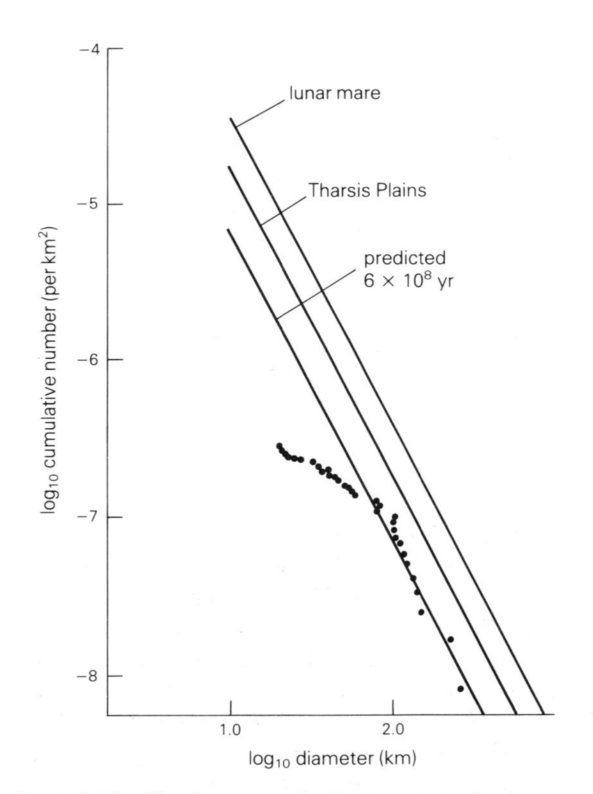

Figure 6.10 Size-frequency distribution of circular features on Venus compared with impact crater distributions on the Moon and Mars, and as predicted for Venus (from Campbell & Burns 1980, *J. Geophys. Res.*, copyright American Geophysical Union).

These implications for ages are not necessarily inconsistent. Except for features such as Atalanta, class I features could be impact basins that have been viscously deformed, yet still preserve blocky ejecta fields or fracture patterns in bedrock that would appear radar-bright on images. The large ejecta blocks associated with this size impact event may not be fragmented owing to atmospheric deceleration in the final stages of their trajectory. Smaller craters, represented by class II features, would be less affected by viscous relaxation but could be obliterated (or at least made radar "invisible") by processes such as burial by lava flows, pyroclastic deposits or aeolian sediments. Thus, while parts of the venusian *crust* may be relatively ancient (preserving larger impact basins), some *surfaces* (e.g. lava flows and sedimentary mantles) may be relatively young.

6.4 Volcanism

The similarity of Earth and Venus in mass, density and diameter would suggest that volcanism and tectonism should be as important on Venus as on Earth. Although, as with the question of impact craters, definitive evidence for volcanism is lacking, radar images reveal various landforms that resemble volcanoes.

Furthermore, surface compositions determined for five sites on Venus (Fig. 6.1) are suggestive of volcanic rocks. Some Venera missions (Table 6.1) consisted of a two-part vehicle – a landing probe released prior to arrival at Venus and a flyby bus which served as a communications link between the lander and Earth. Veneras 8, 9 and 10 measured the abundances of radioactive uranium, thorium and potassium using gamma-ray spectrometers. Venera 8 landed in rolling plains and yielded compositions similar to alkali basalts. Veneras 9 and 10 are located on the slopes of Beta Regio – a suspected volcano – and yielded results (Surkov 1983) suggestive of alkali-olivine basalts and tholeiitic basalts, respectively (Fig. 6.11). Veneras 13 and 14 used drills to obtain small soil samples, then determined the chemical compositions using X-ray fluorescence techniques (Moroz 1983). Venera 13 landed in rolling plains and found materials similar to leucite basalt, whereas Venera 14, located on smooth plains, found materials similar to tholeiitic basalts. The plains units are within 1 km of the median elevation and are near the flanks of large domical features which may be associated with volcanic activity.

In addition to estimates of composition, Venera 10 assessed the density of surface materials and obtained values of 2.8 ± 0.1 g cm^{-3}, comparable to terrestrial basalts. Thus, igneous processes appear to have operated on Venus with mafic volcanism being domi-

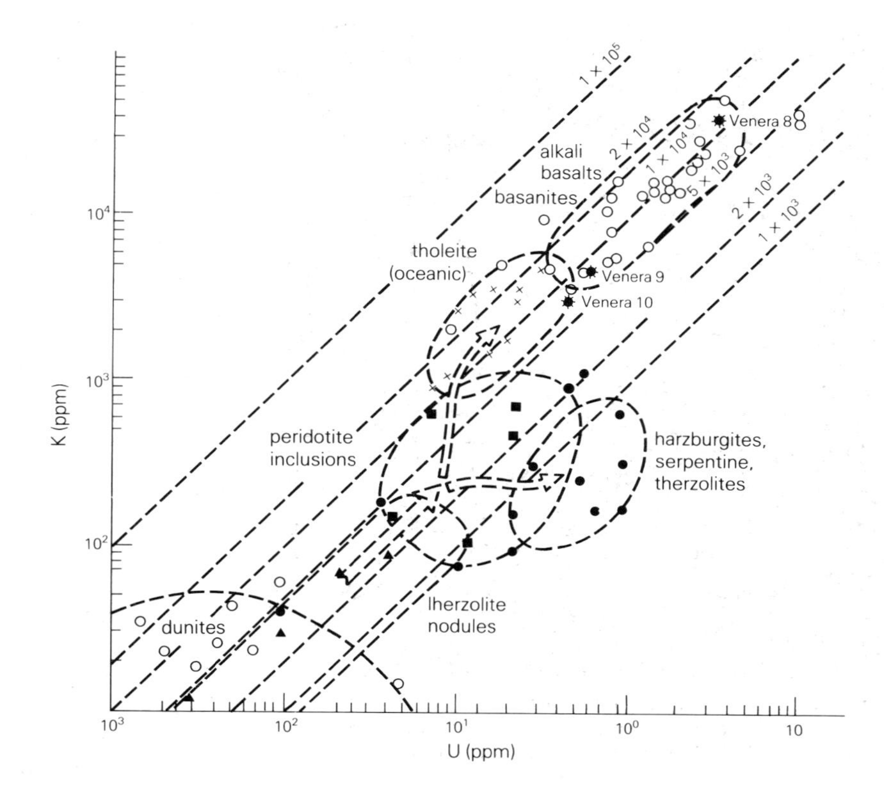

Figure 6.11 Diagram for the content of potassium and uranium in rocks on Earth compared with results obtained by Veneras 8, 9 and 10 (from Surkov 1983).

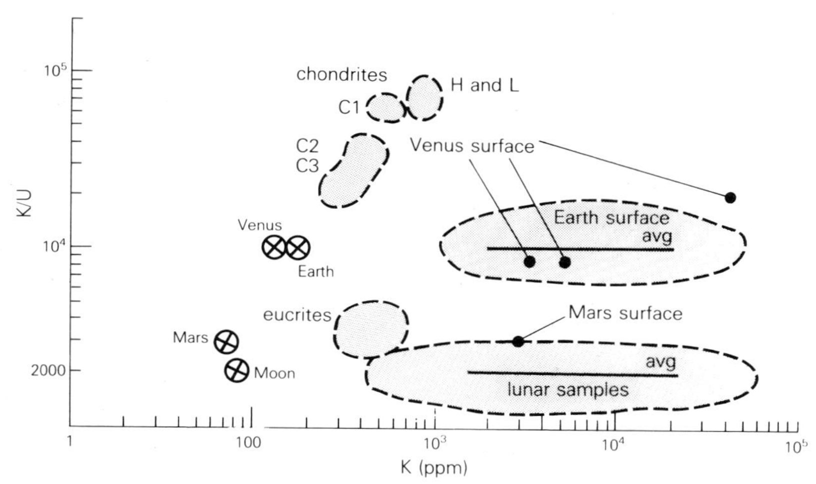

Figure 6.12 Diagram of potassium abundances and K/U ratios for Venus, Earth, Mars, Moon and selected meteorites, showing the similarity between Venus and Earth (from Taylor 1982).

nant at the sites analyzed. Although differentiation apparently has occurred, Venus may not have differentiated to as great an extent as Earth, as suggested by Figure 6.12.

6.4.1 Volcanic features

Possible volcanic features on Venus include central volcanoes, volcanic plains and ring complexes. Beta Regio (Fig. 6.1) is a large, dome-shaped area displaying a variety of radar brightnesses, elevations and r.m.s. slopes and is surmounted by two mountains, Theia Mons and Rhea Mons. Both of the mountains have slopes that average 1°, consistent with shield volcanoes formed by very fluid lavas. Theia Mons, discussed by Saunders and Malin (1977) as a volcano (Fig. 6.13), contains a summit depression which could be a caldera. Arecibo radar images reveal bright, ray-like patterns radiating from Rhea Mons and Theia Mons (Fig. 6.3) which Masursky and colleagues (1980) suggest to be rough lava flows that were extruded onto older, smoother surfaces. One of the first high-resolution images (Fig. 6.14) released by the Soviets from the Venera 15 mission shows a circular feature containing a central

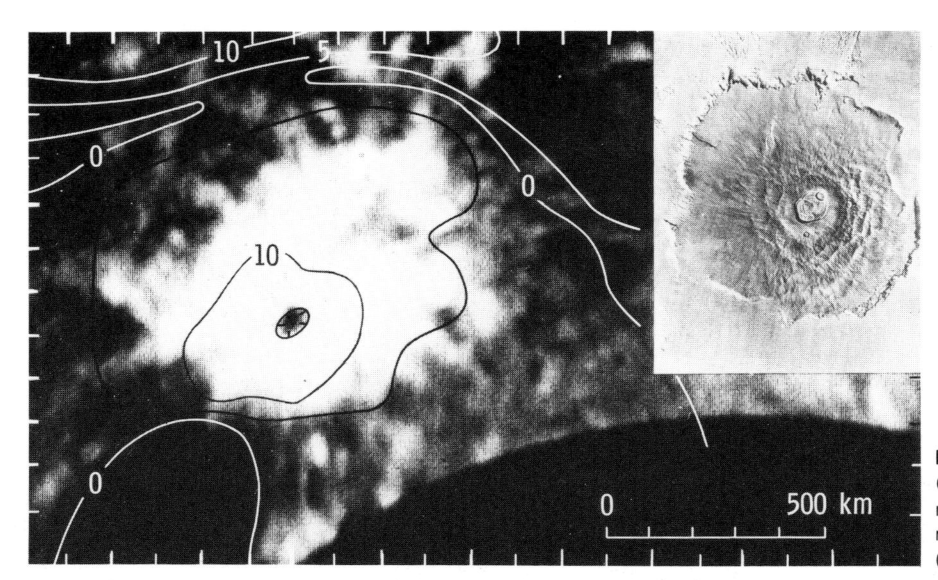

Figure 6.13a Topographic contours (in km) superimposed on Theia Mons, note the central depression; inset is the martian shield volcano Olympus Mons (from Phillips & Ivins 1979).

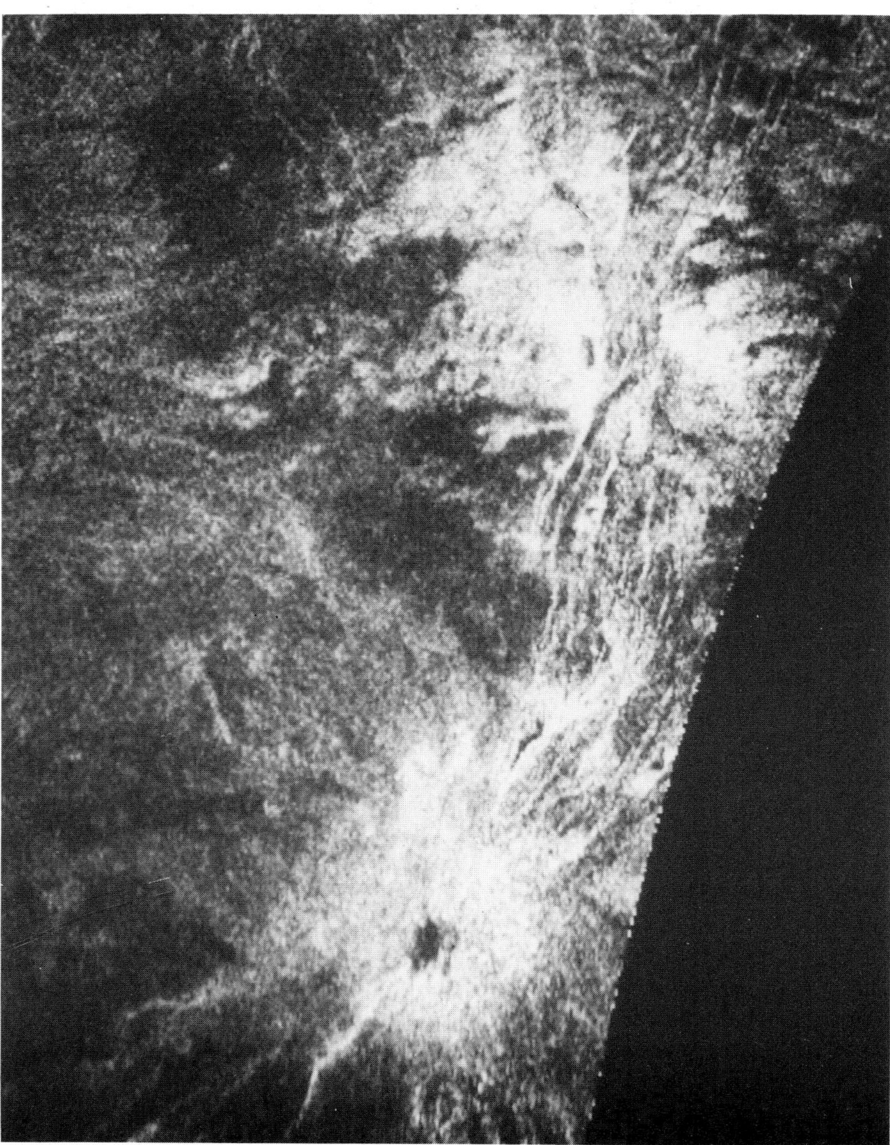

Figure 6.13b High-resolution (~ 4 km) Arecibo radar image of Beta Regio; the northern radar-bright area is Rhea Mons, the southern Theia Mons is commonly interpreted to be volcanic (courtesy D. B. Campbell, Arecibo Observatory).

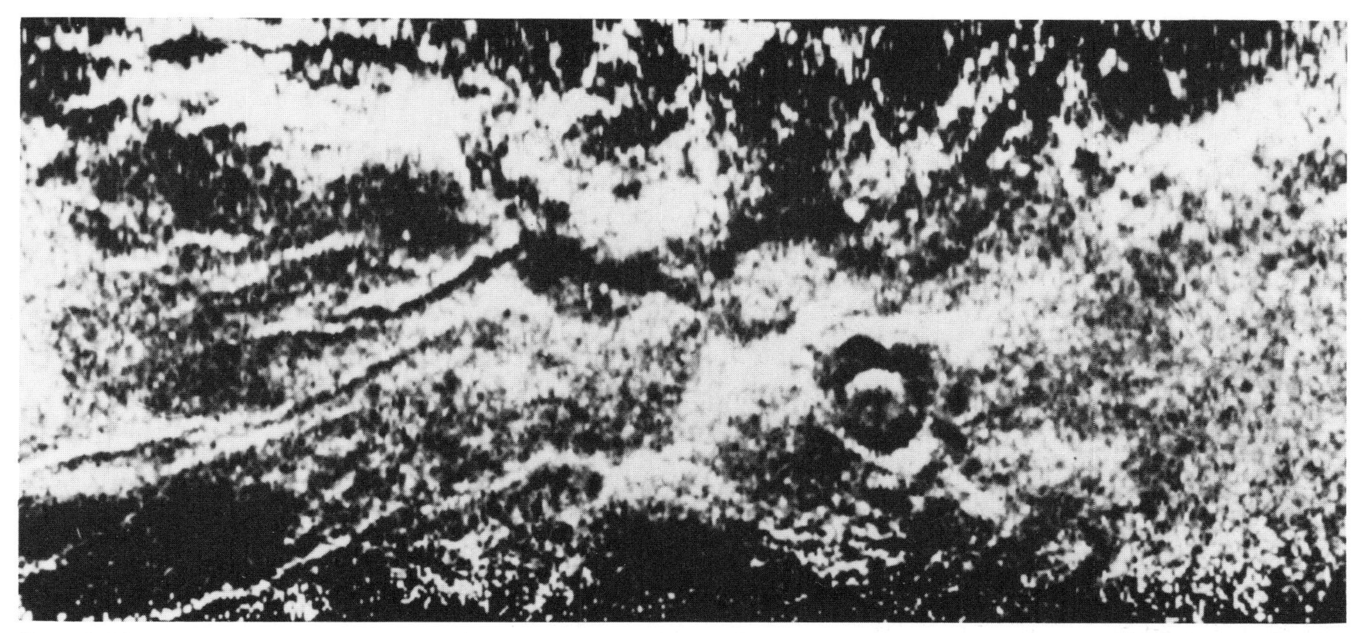

Figure 6.14 High-resolution image from the Venera 15 mission of the area thought to be Metis Regio at about 72°N; the circular dome-shaped feature on the right has a summit depression and appears to be a volcano; bands on the left may represent ridges and valleys resulting from tectonic processes; image is about 140 by 300 km (Tass/Wide World photograph).

depression. This ~35 km diameter volcano-like feature was observed in the Metis Regio highland at about 72°N. Other possible volcanic constructs are found associated with Akna, Maxwell and Freyja Montes, all in Ishtar Terra, and possible volcanic features are suggested in tectonically deformed areas (Schaber 1982c).

Some lowlands may have surfaces covered with lava flows. Venera 14 landed in lowlands southeast of Beta Regio and, as reported by Surkov *et al.* (1983), yielded a surface composition suggestive of basaltic lava flows. Moreover, radar data indicate relatively smooth, homogeneous surfaces and Masursky *et al.* (1980) suggest that the area may be similar to mare regions on the Moon.

Although described in the section under craters, some ring-shaped features may well be volcanic or volcano-tectonic structures. These and other geomorphic features must await higher-resolution images before their origins can be unambiguously determined.

6.4.2 Effect of the environment

As with impact cratering processes, the hot, dense atmosphere on Venus also may affect volcanic processes. For example, the high atmospheric pressure at the surface may retard explosive eruptions and the dominant volcanic style on Venus may involve effusion (flow) activity. Wilson and Head (1983) reason that in the lowlands, the atmospheric pressure is so great that magma cannot be disrupted by growth of gas bubbles – the primary driver in explosive eruptions – thus preventing or at least retarding pyroclastic activity. They calculate that if pyroclastic eruptions do occur, then cloud heights on Venus would be only about one-third as high as produced by comparable eruptions on Earth. As pointed out by Garvin *et al.* (1982), the net effect could be a scarcity of volcanic ash on Venus, and those pyroclastics that are produced might be concentrated near their vents, assuming no reworking by aeolian processes.

Charles Wood (1979) notes that the atmospheric pressure on most of Venus is equivalent to an ocean depth of 1 km on Earth. In this environment on Earth, the prevailing volcanic activity is the eruption of lava flows. In the high-temperature environment on Venus, cooling of such flows would be substantially retarded and, consequently, some flows may extend longer distances on Venus than on Earth, all other factors being equal.

6.5 Tectonism

Venus displays numerous features that can be attributed to tectonic processes, including scarps that are hundreds of kilometers long and huge riftlike valleys. Because of its similarities to Earth, there has been considerable speculation as to whether or not Venus experiences plate tectonism in a style similar to Earth, as reviewed by Phillips *et al.* (1981), Head *et al.* (1981a), McGill *et al.* (1983), Phillips and Malin (1983) and others.

Sean Solomon and Jim Head (1982b) provide a lucid discussion of the role of plate tectonics in general planetary evolution and they note that the mode(s) by which planets transport heat from the interior outward has an important effect on surface morphology. On the solid planets and satellites, transport of lithospheric heat occurs through plate recycling, lithospheric conduction and hot-spot volcanism, or some combination of these processes. What surface features result from these processes and are any of these features observed on Venus? On Earth, hot-spot volcanism is generally manifested through volcanic constructs, typically shield volcanoes. Morgan and Phillips (1983) note that the number of hot spots on Venus, taking into account that virtually all the heat lost is due to conduction through the crust at the hot spot, is 35 to 70. It seems likely that Venus, like Earth, probably gets rid of lithospheric heat by a combination of mechanisms.

Phillips and Malin (1983) review the topographic features associated with plate tectonics on Earth and recognize two general categories: features associated with sea-floor spreading and features related to continental drift. Sea-floor spreading features include mid-ocean ridges, deep-sea trenches and associated island arcs, and fracture zones. On Earth, mid-ocean ridges are prominent global features whose morphology is dependent upon the velocity of spreading and the rate of cooling. Phillips and Malin note that slow-spreading centers display sharply peaked ridges, whereas fast centers form more subdued and fractured ridges.

What would comparable ridges on Venus look like? Figure 6.15 shows a mid-ocean ridge (the East Pacific Rise) on Earth corrected for the venusian environment. Phillips and Malin point out the difficulties in detecting this Earth-type ridge with currently available data for Venus. Although ridge features are present on Venus, they are much narrower and display different morphologies than would be expected for those associated with spreading centers, at least in comparison with Earth.

Deep-sea trenches and associated island arcs, fracture zones and transform faults accompany sea-floor spreading. Although some features on Venus, such as Ut Rupes, a ridge that parallels Ishtar Terra, resemble some elements of sea-floor spreading on Earth, again the lack of adequate resolution data precludes positive identification.

Numerous features associated with continental drift are also indicative of plate tectonics. These include rifts (the continental equivalent of mid-ocean ridges) and linear mountain chains (Fig. 3.17), which on Earth typically result from the collision of crustal plates. Phillips and Malin (1983) note that large, distinctive mountain ranges on Venus occur only in Ishtar Terra and include Akna, Freyja and Maxwell Montes. Campbell and co-workers (1983) have obtained high-resolution radar images of this area (Fig. 6.16) from Arecibo. Much of this complex, several thousand kilometer long series of ranges, is composed of parallel, 10–20 km wide bands of high and low radar backscattering terrain, interpreted to be ridges and valleys. They suggest that this terrain results from either compression (fold mountains; Fig. 3.16) or extension (block faulting; Fig. 3.14), neither of which is necessarily diagnostic for plate tectonics. Akna and Freyja Montes also exhibit these parallel bands.

As is true for large impact craters on Venus, the high ambient temperature would enhance viscous relaxation of large topographic features over geologic time if these high temperatures have existed throughout this time. Weertman (1979) and Stephens and co-workers (1983) demonstrate that the present relief of the Ishtar mountain ranges cannot be older than a few hundred million years, whereas other highland areas of lesser relief are unlikely to exceed 10^9 years in age.

Riftlike valleys occur in several areas on Venus. Masursky *et al.* (1980) describe Artemis, Dali and Diana Chasmae in Aphrodite as possible tectonic rifts (Fig. 6.1). McGill and co-workers (1981) describe features in the Beta Regio to Phoebe Regio area (shown in Fig.

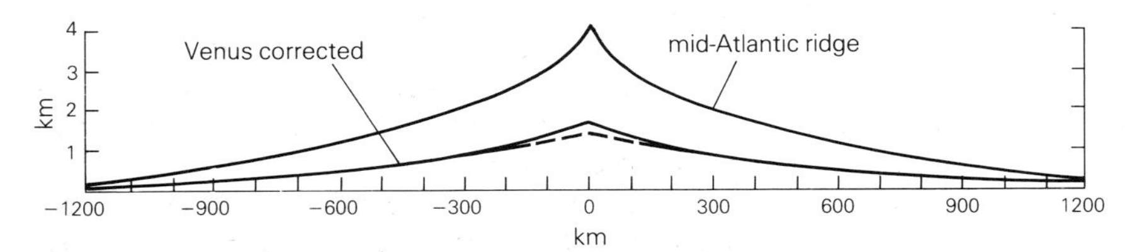

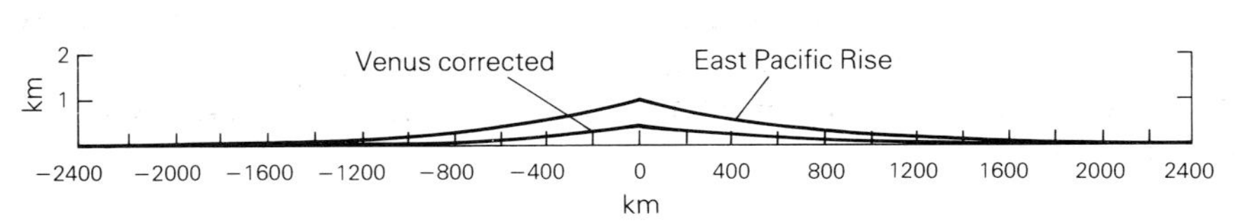

Figure 6.15 Topographic profiles across the mid-Atlantic ridge (a slow-spreading feature) and the East Pacific rise (a rapid-spreading feature) on Earth, and profiles inferred for Venus when the venusian environment is taken into account; the dashed line shows the smoothing that would be caused by the Pioneer Venus altimeter (from Phillips & Malin 1983).

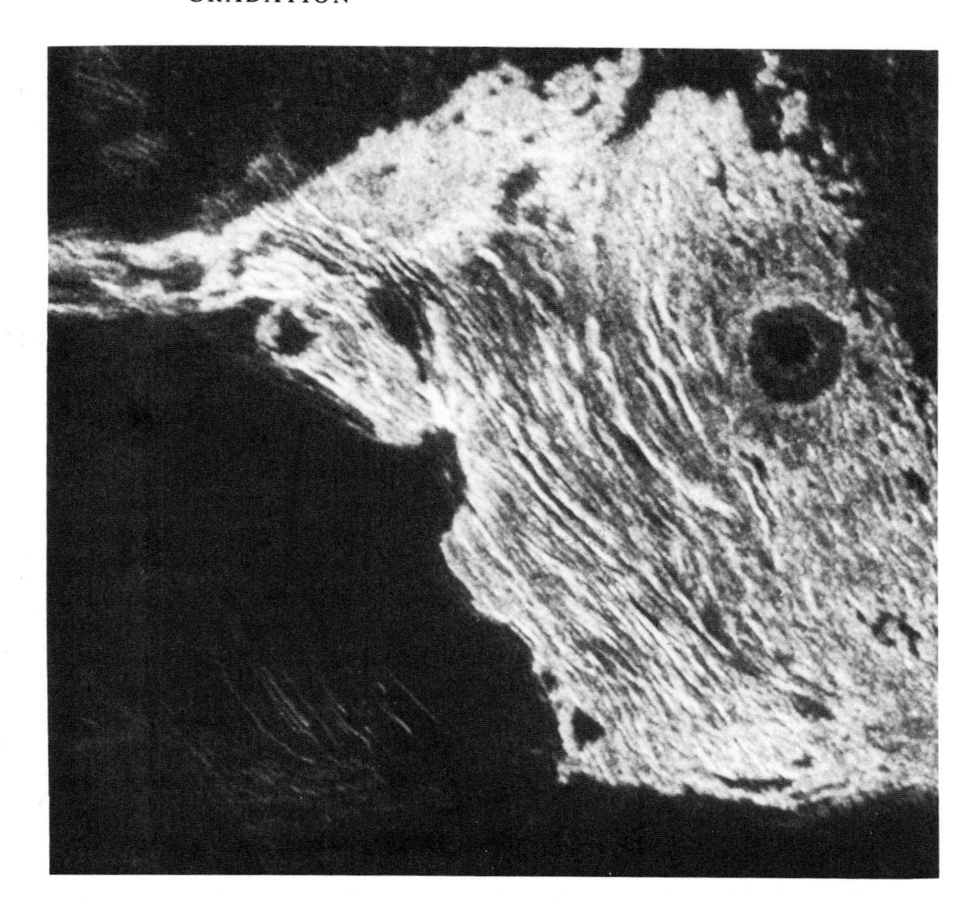

Figure 6.16 High-resolution radar image from Arecibo of Maxwell Montes, showing alternating bands of high and low radar backscattering; bands may be the consequence of folding and/or faulting. Circular feature on the right is more than 100 km in diameter and could be a caldera or impact structure (photograph courtesy of D. B. Campbell, Arecibo Observatory).

6.17) and suggest a supplement to the volcanic interpretation for the area. They compare the size and morphology of the venusian features to the East African Rift zone and to other domes associated with continental rift systems on Earth. They describe a series of elongate, linked depressions that define a trough 0.5–2.5 km below the upland surface of Beta Regio. The southern part of the trough consists of en-echelon depressions. McGill and co-workers suggest that Beta Regio, in places, is a complex, volcano-covered dome associated with a major rift accompanied by dynamic uplift of the crust of Venus.

In summary, numerous venusian features of probable tectonic origin can be identified on radar images and inferred from global topography. However, there is considerable controversy whether these features are indicative of Earth-like plate tectonism. Some investigators feel that plate tectonics of the style seen on Earth may not even be possible on Venus. Anderson (1981) calculates that the high surface temperature would not permit the lithosphere to cool to the point of negative buoyancy with respect to the underlying asthenosphere. If the lithosphere is impossible to subduct into the mantle, such features as deep-sea trenches and island arcs presumably cannot form. Furthermore, attempts to predict the appearance of venusian tectonic features are highly model-dependent and there is little agreement among workers. All investigators do agree, however, on the need for more and better-resolution data.

6.6 Gradation

Gradation includes the weathering, erosion, transportation and deposition of materials on planetary surfaces. On Earth, gradation is dominated by liquid water; on Venus, liquid water would quickly boil away in the high temperature environment. Thus, what processes of gradation might be expected?

Weathering products on Venus might be transported by aeolian processes, sediment-gravity currents in the atmosphere and mass wasting. And, although liquid water cannot presently exist on Venus, models based on the deuterium–hydrogen abundance ratio in the atmosphere suggests that Venus might once have possessed oceans (Donahue *et al.* 1982). Thus, there is the possibility of "fossil" landforms – such as river channels, stream patterns and shoreline features – generated by an ancient venusian hydrologic cycle.

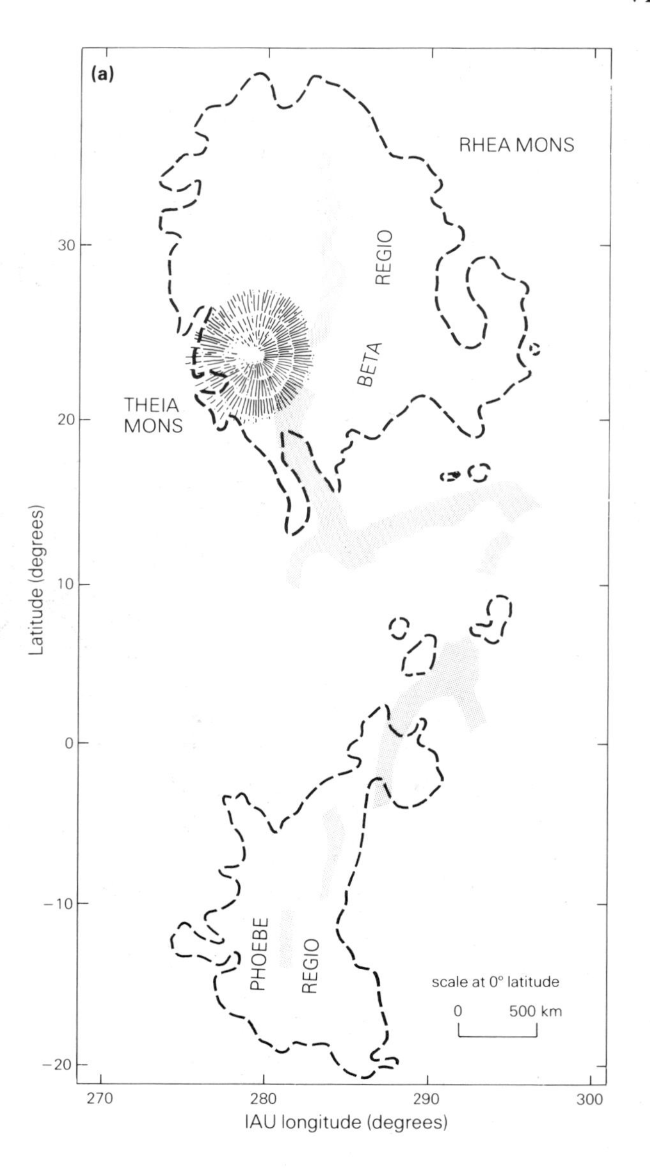

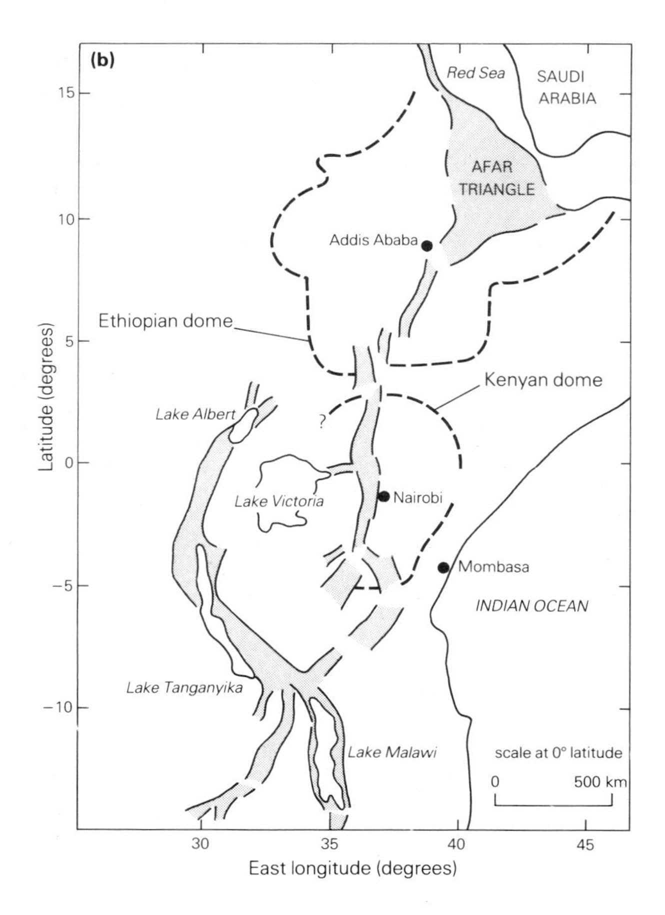

Figure 6.17 Sketch map of (a) the Beta Regio–Phoebe Regio area, based on radar images, compared to (b) the East Africa Rift; similarities in size, form and geometric relationships led McGill *et al.* (1981) to consider this region of Venus to result from mantle upwelling, tectonic deformation of the crust and eruptions to produce superposed volcanoes (from McGill *et al.* 1981, *Geophys. Res. Let.*, copyright American Geophysical Union).

6.6.1 Weathering

Harold Urey (1952) first considered the possible interactions of the atmosphere and lithosphere on Venus and proposed a model in which the CO_2 atmosphere is buffered by reactions with wollastonite (a calcium silicate mineral) to form quartz and calcite. Barsukov and co-workers (1982) reconsidered the Urey model (and subsequent models), taking into account the Venera and Pioneer Venus measurements of surface and atmospheric compositions. They suggest that a pyrite–anhydrite–magnetite assemblage buffers the atmosphere, resulting in reducing conditions in which carbonate minerals are unstable.

Nozette and Lewis (1982) have also derived a chemical model which takes into account potential differences in reactions occurring in the high-temperature/pressure environment in lowland areas and the relatively low-temperature/pressure conditions found in the highlands. They suggest that calcium- and magnesium-rich products form in the highlands from the weathering of igneous rocks and are then transported to the lowlands by aeolian processes. During transport, the weathering products react further with the atmosphere, ultimately forming calcium-rich mineral assemblages in lowland areas.

In addition to various chemical weathering processes, mechanical weathering also probably occurs on Venus. Volume changes associated with chemical weathering, fracturing associated with tectonic, volcanic and impact processes, and erosion during transport by wind and gravity could all produce fragmental debris.

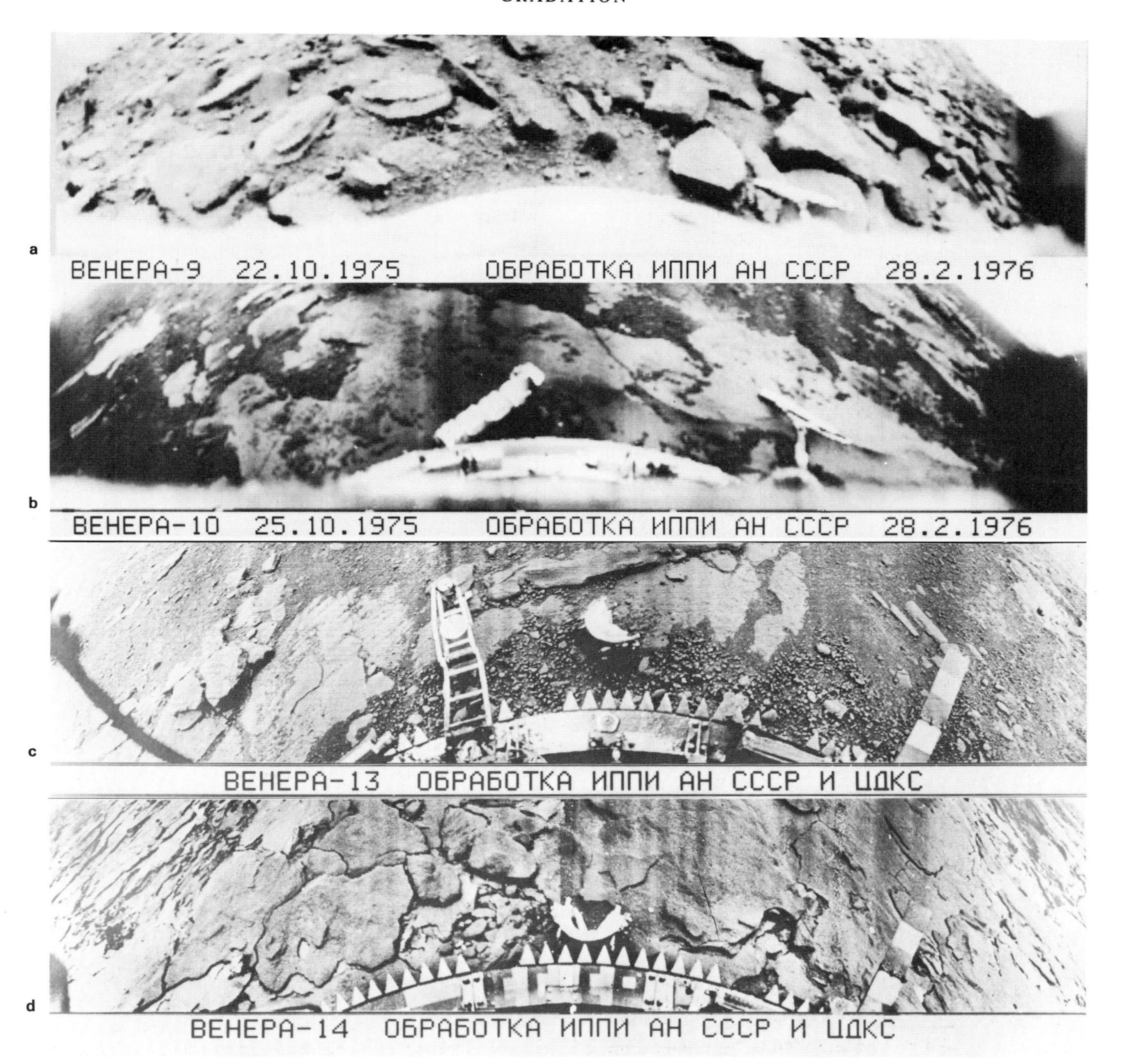

Figure 6.18 Venera images: (a) Venera 9, (b) Venera 10, (c) Venera 13 and (d) Venera 14 (images courtesy of Jet Propulsion Laboratory).

6.6.2 *Venera lander images*

Pictures taken on the surface by Venera landers provide clues to various processes of gradation. Although limited in areal coverage and quality, images have been obtained for four sites (Fig. 6.1), each of which appears somewhat different (Fig. 6.18). The Venera 9 and 10 sites have been described by Florensky *et al.* (1977) who note that Venera 9 landed on a slope composed of slabs and fine-grained material while the Venera 10 site is on a plain displaying scattered rock outcrops set in a matrix of fine-grained material. They consider six possibilities for the materials observed: (a) lavas, (b) exposed igneous intrusive rocks, (c) pyroclastic deposits, (d) impact-lithified material, (e) sedimentary rocks and (f) materials lithified by some atmospheric process. In addition, they consider the rocky slabs at the Venera 9 site to have been mass wasted down hill and suggest that the rocks at the Venera 10 site have been rounded through chemical weathering.

The Venera 13 and 14 images have been described by Moroz (1983), Florensky *et al.* (1983) and Basilevsky *et al.* (1984). These images were improved over those from the Venera 9 and 10 missions in resolution, coverage and by the use of filters to produce a color image for the Venera 14 site. Venera 13 landed between two gently sloping hills which display subhorizontal rock outcrops. Rocky slabs are visible near the spacecraft and are separated by loose material of lower albedo, similar in appearance to the Venera 10 site. In contrast to the Venera 9, 10 and 13 sites, Venera 14 generally lacks the matrix of loose, fine-grained material, although small accumulations of loose material are visible in shallow depressions. Large fragments appear slabby, while those smaller than 5 cm appear to be rounded. Outcrop units appear to be thinly layered.

From their analyses of images from all four Venera sites, Basilevsky and colleagues conclude that two materials are present: consolidated, mostly layered rock units and dark, loose materials which occur in greater or lesser amounts at each site. Garvin and co-workers (1984) concluded that the slabby bedrock forms seen in the Venera images originated from cooling of lava flows similar to pahoehoe basalts. At least part of the fine-grained material is considered to result from *in situ* disintegration of the rocky material as a result of deep chemical weathering.

6.6.3 *Modes of transportation*

In the absence of liquid water, material on Venus can be transported by the atmosphere, thrown ballistically by ejection from volcanic or impact processes, or carried by mass wasting. Although winds near the surface are rather sluggish (less than $2\ \mathrm{m\,s^{-1}}$ as measured by Venera spacecraft) the velocities are well within threshold speeds, as shown in Figure 3.36. Estimates based on wind tunnel simulations (Greeley *et al.* 1984) show that substantial quantities of sediment could be transported by even these gentle winds on Venus. While direct abrasion of fresh rocks by windblown particles may be minimal given the low velocities, removal of chemical-weathering products by wind and windblown grains may be an important process. Furthermore, aeolian processes may play a key role in buffering of the atmospheric composition as discussed in the Nozette and Lewis model.

The dense venusian atmosphere may allow sediment-gravity flow to be an important mode of erosion and transportation on Venus (Basilevsky *et al.* 1984). Candidate sites with sufficient topography for gravity flow include the southern scarp of Lakshmi Planum, slopes on Akna, Freyja and Maxwell Montes, and rift valleys of Aphrodite Terra. Slow-moving slope winds generated in these areas may also produce turbid, sediment-gravity flows (Williams & Moore 1984) that could carry enormous quantities of sediments and scour the surface in a manner somewhat analogous to gravity flows in oceanic environments on Earth.

6.7 Conclusions

Venus remains one of the most exciting objects for study for comparative planetology because of its potential Earth-like qualities. Questions of active or previous volcanism and plate tectonism, the possibility of former oceans and of features related to the hydrologic cycle, and the preservation of impact scars all remain unanswered. Most of these questions, however, can be addressed through the analysis of landforms and will require radar imaging of better resolutions than currently available.

Such imaging is promised through continued improvements in Earth-based radar systems and the Venera 15 and 16 radar orbiters which began operation in late 1983. In addition, the US Venus Radar Mapping mission, proposed to be flown in the late 1980s, will yield images for 70 percent of the planet at spatial resolutions better than 500 m.

7 Mars

7.1 Introduction

It is no wonder that the planet Mars is named for the Roman God of War. Its bright, blood-red color is in stark contrast to the night sky and must have inspired visions of battle to ancient observers. In the last hundred years or so, the Red Planet, like Venus, has evoked speculations on bizarre lifeforms, advanced civilizations, and Earth-like climates ranging from cold and wet to hot and dry.

These speculations were based on observations combined with fertile imaginations. When seen through telescopes, Mars displays distinctive surface patterns (Fig. 1.1a), some of which change with martian seasons. For example, white patches in the polar regions shrink toward the poles in the summer and a dark zone, called the "wave of darkening", extends toward the equator. This sequence was interpreted to represent the melting of ice caps and the subsequent release of water. In all, this is not an unreasonable interpretation based on pre-space age knowledge – the white patches are, in fact, polar caps, but they are composed mostly of CO_2 frost which ablates during the summer. The wave of darkening, however, may simply be a photometric effect of differences in illumination and viewing angles from season to season. Alternatively, it may be the result of contrast differences in which some areas become brighter, perhaps as a result of atmospheric dust.

Many imaginative science fiction themes for Mars arose from misuse of the term **canale** – the Italian word for *channel*. Apparently introduced in the 1860s by the Italian observer, Secchi, for some linear markings that he saw on Mars, Schiaparelli later used the term during his observations in 1877 and recorded them on his well known map of 1878. *Canale* was transliterated by English-speaking workers to *canal* and touched off the notions of intelligent life. Enamored by this possibility, Percival Lowell, of a wealthy Boston family, built an observatory at Flagstaff, Arizona, for the primary purpose of observing Mars. He drew maps showing extensive canals and, around the turn of the century, Lowell published three books in which he proposed the existence of advanced civilizations on Mars and described the construction of elaborate irrigation network canals for the distribution of water over the whole planet. These and earlier reports inspired science fiction writers, including Edgar Rice Burroughs and H. G. Wells, who entertained – and, in some cases, terrified – many a reader. The ideas of intelligently constructed canals and notions of advanced lifeforms were not put to rest until well into the space age.

Despite several successful early missions to Mars (Table 7.1), it was not until Mariner 9 that global knowledge of the surface, including the absence of canals, was obtained. Nonetheless, it was the quest for extraterrestrial life that stimulated the subsequent Viking missions. Involving two orbiters and two landers, all operating simultaneously, the Viking project remains the most complex unmanned mission in Solar System exploration. The primary objective of the pro-

Table 7.1 Successful missions to Mars.

Spacecraft	Encounter date	Mission	Event
Mariner 4	14 July 1965	flyby	Closest approach 9912 km; 22 images.
Mariner 6	31 July 1969	flyby	Closest approach 3330 km; 74 images.
Mariner 7	5 Aug. 1969	flyby	Closest approach 3518 km; 125 images.
Mariner 9	13 Nov. 1971	orbiter	6900 images; ultraviolet and infra-red spectrometers; infra-red radiometer.
Mars 5*	2 Feb. 1974	orbiter	Transmitted images and served as relay for lander.
Mars 6*	12 Mar. 1974	lander	Soft landed at 24°S, 25°W; returned atmospheric data during descent.
Viking 1	20 July 1976	lander and orbiter	Lander obtained images, measured wind speeds, temperatures and directions; measured chemical and physical properties of the surface.
Viking 2	3 Sept. 1976	lander and orbiter	Orbiters returned 55 000 images showing surface details as small as 10 m, collected gravity field data, monitored atmospheric water levels, thermally mapped selected surface sites.

*Soviet missions.

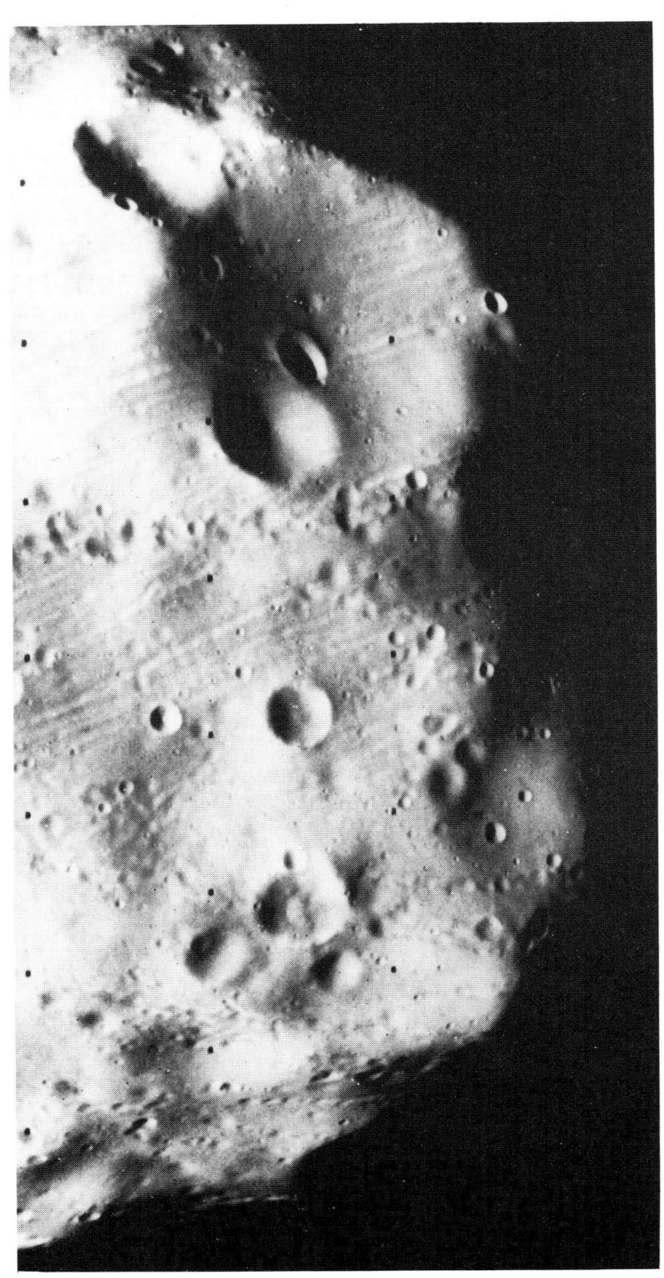

Figure 7.1 (opposite) High-resolution image of the martian moon, Deimos, taken by Viking Orbiter 2 from a distance of only 50 km. Rocks as small as 3 m across are visible and are thought to be impact ejecta; illumination from the left (JPL P-19599).

Figure 7.2 (below left) View of the martian moon, Phobos, showing grooved and cratered surface (Viking Orbiter 39B84).

ject was to search for life on Mars. Although results from the biological experiments were essentially negative or at best ambiguous (Klein 1979 provides an excellent review), the Viking mission provided a wealth of data on the geologic, geophysical and atmospheric characteristics of Mars.

Mars has turned out to be a planet of superlatives: it has the biggest volcanoes, the largest impact basin, the biggest canyons and the longest channels of any planet or satellite that we have seen in the Solar System. Thus, while the science fiction enthusiasts may have been disappointed in the biological results, the geologic aspects have been exceedingly rewarding.

7.2 Phobos and Deimos

Mars has two small moons, Phobos (21 km in diameter) and Deimos (13 km in diameter). Both are heavily cratered objects, having crater frequency distributions similar to the Moon (Veverka & Thomas 1979). Viking Orbiter 2 flew as close as 28 km to Phobos and returned the highest resolution (~ 1 m) pictures from the mission. Similar high-resolution images of Deimos show tiny craters and blocks of presumed ejecta (Fig. 7.1).

Images of Phobos (Fig. 7.2) show a curious pattern of grooves. As mapped by Thomas *et al.* (1978), these grooves radiate from a 10 km crater, Stickney, and converge on the opposite side of the satellite at the antipode to the crater, suggesting that the grooves are somehow related to the impact event.

The albedo, spectral reflectance and low density (<2.0 g cm^{-3}) of both moons suggest that they have a composition similar to carbonaceous chondrites (meteorites rich in water and organics). Because carbonaceous chondrites are considered to have formed in the asteroid belt, Pollack *et al.* (1979) suggested that Phobos and Deimos are captured asteroids.

7.3 Physiography

Mars exhibits a wide range of terrains (Fig. 7.3) and has considerable topographic relief. Place names are derived from classical mythology and from a system adapted by the International Astronomical Union

(Blunck 1977). In general, large martian craters are named for famous people, small craters (5–100 km) are named for small towns on Earth, sinuous valleys are named after "Mars" in various languages, and other landforms are named from regional albedo features in combination with a generic term, such as **mons** (mountain).

As shown by its hypsographic curve (Fig. 6.5), Mars has a three-fold topographic distribution: a mid-level terrain which has the greatest areal extent, the lowland plains and the mountains. Elevations are referenced to a 0-elevation plane which is based on atmospheric pressure. The pressure selected for datum is 6.1 mbar, which is the triple point for water (at lower pressures, liquid water is unstable) and is near the present mean pressure. The summit of Olympus Mons at 27 km marks the highest elevation on Mars while the floor of the Hellas basin is about – 4 km elevation. Thus, the total relief on Mars exceeds 30 km, over a third greater than the total relief on Earth.

There is a general asymmetry in the distribution of the major terrains on Mars. The sparsely cratered lowland plains of the northern hemisphere are separated from the southern cratered highlands by a great circle inclined about 30° to the equator in the western hemisphere. Superimposed on the young, lowland plains are the elevated Tharsis and Elysium regions and their volcanoes. The reason for the lowland–highland dichotomy remains one of the major unsolved problems on Mars.

Data from the Mariner and Viking missions enabled the general physiographic provinces and terrain types on Mars to be defined, as shown in Figure 7.4 (terrain units are indicated by abbreviations in the text and on the map). The general geology of Mars prior to the Viking mission is given by Mutch and colleagues (1976) and summarized on a geologic map by Scott and Carr (1978). Excellent post-Viking discussions of the surface of Mars are provided by Arvidson *et al.* (1980) and Carr (1980 & 1981) and outlined on maps by the US Geological Survey.

7.3.1 Basins

Circular basins of presumed impact origin constitute the oldest recognizable features on Mars. In addition to the Hellas, Argyre and Isidis basins, Wood and Head (1976) and Schultz and co-workers (1982) recognize numerous additional basins (Table 7.2; Fig. 7.4). Centered at 295°W, 40°S, the Hellas basin measures 1600 by 2000 km and has a rim that is 50–400 km wide, making it the largest impact basin in the Solar System. Concentric patterns of weakness extend as far as 1600 km from the center of Hellas. The floor of the Hellas basin appears to be filled with volcanic plains and is mantled

Table 7.2 Impact basins on Mars (from Wood & Head 1976 and Schultz *et al.* 1982).

Designation	Location	Ring(s) diameter (km)
1. (overlapped by Newcomb Crater)	3°W, 23°S	380, 800
2. Aram Chaos	21.5°W, 3°N	140, 250, 440, 550
3. Ladon basin	29°W, 18°S	270, 470/580, 975
4. Holden basin	32°W, 25°S	260, 580
5. Chryse basin	47°W, 19°N	800, 1400, 2000, 2750, 3600, 4300
6. Mangala basin	147°W, 0°N	300, 570
7. Sirenum basin	166.5°W, 44°S	500, 1000
8. (arc of massifs in Arcadia)	167°W, 37°N	380, 600
9. (south of Hephaestus Fossae)	180°W, 30°S	180, 340, 1000
10. Al Qahira basin	190°W, 20°S	300, 780, 1200
11. (south of Hephaestus Fossae)	233°W, 10°N	500, 1000
12. (south east of Hellas)	273°W, 58°S	225, 500
13. Nilosyrtis Mensae	282.5°W, 33°N	380
14. (south of Renaudot)	297°W, 38°N	600
15. (south of Lyot)	322°W, 42°N	60, 145, 260, 400, 480, 570
16. Cassini	328°W, 24°N	220, 450, 850
17. Deuteronilus B	338°W, 43°N	55, 201
18. Deuteronilus A	342°W, 44°N	44, 80, 220, 280
19. (overlapped by South Crater)	213°W, 73°S	300, 680
20. (overlapped by Schiaparelli)	346.5°W, 5°S	140, 560
21. (west of Le Verrier)	356°W, 37°S	430
22. Liu Hsin	171.5°W, 55°S	60, 135
23. Gale	222.0°W, 5°S	85?, 150
24. near Columbus	164.5°W, 25°S	70, 145
25. Ptolemaeus	157.5°W, 46°S	65?, 150
26. Phillips	44.5°W, 67°S	90?, 175
27. Molesworth	210.5°W, 28°S	80?, 180
28. Lowell	81.3°W, 52°S	90, 190
29. Lyot	330.5°W, 50°N	100, 200
30. Kaiser	340.5°W, 46°S	100?, 200
31. Kepler	218.5°W, 47°S	110, 210
32. Galle	31.0°W, 51°S	100, 220
33. Herschel	230.2°W, 15°S	160?, 290
34. Antoniadi	299.1°W, 22°N	200?, 400
35. Huygens	304.2°W, 14°S	250, 460
36. Schiaparelli	343.6°W, 3°S	230, 470
37. South Polar	267.0°W, 83°S	?, 850
38. Argyre	42.0°W, 50°S	560?, 1200
39. Isidis	272.0°W, 16°N	1100, 1900
40. Hellas	291.0°W, 43°S	?, 2000

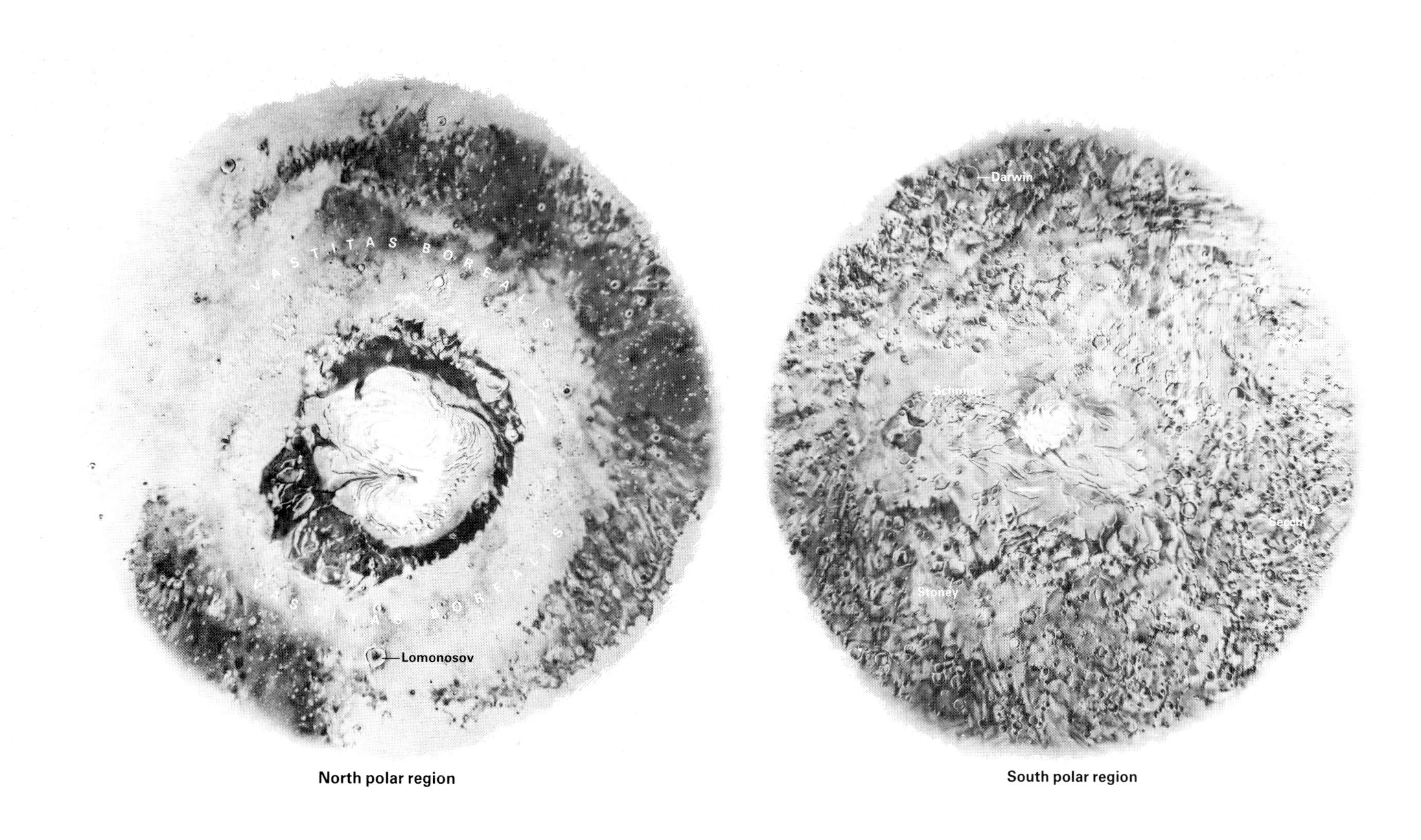

North polar region

South polar region

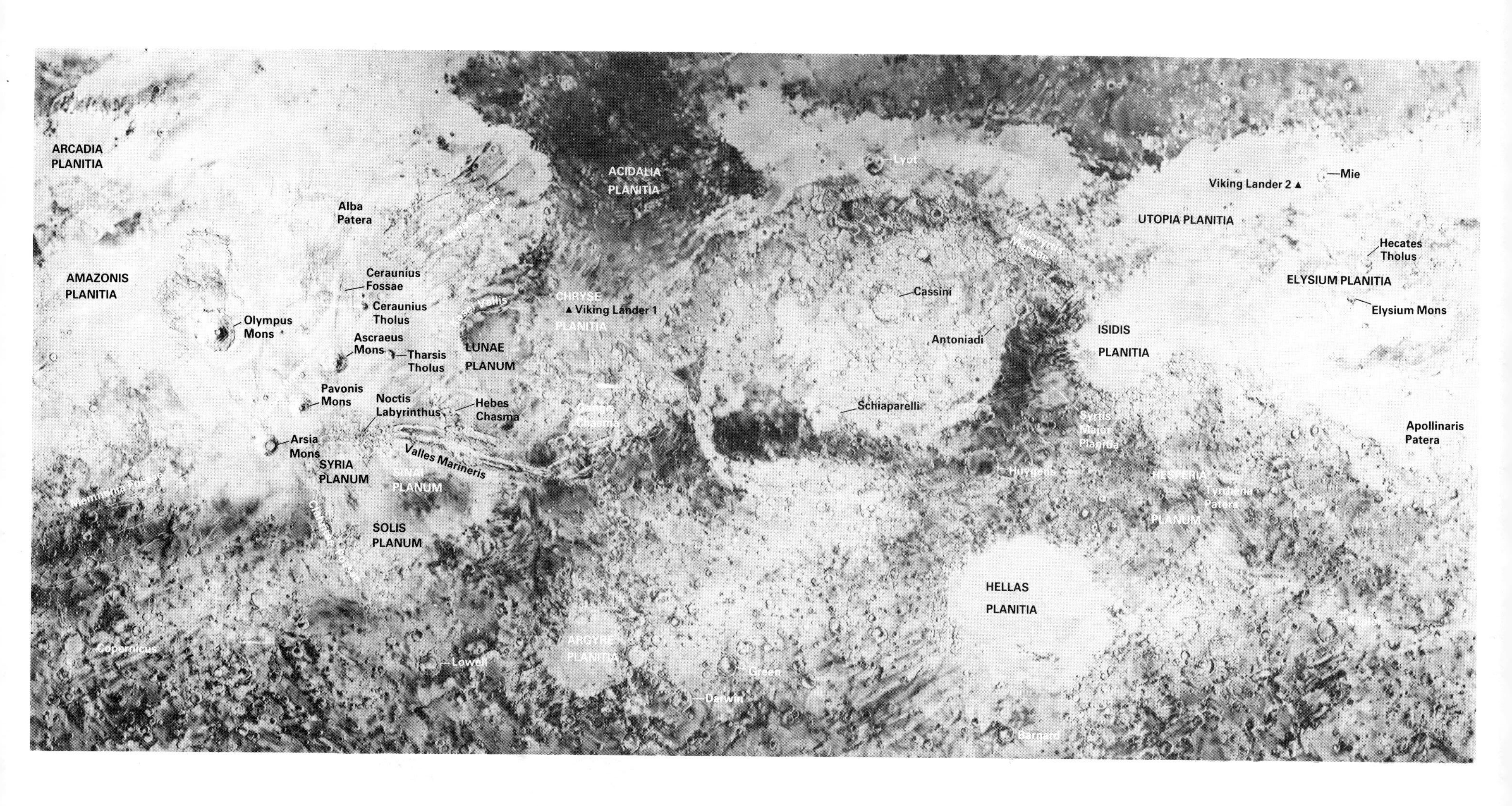

Figure 7.3 Shaded airbrush relief map of Mars showing prominent features, selected place names, and Viking Landers 1 and 2 sites (courtesy US Geological Survey).

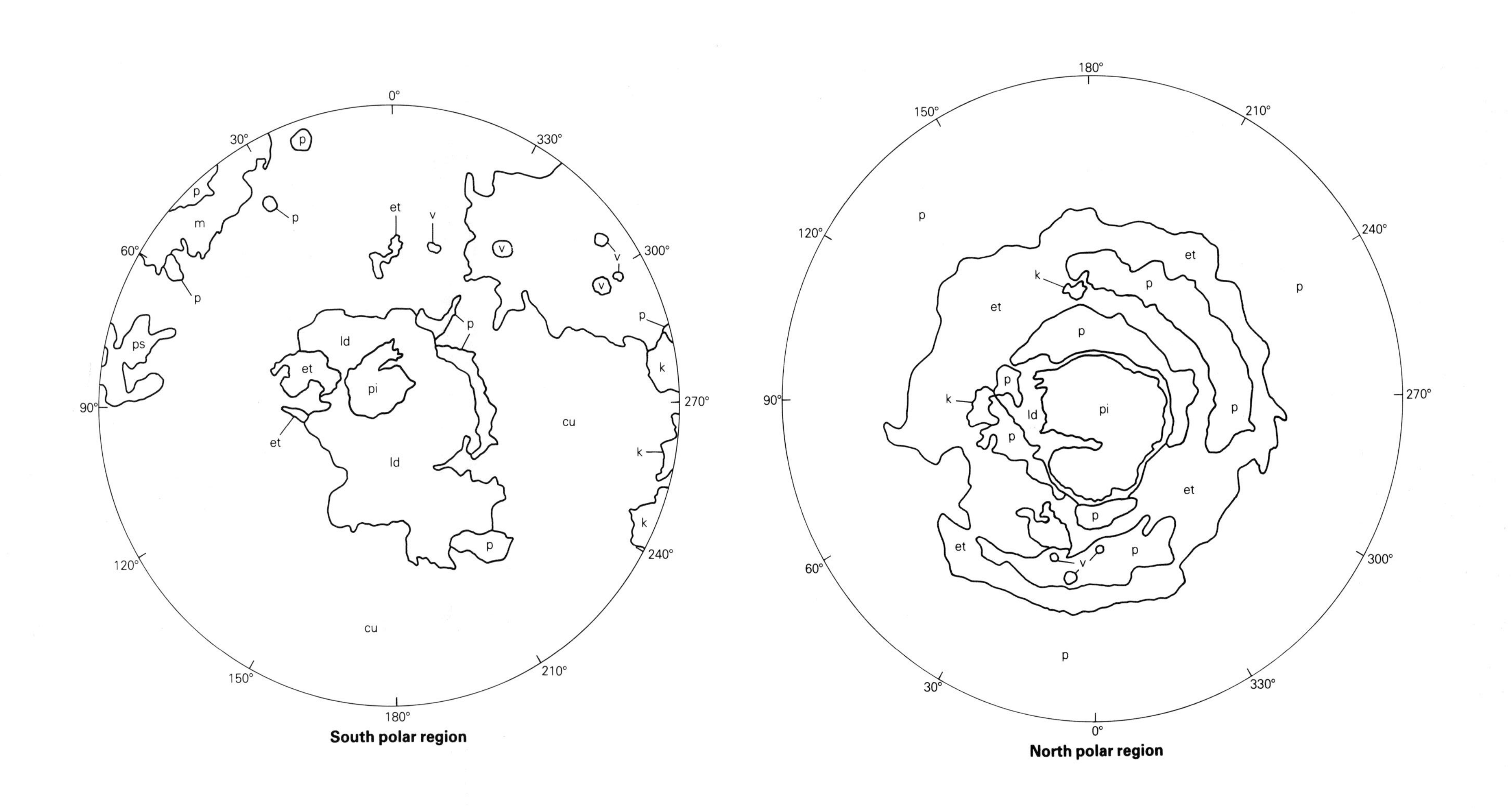

South polar region

North polar region

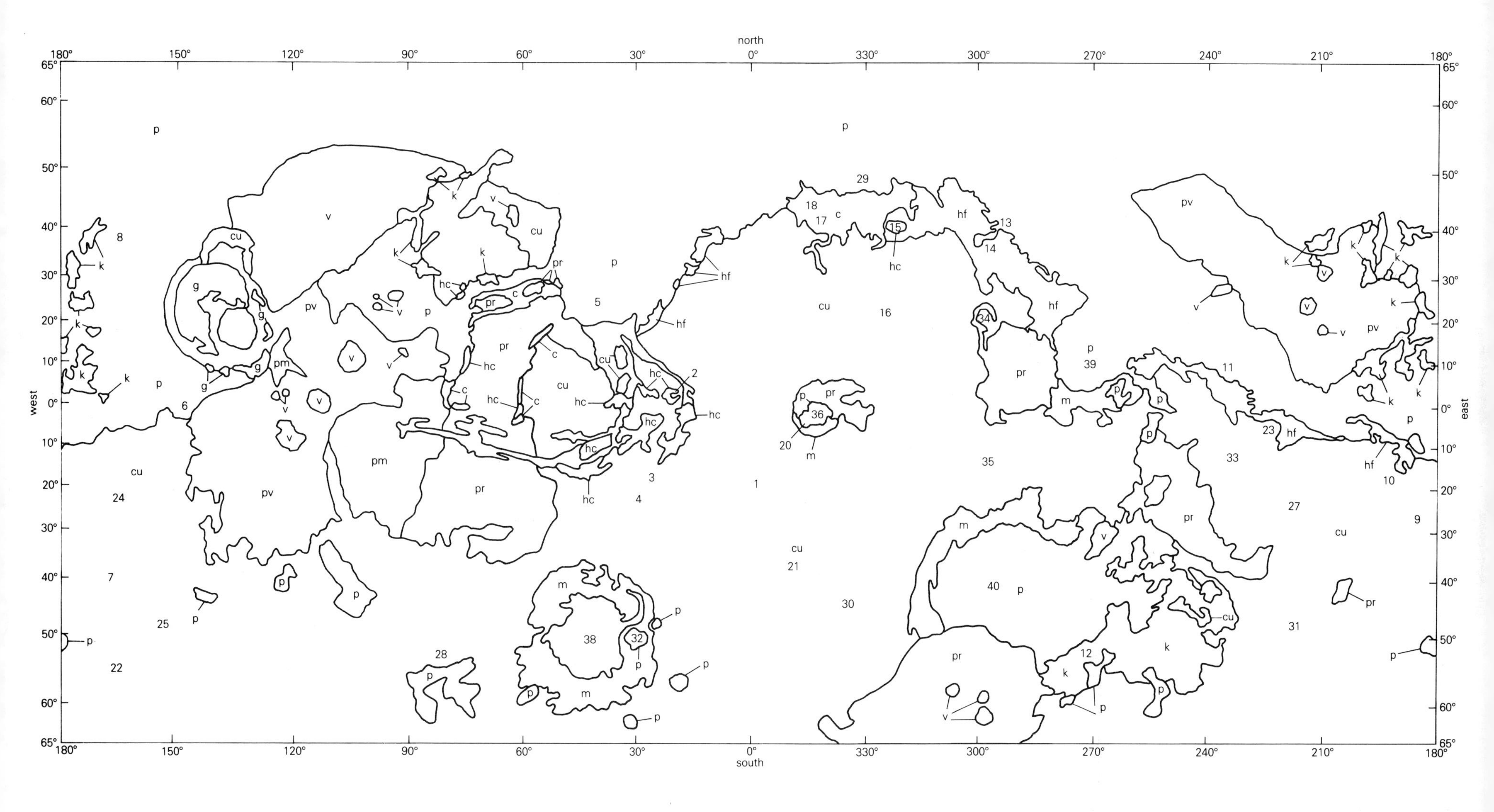

Figure 7.4 Geological terrain map; massifs (m), heavily cratered terrain (cu), undivided plains (p), moderately cratered plains (pm), ridged plateau plains (pr), volcanic plains (pv), volcanoes (v), ice caps (pi), layered terrain (ld), etched terrain (et), canyonlands (c), channeled terrain (ch), chaotic terrain (hc), fretted terrain (hf), knobby terrain (k), and grooved terrain (g) (modified from Scott & Carr 1978). Numbers indicate the position of impact basins, keyed to Table 7.2.

Figure 7.5 Oblique view obtained by Viking orbiter across the Argyre basin showing the rugged massifs that define the rim. The floor of the basin appears to have been flooded with lavas. Horizon in the background shows haze layers 25–40 km high, thought to be crystals of carbon dioxide (JPL P-17022).

with aeolian deposits. The 900 km diameter Argyre basin (Fig. 7.5) is also in the southern hemisphere and appears to be mantled. Much better images exist of the Isidis basin, a ~1100 km feature in the northern hemisphere. The floor of this feature has been buried by flood lavas and, later, larger outpourings of lava occurred to the east of the basin.

All three basins are defined principally by rugged **massifs** or mountain blocks (unit m), 20 km or smaller across and inward-facing scarps. The massifs tend to be closely spaced near the basin and become more widely spaced and subdued away from the rim. They are probably analogous to basin-related massifs on the Moon (Fig. 4.4) and Mercury (Fig. 5.25) and represent uplifted blocks of crust. However, radial ejecta deposits are lacking around most martian basins.

7.3.2 Heavily cratered terrain

Heavily cratered terrain (unit cu) occurs mostly in the southern hemisphere. Detailed photogeologic mapping shows that this unit can be further subdivided based on the degree and style of modification. At first appearance, this terrain is similar to the lunar highlands since it is dominated by large (>20 km), degraded impact craters. Closer inspection shows some important differences in comparison with the Moon. Most of the large craters are relatively shallow; ejecta deposits are degraded or lacking, the rims are degraded and the floors are flat and may have been filled with deposits of various origins.

Furthermore, the large craters are more widely spaced than those on the Moon and are separated by complex intercrater plains. The plains are superposed on most of the craters and cover the crater ejecta deposits. Some crater rims and plains display extensive valley networks (Fig. 7.6) of presumed fluvial origin, while other plains show evidence of aeolian processes; thus, some of the plains are probably sedimentary deposits. Most of the plains within the cratered terrain, however, are thought to be volcanic (Greeley & Spudis 1978b), and it is likely that their history reflects a complex interplay of fluvial,

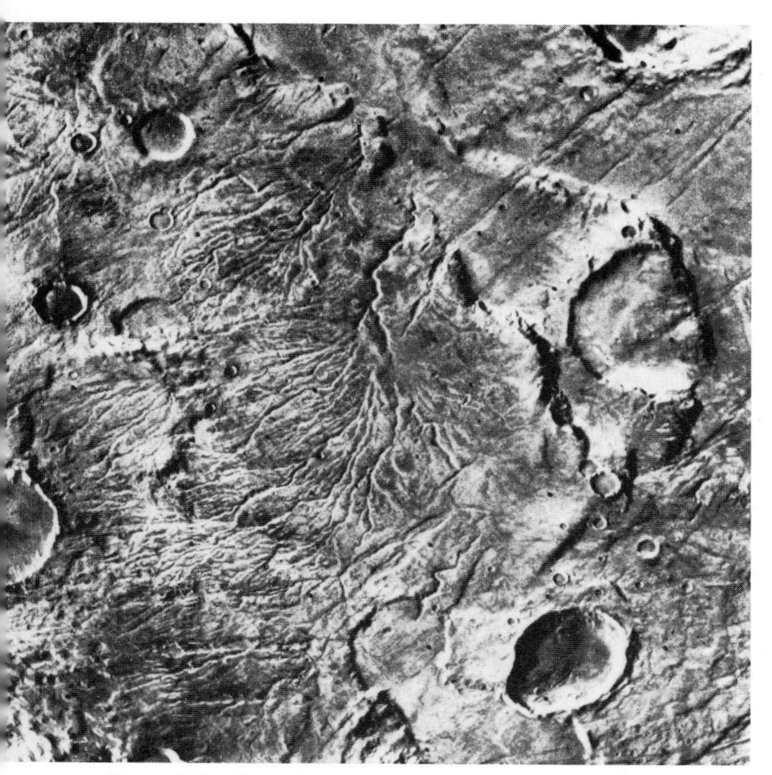

Figure 7.6 View of cratered terrain in the southern hemisphere showing valley networks and fracturing. View is 200 km across; north is toward lower left (Viking Orbiter 63A09).

Figure 7.7 Pedestal craters in the northern lowland plains. Pedestals are considered to be surfaces of ejecta deposits that have been armored with rocks to prevent erosion while the surrounding plains have been lowered by deflation. Many of the small mounds with central craters may be small pedestal craters, rather than volcanic cones, as has been proposed for similar features in other areas of Mars. Area shown is 45 by 65 km (Viking Orbiter 60A53). ▷

aeolian and volcanic processes. Superimposed on the large craters and the intercrater plains is a population of small (<20 km), relatively fresh impact craters, most of which have ejecta patterns suggestive of fluidized flow (Fig. 3.4).

7.3.3 Plains and plateaus

In addition to the intercrater plains described above, numerous other plains occur on Mars. These have been mapped as undivided plains, moderately cratered plains, ridged plateau plains and volcanic plains.

Plains (undifferentiated; unit p) occur principally in the northern lowlands and on the floors of large craters and basins. The northern lowland plains exhibit a wide range of landforms when seen at high resolution, some of which are attributed to periglacial processes; others resemble small volcanic features, but could also be eroded impact craters (Fig. 7.7). Although both Vikings landed in the north (Fig. 7.3), it is not known how representative these sites are for the plains as a whole. Nonetheless, images from the landers provide clues to the processes that have shaped the surface (Fig. 1.4). Most of the rocks are thought to be impact ejecta, although some may be derived from frost-heave or other endogenic processes. Aeolian features, including drifts of windblown sediments and possible ventifacts (wind-sculptured features), are abundant at both sites.

In general, these plains of Mars represent diverse origins and processes of modification. The relative lack of large impact craters indicates the relative youth of the surface. As detailed photogeologic mapping progresses, it is likely that these plains will be subdivided by both type and age.

Moderately cratered plains (unit pm) occur as a plateau centered in the Syria Planum–Sinai Planum area and around Alba Patera. As the unit name implies, impact crater frequencies indicate an intermediate age. Extensive grabens and other fractures show that this unit has been subjected to tectonic deformation, mostly crustal extension (Fig. 7.8).

Ridged plateau plains (unit pr) are found primarily in four areas: within the Isidis basin, as a north–south band (Lunae Planum) centered at 70°W, and in two areas around the Hellas basin. Their primary characteristic is the presence of widely spaced ridges that are very similar to lunar mare ridges, as described by Lucchitta and Klockenbrink (1981) (Fig. 7.9). The unit is relatively smooth on 100 m resolution images but is peppered with small, fresh craters when seen at higher resolution. Ridged plateau plains are considered to be volcanic flood lavas.

Volcanic plains (unit pv) display flow lobes (Fig. 7.10) and other features diagnostic of lava flows. Volcanic plains occur in association with the two principal volcanic provinces, the Tharsis region and the Elysium region (Fig. 7.3). Source vents for most volcanic plains on Mars remain hidden, similar to the mare flows on the Moon. Like many volcanic plains, the martian units probably originated from fissure vents which were subsequently buried by their own flows or by flows from other sources. By analogy with Earth, the martian volcanic plains probably involve a combination of flood eruption and "plains" eruptions (Fig. 3.27). Some of the volcanic plains appear to have originated on the flanks of the large shield volcanoes, then flowed beyond the margin of the constructs.

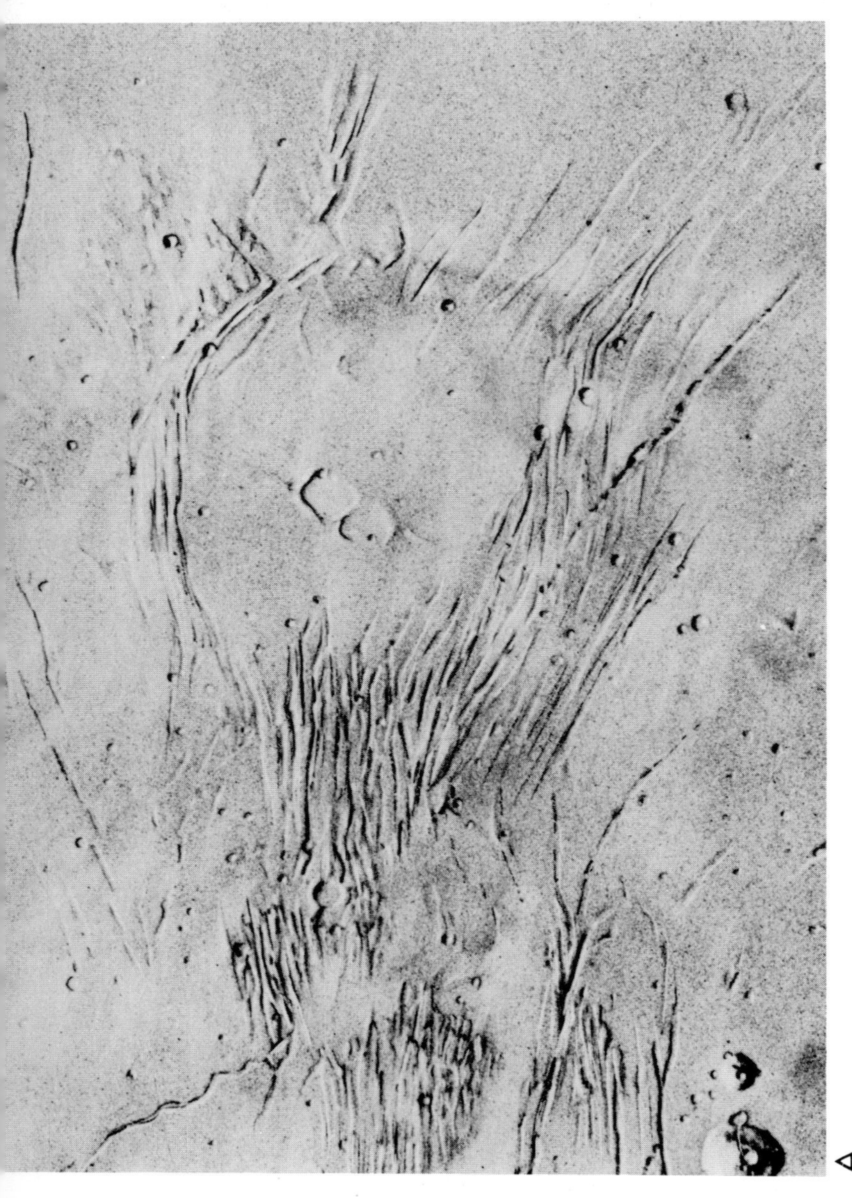

Figure 7.9 Mosaic showing ridged plateau plains east of Mellis Dorsa in the southern hemisphere of Mars. Ridges resemble mare ridges on the Moon. Plains have been fractured and the feature at A appears to have been offset laterally. Crater in center is a typical ejecta flow crater showing multiple flow lobes (Viking Orbiter 610A01-3, 608A409).

◁ **Figure 7.8** Shaded airbrush relief chart showing Alba Patera, a unique volcanic structure, and the set of fractures and grabens surrounding the caldera region. Area shown is 1000 by 1500 km (courtesy US Geological Survey).

Table 7.3 Classification of large martian volcanoes.

Name	Location	Type	Area (10^4 km^2)	Note
Alba Patera	110°, 40°	Alba	113.0	1
Albor Tholus	210°, 19°	dome	1.94	2
Amphitrites Patera	299°, −59°	highland patera	6.55	3
Apollinaris Patera	186°, −8°	shield	5.35	4
Arsia Mons	121°, −9°	shield	33.3	5
Ascraeus Mons	105°, 11°	shield	14.0	5
Biblis Patera	124°, 2°	shield	1.18	6
Ceraunius Tholus	97°, 24°	dome	1.01	2
Elysium Mons	214°, 25°	shield	13.1	7
Hadriaca Patera	267°, −30°	highland patera	9.24	3
Hecates Tholus	210°, 32°	shield	2.59	6
Jovis Tholus	118°, 18°	shield	0.29	6
Olympus Mons	133°, 18°	shield	37.4	5
Pavonis Mons	113°, 0°	shield	15.0	5
"Tempe" Patera	63°, 44°	highland patera	1.54	3, 8
Tharsis Tholus	91°, 13°	dome	1.68	2
Tyrrhena Patera	254°, −22°	highland patera	3.94	3
Ulysses Patera	122°, 3°	shield	0.85	6
Uranius Patera	93°, 26°	shield	2.71	6
Uranius Tholus	97°, 26°	dome	0.38	2

[1] Unique volcanic structure on the terrestrial planets.
[2] May represent buried older shields or different style of volcanic activity.
[3] Interpreted as ash shields with less lava present.
[4] Oldest lava shield recognized on Mars; occurs on plains/uplands boundary.
[5] Main shields of the Tharsis complex.
[6] Shields that have been partly buried by surrounding plains lavas.
[7] Slightly steeper sides than the Tharsis shields.
[8] Informal name, see Plescia and Saunders (1979).

Figure 7.10 High-resolution image showing flow lobes of presumed volcanic origin (Viking Orbiter 806A60).

7.3.4 Central volcanoes

Martian central volcanoes (unit v) are characterized by localized source vent(s) and occur as shield volcanoes, dome volcanoes, highland patera and as a unique feature – Alba Patera. In addition are numerous small volcanoes including cones, low shields and possible composite cones which are discussed in Section 7.5.

Shield volcanoes provided the first clear evidence that volcanism was important on Mars. Shield volcanoes (Figs 7.11 & 7.12) in the Tharsis and Elysium regions are composed of thousands of individual flows, many of which were emplaced through lava tubes and channels, typical of shield-forming flows on Earth (Fig. 3.18). The summit areas are characterized by calderas (Fig. 7.13), most of which are complex and exhibit repeated episodes of eruption, collapse and tectonic modification. Many of the martian shield volcanoes have low slopes on the outer flanks and a steeper summit region which may reflect a change in the style of volcanism.

Dome volcanoes (tholii) are smaller and have steeper slopes (Fig. 7.14) than shield volcanoes. The steeper slopes may be attributed to more viscous lavas, incorporation of pyroclastic deposits with lava flows, lower rates and volumes of effusion, or some combination of these factors, any one of which could result in the development of dome-shaped volcanoes. Many of the features on Mars first identified as domes are now considered to be partly buried shield volcanoes, in which the steeper summit area remains exposed (Table 7.3). Although rare, some features have been found on Mars (Fig. 7.15) that may be similar to composite cones.

Highland patera, as defined by Jeff Plescia and Steve Saunders (1979) of the Jet Propulsion Laboratory, are characterized by low profiles, radial channels and complex central calderas (Fig. 7.16). Some of the features first identified as patera on Mariner 9 are now seen to be of non-volcanic origins. With the exception of Tempe Patera in the northern hemisphere, all the highland patera (Table 7.3) occur near the Hellas basin. Because these features are about the same age and appear to be situated over ring-fractures associated with

Figure 7.11 Olympus Mons, one of the prominent shield volcanoes in the Tharsis region, measures 600 km across and is surmounted by a complex summit caldera (see Fig. 7.13). The volcano is surrounded by a prominent scarp several kilometers high. Flows derived from the volcano can be traced onto the surrounding plains at (A). North is toward upper left (Viking Orbiter 646A28).

Figure 7.12 Viking orbiter mosaic of Hecates Tholus, one of the prominent volcanoes in the Elysium region, measuring more than 170 km across. Various radial channels have been interpreted as erosional ash channels, lava channels or channels eroded by fluvial processes.

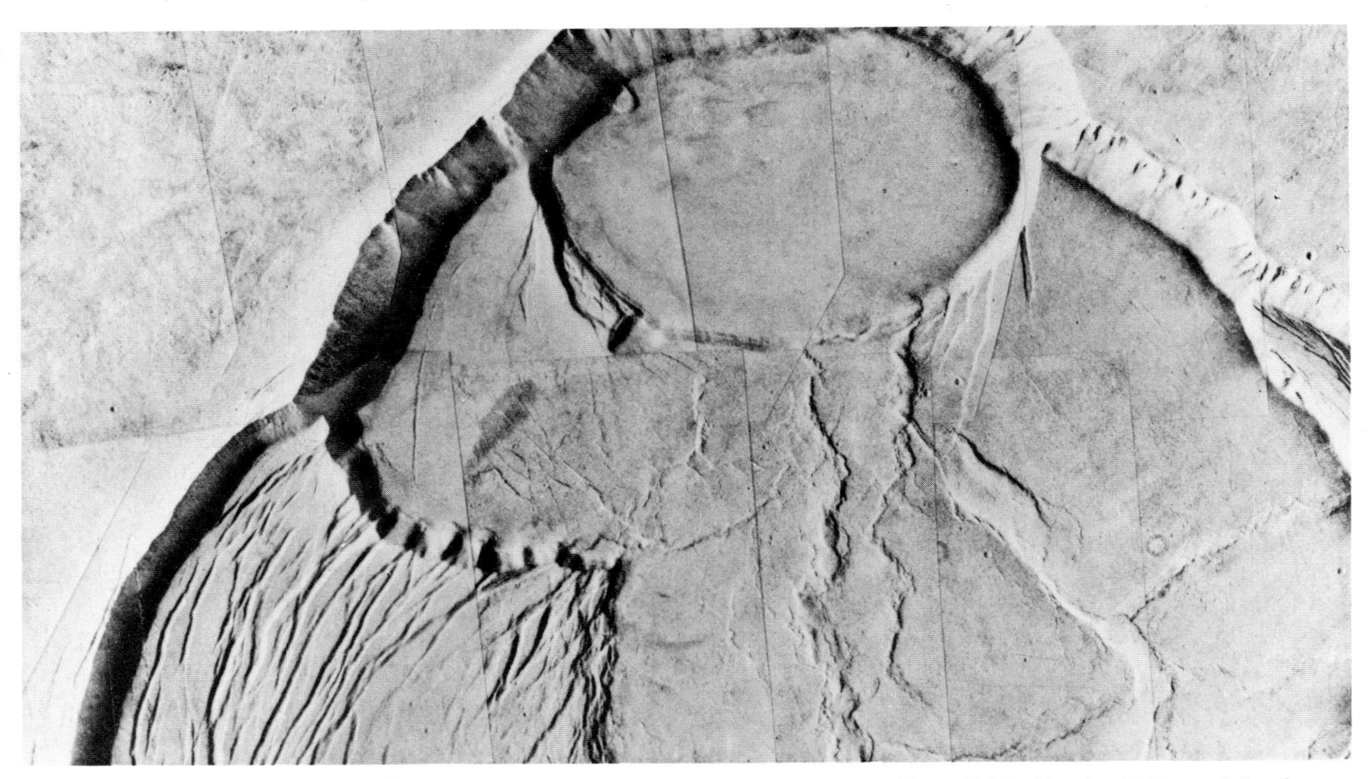

Figure 7.13 Mosaic of high-resolution frames of part of the Olympus Mons summit caldera, showing extensive wrinkle ridges in one of the floor units. These ridges are typical in morphology to those that occur on lunar maria. Area shown is 75 km across (JPL P-19381B).

Figure 7.14 Tharsis Tholus, measuring 110 by 170 km, has steep flanks and is classified as a dome volcano (Viking Orbiter 858A23).

Figure 7.15 This 2 km high feature, located at 21°S, 188°W, may represent a composite cone (Greeley & Spudis 1978b), suggested by slopes 10° to 20° and its rugged slopes. The summit caldera is about 8 km across and appears to be filled with lavas. Numerous channels and possible flows radiate from the caldera. Apron-like flows (arrows) could be either landslides or lava flows (JPL P-24656).

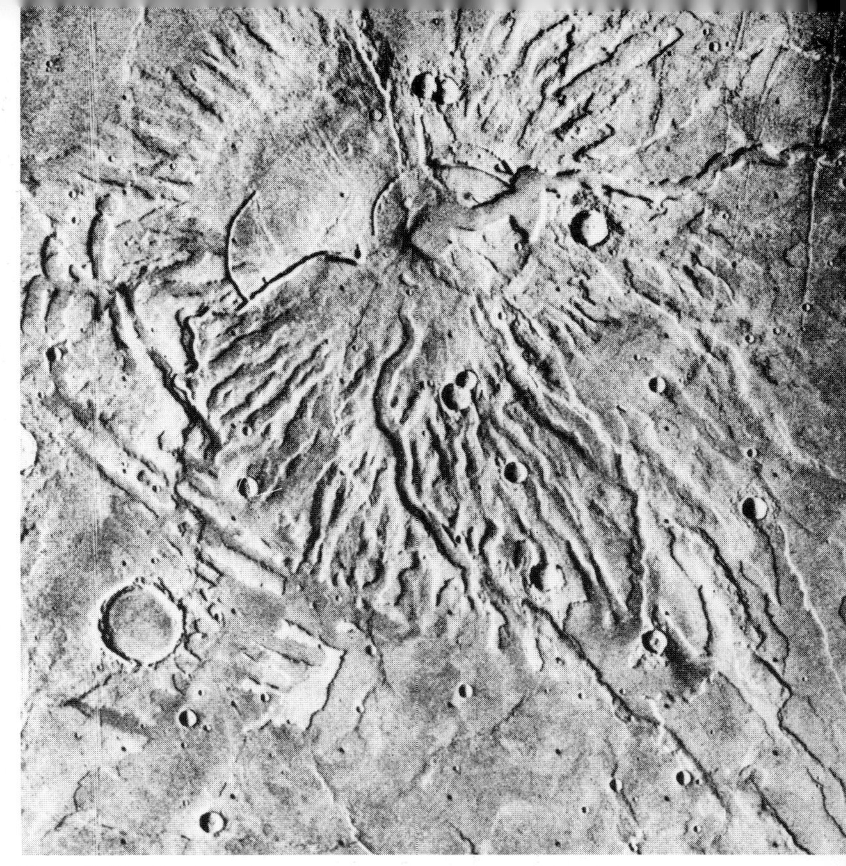

Figure 7.16 Tyrrhena Patera, one of the highland patera associated with the Hellas basin in the southern hemisphere. The area shown is about 280 km across (Viking Orbiter 87A14).

Hellas, Peterson (1978) suggested that they may represent early-stage volcanism associated with post-impact adjustments of the Hellas basin.

Alba Patera represents a volcanic feature that may be unique. Stretching more than 1000 km across, it is the largest central volcano seen anywhere in the Solar System. It is characterized by a set of complex ring fractures surrounding a 100 km caldera centered at 110°W, 40°N (Fig. 7.8). Sheet lavas, channel-fed flows and complex tube-fed lavas can be traced westward from the central caldera for hundreds of kilometers. Small domes that may be either parasitic vents or "rootless" vents occur in association with some of the flows at the western extremity of the volcano.

7.3.5 Polar units

The polar regions incorporate several distinct terrains, including *ice caps* (unit pi), *layered terrain* (unit ld) and *etched terrain* (unit et). The ice caps form bright zones around the north and south poles. They persist through the late spring and summer and, at least in the north, are considered to contain both water ice and frozen CO_2, as inferred from infra-red data obtained from the Viking orbiters (Kieffer & Palluconi 1979, Farmer *et al.* 1977). As the caps retreat with the warm seasons, outlying patches of CO_2 frost often remain in low-lying and shaded areas, such as behind crater rims.

Layered terrain (Fig. 7.17) occurs in both polar areas. In the north it is superposed on plains units and in the south it overlies cratered terrain. Layering is visible as fine bright and dark bands exposed on slopes. Thicknesses of individual bands have been estimated at 10–50 m, but they probably exist as thinner units which are simply too small to be seen. Many of the bands can be traced hundreds of kilometers and most of them tend to follow the topography, suggesting that the bands represent nearly horizontal layers. Some bands transect (cross-cut) underlying sets of bands, indicating discontinuities in deposition. The origin of the layers has evoked considerable controversy, as reviewed by Cutts and Lewis (1982). Most investigators suggest that they are alternating deposits of dust, water-ice, CO_2 ice or some combination of materials that accumulate on an annual or longer-term cycle. A major paradox is the seemingly young age of the polar layered terrains. The number of layers indicate an age <1 my. The number of surrounding pedestal craters (see Section 7.7.1)

Figure 7.17 (top) View of layered terrain in the north polar area. Area shown is 65 km across, centered at 80°N, 347°W (Viking Orbiter 56B84).

Figure 7.18 (above) Etched terrain in the south polar region, centered at 76°S, 74°W, showing characteristic pitted surfaces that may be the result of deflation and/or ablation of ice. Area shown is 200 km across (Viking Orbiter 390B90).

indicate an age younger than Olympus Mons (P. Schultz, personal communication).

Etched terrain consists of areas that have been eroded to form pits and valleys (Fig. 7.18). First described by Sharp (1973), this unit appears to involve layered terrain that has been eroded by some processes such as wind deflation. For example, ablation of ice layers may leave behind loose grains of sand which can then be removed by the wind.

7.3.6 Canyonlands

Stretching 4000 km along the equatorial zone eastward from the Tharsis Montes are grabens, canyons, pit craters and channels which form vast *canyonlands* (unit c). Its length is the equivalent of a canyon system extending across the United States from the Pacific to the Atlantic Oceans! The region, including the main canyon system, was named Valles Marineris in honor of the Mariner 9 spacecraft which returned the first pictures of this remarkable area. Individual canyons are as wide as 200 km and as deep as 7 km. As shown in Figure 7.19, the canyonlands can be subdivided into three sections, Noctis Labyrinthus in the west, the main section of canyons in the center and a complex eastern section. Noctis Labyrinthus consists of short, narrow canyons that create a mosaic of crustal blocks (Fig. 7.20). The area is slightly east of the *Tharsis bulge*, a prominent "swelling" on Mars which rises more than 11 km above datum. Most of the canyons appear to be grabens that formed in response to extension associated with the uplift of the Tharsis bulge.

The central part of Valles Marineris is more than 2400 km long and consists of multiple, parallel canyons, all trending east–west, with smaller grabens and chains of coalescing pits. The walls of the canyons exhibit layering which could represent the sequences of the lava flows that are thought to make up the plains into which the canyons have cut.

The eastern part of Valles Marineris loses the prominent east–west fabric and consists of loosely connected canyons and depressions, some of which are as large as 150 km across. These grade into enormous channels, most of which eventually empty into Chryse Planitia to the north. The general area is classified as **chaotic terrain** (discussed below) and is characterized by its jumbled, irregular appearance (Fig. 7.21), suggestive of extensive mass wasting.

Most investigators agree that tectonic processes, somehow related to the Tharsis region, played a key role in the origin of the canyonlands. Crustal extension generated fractures and grabens that were enlarged by mass wasting (Fig. 7.22) and further eroded by aeolian and other processes. From analysis of deposits emplaced within some canyons, Jack McCauley (1978) of

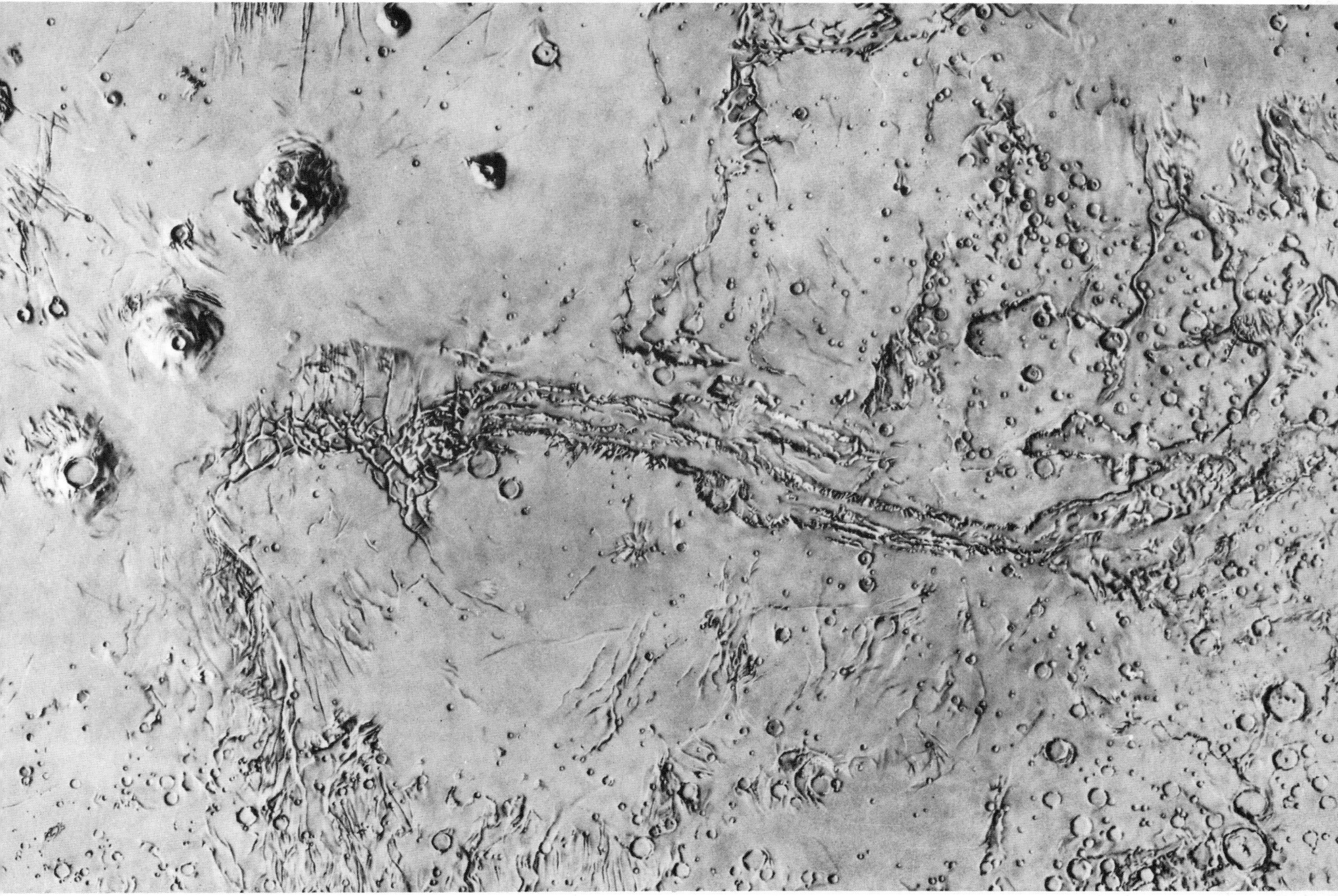

Figure 7.19 Shaded airbrush relief chart showing the Valles Marineris region. This area, known collectively as the canyonlands, is composed of three elements, Noctis Labyrinthus in the west, the central canyon area, and an eastern area made up of chaotic terrain and outflow channels. Area shown is about 3500 by 6000 km (courtesy US Geological Survey, Flagstaff).

the US Geological Survey proposed that vast lakes were once contained within the canyons. He suggests that as dams within the canyonlands were eroded, the lakes drained eastward through the chaotic terrain.

Groundwater processes and subsurface ice are also involved in most models of the evolution of Valles Marineris. Some canyonland elements, such as Hebes Chasma, are closed systems and ablation of ice to the atmosphere and subsurface erosion by flow of groundwater through "piping" may account for some of the material removed from these areas.

7.3.7 Channeled terrain

"Dry riverbeds were one of the most unexpected finds of the Mariner 9 mission to Mars": so wrote David Pieri in his 1983 review of the valleys and channels on Mars. Why unexpected? Because by the Mariner 9 mission, it was clearly established that liquid water can exist on the surface of Mars for only short periods of time (~minutes) – yet the features revealed by the Mariner 9 cameras clearly spoke of erosion by water. This implied either that environmental conditions on Mars were quite different in the past or that the features were formed by processes other than free-flowing water.

Two types of features have been identified: *outflow channels* (Fig. 7.21) and *small valleys* (Fig. 7.6). Small valleys and valley networks are discussed in Section 7.7.2. The *channeled terrain* (unit ch) refers to areas that have been modified by outflow channels and includes terrain modified by the "fretted" channels of Sharp and Malin (1975). Outflow channels generally begin in terrain that has been modified by collapse and mass wasting. Most of these channels originate fully developed in uplands and flow northward across diverse

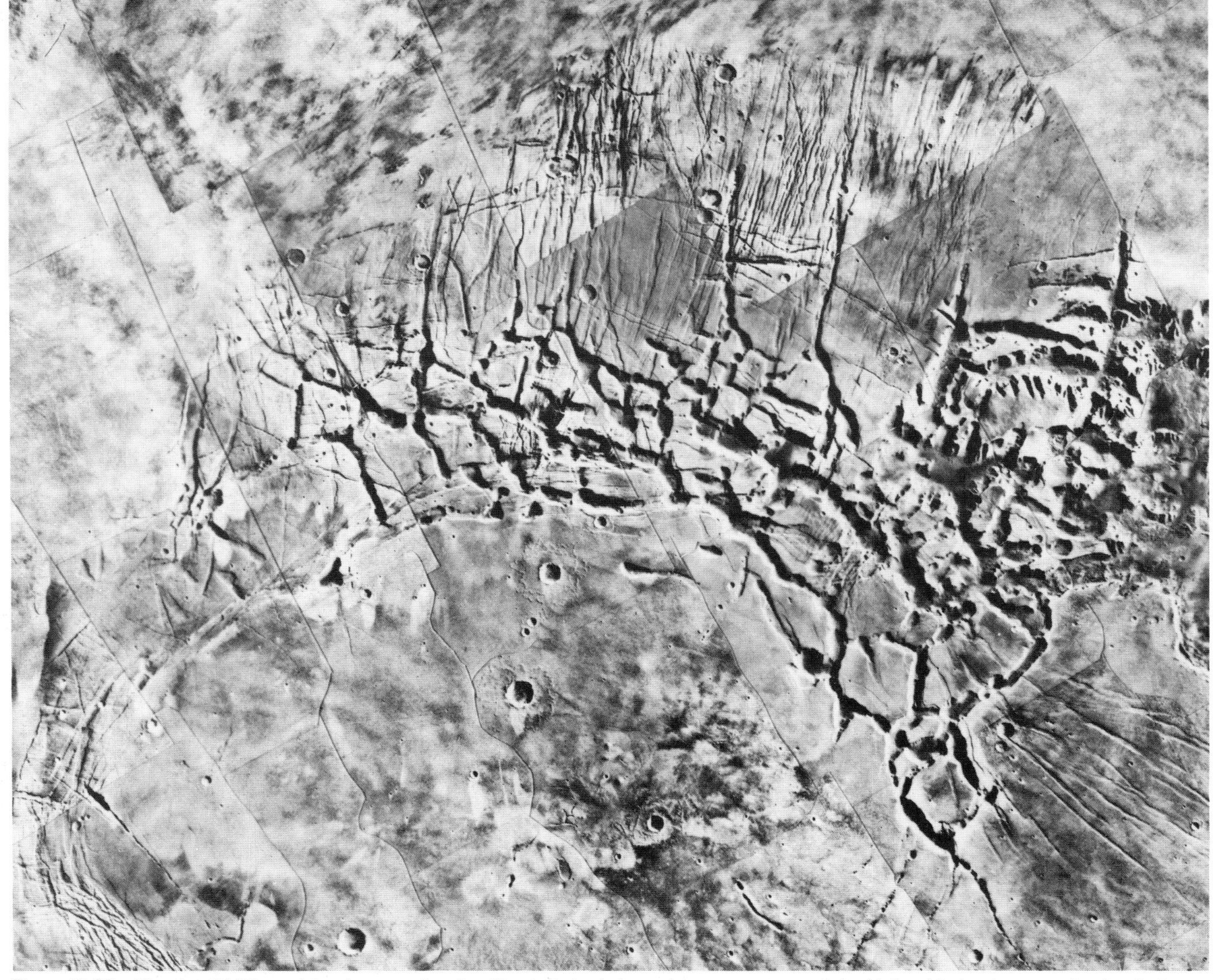

Figure 7.20 Mosaic of Viking orbiter images showing Noctis Labyrinthus, the western part of the Valles Marineris canyonlands. Grabens and other fractures resulted from the uplift of the Tharsis bulge (from US Geological Survey MC17 NE).

terrains onto the lowland plains. They are typically 800–1200 km long, although some, such as Kasei Vallis, may reach 3000 km in length. Some outflow channels are 2.5 km deep and exceed 150 km in width. In general, channel depth is inversely related to width. Gradients for outflow channels have been estimated from a minimum of 0.5 m km^{-1} for Simud Vallis to a maximum of 20 m km^{-1} for parts of Maja Vallis.

Many outflow channels exhibit terracing or layering along the walls which may reflect differential erosion, or repeated episodes of channel formation, or both. Episodic channeling is clearly documented in Chryse Planitia by stratigraphic relations, shown in Figure 7.23. Evidence for scouring, grooving and erosional "plucking" are seen in association with many outflow channels which sweep across wide areas (Fig. 7.24).

7.3.8 Modified terrain

Several areas of Mars have been so severely modified that they have been singled out as distinctive terrain units. **Chaotic terrain** (unit hc; Fig. 7.21) is characterized by disordered, jumbled blocks of terrain that appear to have been subjected to extensive mass wasting. In addition to the eastern part of Valles Marineris, chaotic terrain is found along the western boundary of Lunae Planum, south of Apollinaris Patera and north of Hecates Tholus, most of which is on the rings of ancient impact basins.

Fretted terrain (unit hf) is found primarily along the boundary of the cratered terrain and the northern lowland plains. First described by Sharp (1973), it typically consists of two parts, an upper section of relatively undisturbed terrain and a lower section of relatively smooth plains that appears to have formed by scarp retreat at the expense of the upper section (Fig. 7.25). The lower section often grades imperceptibly into the lowland plains. The "fretting" is clearly controlled by fracture patterns in some areas and by channels in other areas.

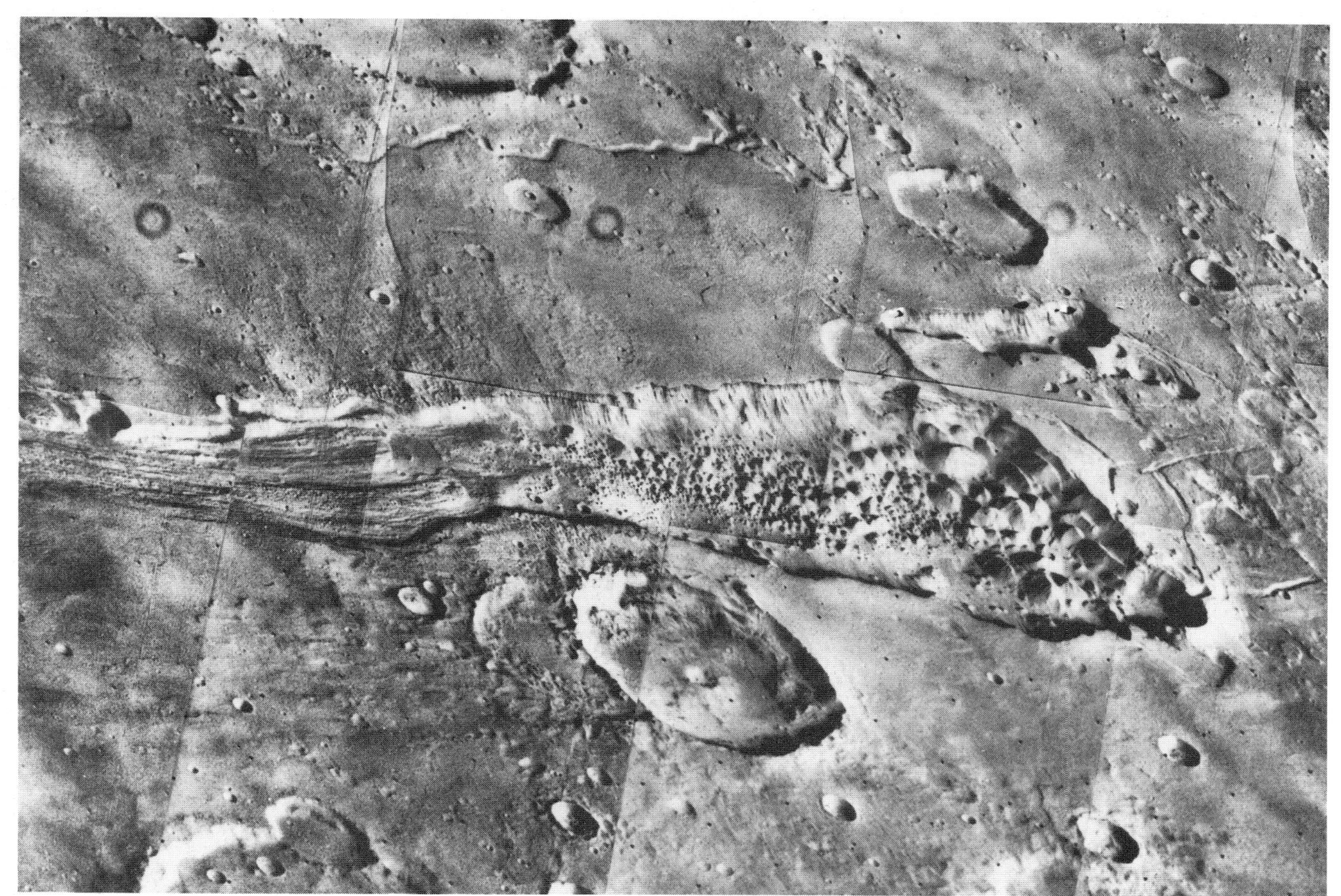

Figure 7.21 Mosaic of Viking orbiter images showing a 20 km wide channel emerging from the chaotic terrain. This channel eventually connects with Simud Vallis and flows northward into Chryse Planitia (JPL P-16983).

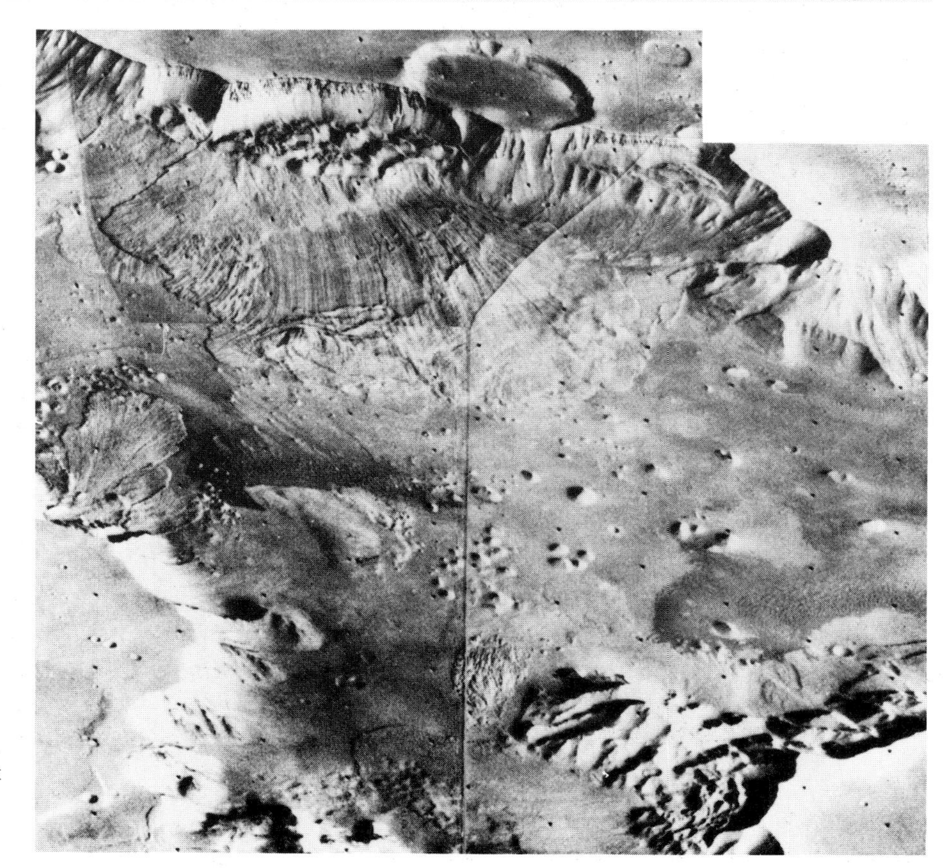

Figure 7.22 Oblique Viking orbiter view across Gangis Chasma in the canyonlands. The landslide on the far wall extends as far as 50 km from the canyon wall and is one of several landslides that have enlarged the canyon. Visible in the lower right is a dark deposit which consists of sand dunes, demonstrating aeolian activity (JPL P-16952).

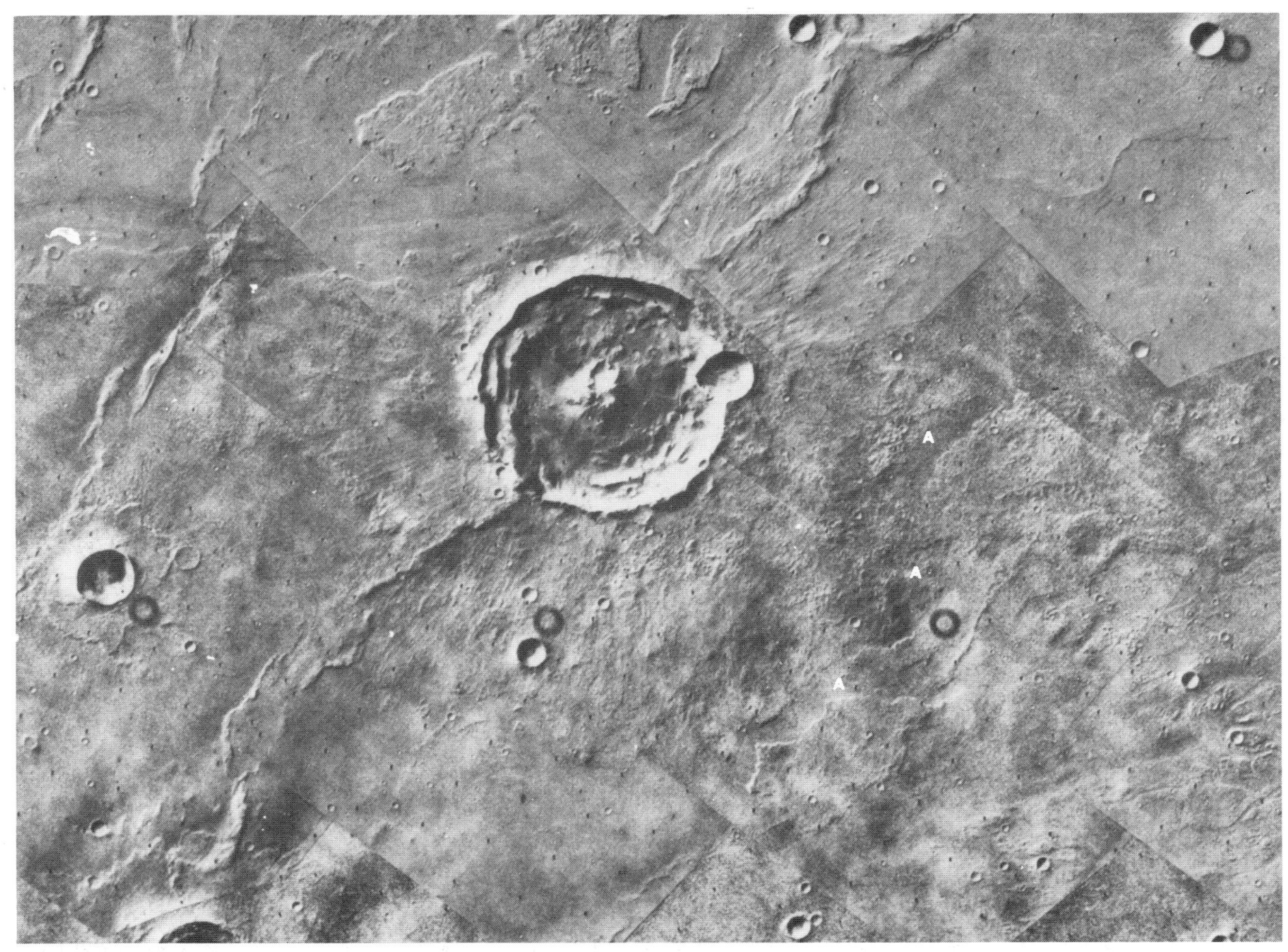

Figure 7.23 Viking orbiter mosaic of a 21 km diameter impact crater in western Chryse Planitia showing multiple channeling episodes. A set of degraded channels (A) is overlain by ejecta deposits from the crater showing that channel formation pre-dates the impact. On the north (upper half of figure) side of the crater the ejecta deposits have been eroded by a later episode of channeling. The regularly spaced dark rings are artifacts of the camera system (from Greeley *et al.* 1977).

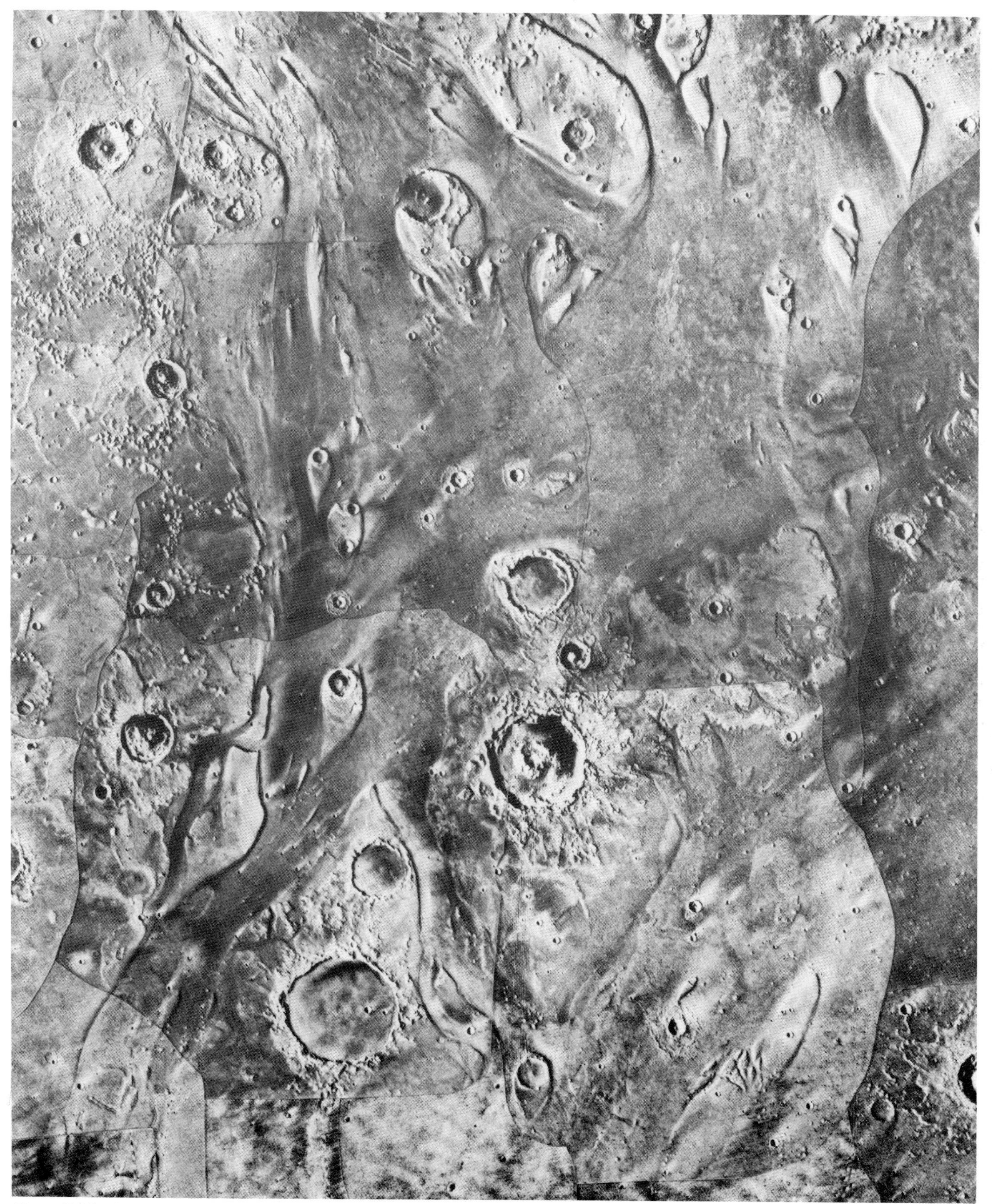

Figure 7.24 Part of channel system in Chryse Planitia showing channel segments and teardrop-shaped islands. Area shown is 600 by 750 km, centered at 23°N, 33°W (mosaic part of MC 11-NW, US Geological Survey).

A large area northwest of Olympus Mons shield volcano has been extensively modified to produce **grooved terrain** (unit g) which extends as lobes from the base of Olympus Mons several hundred kilometers (Fig. 7.26). The origin of this terrain, known also as the Olympus Mons aureole, has evoked considerable debate and includes models involving erosion of ash sheets (King & Riehle 1974, Morris 1982), eruptions through ice sheets (Hodges & Moore 1979) and mass wasting processes (Harris 1977, Lopes *et al.* 1980 & 1982).

7.3.9 Knobby terrain

Knobby terrain (unit k) is found mostly in the northern hemisphere where it occurs east of Elysium, within the northern plains (Guest *et al.* 1977), and along the cratered uplands–northern lowland plains border where knobs are often gradational with fretted terrain. Individual knobs range up to 10 km across and appear to be of diverse origins, including erosional remnants of older terrain such as highly degraded crater rims (Fig. 7.27). Some knobs stand higher than the upper surfaces of fretted terrain, indicating that these knobs are more than simply erosional remnants. In other areas, many of the knobs have small summit craters, suggesting that they may be small volcanoes.

Figure 7.25 View of fretting along the border of the cratered terrain and the northern lowlands. "Fretting" results from scarp retreat and the breakup of the uplands to form small, outlying mesas. Area shown is 250 by 300 km (US Geological Survey, MC5-SC, part).

Figure 7.26 Mosaic showing Olympus Mons and the surrounding aureole material. Northern lobe (A) is bounded on the eastern side by scarp (C) and partially overlain by the northwestern lobe (B) (from Lopes *et al.* 1982; mosaic courtesy of E. C. Morris, US Geological Survey).

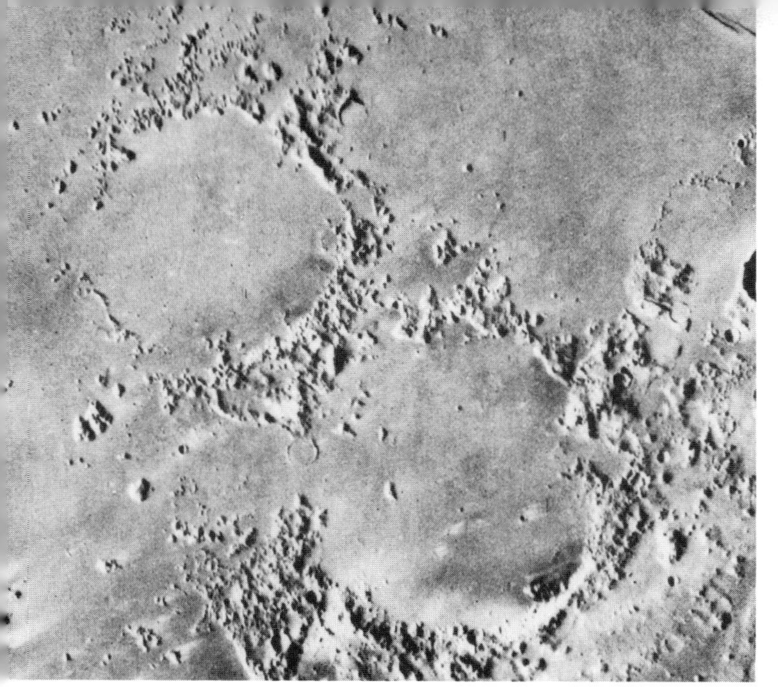

Figure 7.27 Knobs east of the Elysium region which represent rim of degraded crater about 280 km across (Viking Orbiter 672A64).

Figure 7.28 Impact crater 50 km in diameter in the Noachis region. The central peak in this crater has a pronounced pit which may have formed by collapse during the final stages of crater formation. The abundance of central pits on Mars suggests that they may be related to volatiles, such as subsurface water or ice contained at the time the impact occurred (Viking orbiter mosaic 211-5735 part).

7.4 Craters

When Mariner 4 returned the first spacecraft images in 1964, Mars was seen to be a cratered object similar to the Moon – an impression that did not change substantially with the later flybys of Mariners 6 and 7. As fate would have it, all three spacecraft imaged essentially the same physiographic region of Mars, the heavily cratered terrain. Yet, even though these images were crude by today's standards, the craters appeared to be somewhat different from those on the Moon. Mariner 9, carrying both wide angle and telescopic cameras, revealed not only the diversity of terrains on Mars but also provided additional details on martian craters – details that were further enhanced by Viking images.

7.4.1 Crater morphology

Even on earliest spacecraft images, martian craters were seen to be shallower than lunar craters. This was attributed to erosion of the rims and partial filling of the craters by windblown sediments and other deposits (Fig. 3.35). We know that lunar craters are enlarged by slumping, and if volatiles were present in the "target" materials on Mars, this process would probably be enhanced and could lead to a lower depth-to-diameter ratio.

The morphology of craters might provide some insight into the environment (including "target" properties) at the time the craters were formed. For example, from analysis of Viking orbiter images, Eugene Smith (1976) and other investigators noted that many martian craters have central pits (Fig. 7.28). Wood and his colleagues (1978) attributed this difference to the presence of subsurface ice. They suggested that during impact the transient cavity penetrated a zone containing ice which was volatilized by impact-generated heat and vented to form the central pit.

One of the more striking results of the Viking mission was the discovery of ejecta deposits (Fig. 3.4) which appear to have been emplaced at least partly by flow of fluidized materials (Carr *et al.* 1977). In many cases, the final stages of ejecta emplacement involved ground-hugging flow, rather than ballistic emplacement (Fig. 7.29). These craters, variously termed "splosh" craters, ejecta-flow craters or rampart craters, are not seen on Mercury or the Moon, and it has been suggested that water or the atmosphere of Mars, or both, may have been responsible for the form of the ejecta.

Laboratory impact experiments using viscous targets to simulate impacts on Mars led to a model which accounts for many of the features observed on martian craters (Fig. 7.30). Schultz and Gault (1979) conducted other experiments to assess atmospheric effects on ejecta emplacement and found that ejecta mixed with

Figure 7.29 Viking orbiter image of an ejecta-flow crater (upper right). Emplacement of ejecta must have involved material flowing very close to the surface rather than as a ballistically emplaced deposit, indicated by the control of the small impact crater (A). Area shown is 50 km across (Viking Orbiter 32A28).

even a low-density atmosphere could produce distinctive ramparts, flowlike features and radial grooving, depending on the size of the ejecta particles. Thus, it would appear that volatiles in the form of subsurface water (or perhaps ice) or an atmosphere could result in ejecta flow patterns.

Peter Mouginis-Mark (1979 & 1981) classified martian craters by their morphology (paying particular attention to ejecta deposits) and searched for correlations with terrain, elevation and latitude. In general, he found that ejecta deposits extend farther from the crater rim at higher latitudes and at lower elevations, perhaps indicating that the fluidity of the ejecta was greater during emplacement. This would suggest that a higher abundance of volatiles was present in those areas at the time of impact.

7.4.2 Crater statistics

Crater counts on Mars have generated considerable controversy. For some areas, the shape of the size-frequency distribution curve is remarkably similar to that of the Moon. Other areas are different from the Moon, depending upon the size range of craters counted. Of course, it is reasonable to expect some differences between the Moon, where erosion by non-impact processes is negligible, and Mars, where wind and water have modified the surface. Thus, like impact craters on Earth, many craters on Mars may be obliterated or so degraded that they are impossible or difficult to identify for counting. For example, Wood and Head (1976) noted that the Moon appears to have more basin or basin-like craters than Mars. However, more recently, Schultz and co-workers (1982) have found evidence for 21 possible additional basins (Table 7.2). The features they describe have been severely degraded (Fig. 7.31), but many still display geomorphic features indicative of impact.

Furthermore, processes of gradation have not been uniform through time on Mars. Evidence in the old, heavily cratered terrain suggests that running water may have carved drainage networks early in martian history. Younger terrains do not show this evidence, suggesting a change in climate. Thus, the erosion and obliteration of craters may not be taken into account easily as a function of time.

Attempts to relate crater counts to "absolute" ages of surfaces – typically using the cratered surfaces of the Moon that have been sampled and dated as a calibration – are fraught with difficulties (Neukum & Hiller 1981, Chapman & Jones 1977). In general, martian crater counts exhibit more small craters in proportion to large craters than is typical for the Moon. Soderblom and his colleagues (1974) attribute this to a greater proportion of secondary craters on Mars (perhaps resulting from the break-up of ejecta in the martian atmosphere to produce more secondaries); however, Neukum and Wise (1976) suggest it may be due to a fundamental difference in the distribution of impacting bodies in the vicinity of Mars.

Despite these difficulties, crater counts on Mars provide unique and fundamental information for *relative* dates of surfaces that have not been extensively modified.

7.5 Volcanism

The history of the martian surface is dominated by volcanism. A study based principally on the morphology of surface features shows that a substantial part of the surface of Mars involves various volcanic units (Greeley & Spudis 1981). Although the enormous shield volcanoes are impressive, they constitute less than 1 percent of the total surface. In contrast, plains of probable volcanic origin cover more than 35 percent of Mars. Estimates of the composition of volcanic materials on Mars are derived from analyses of surface materials at the Viking lander sites using X-ray fluorescence techniques, from remote-sensing data based on spectral characteristics, and from theoretical modeling based primarily on geophysical and petrological considerations. Together, the results suggest igneous rocks (such as basalts) that are mafic to ultramafic in composition.

7.5.1 Styles of volcanism

Mars exhibits a wide variety of volcanic features, probably reflecting many different styles of volcanism. Some of the vast plains (Fig. 7.9) may have formed by flood eruptions of mafic lavas. Lack of large channels and collapsed lava tubes in these areas may signal rapid emplacement and ponding of thick sheets of very fluid lavas, analogous to some mare units on the Moon and possibly some of the mercurian smooth plains. Other plains exhibit hummocky surfaces and small channels that could be volcanic (Fig. 7.32). These plains may have been emplaced by lavas erupted at lower rates (Fig. 3.27). Finally, some plains, such as those of Syrtis Major, have central vents and may be very low-profile shield volcanoes (Schaber 1982a).

Some plains may not be lava flows but could be vast volcanic ash flows. Water almost certainly has been present throughout most of the history of Mars, if not on the surface then in the subsurface as groundwater or ground ice (Fanale 1976). With widespread effusive volcanism, it is likely that phreatomagmatic eruptions would have generated abundant ash deposits, as Dave Scott (1982) has suggested for the mantling deposits seen in the Memnonia region (Fig. 7.33). However, Francis and Wood (1982) argue by analogy with terrestrial volcanism that such eruptive styles may not be present to this degree on Mars, and as an alternative,

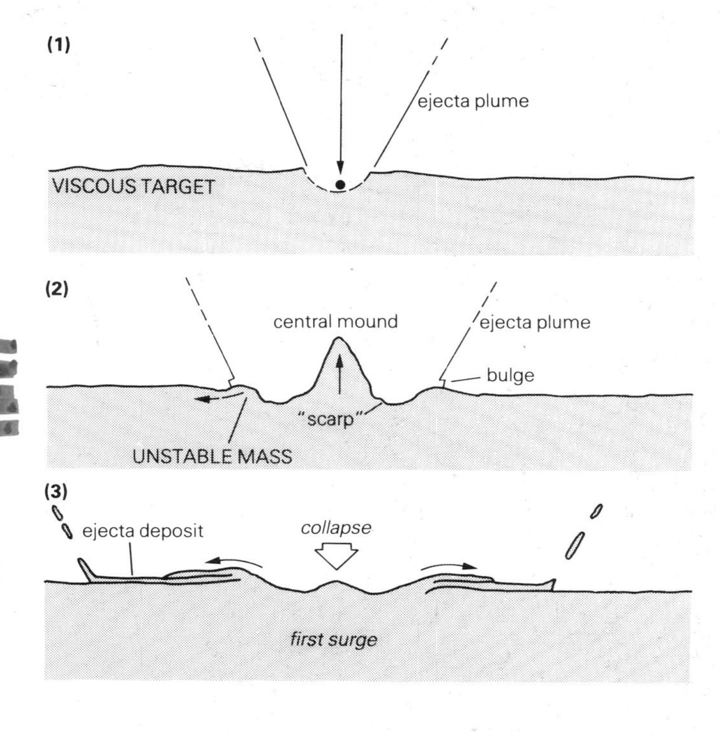

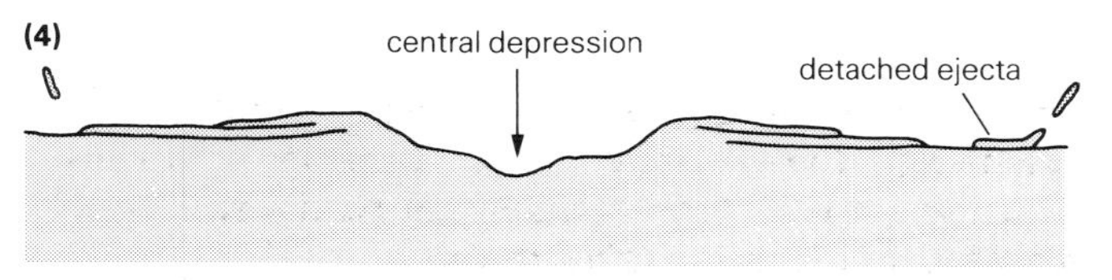

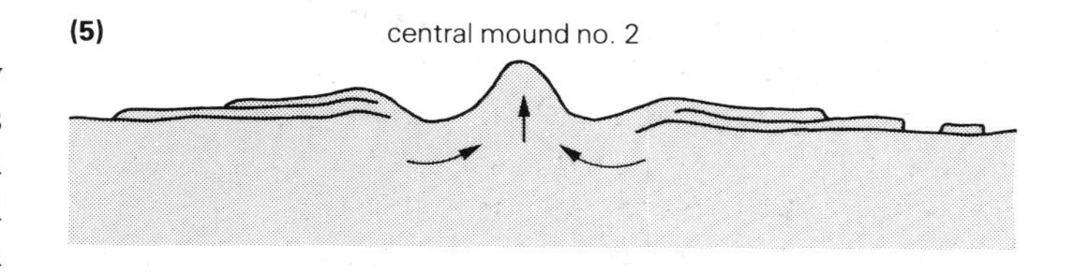

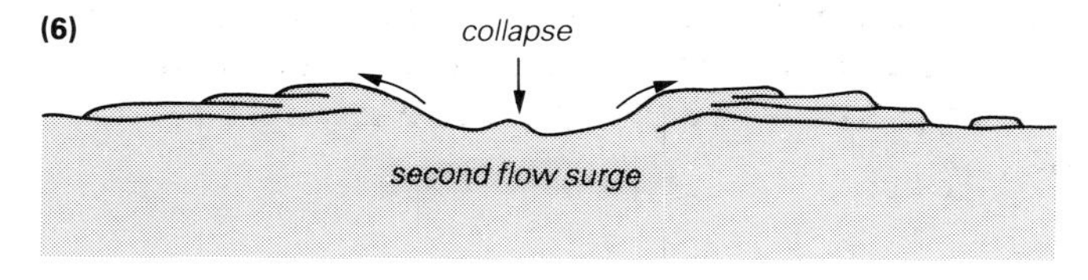

Figure 7.30 Sequence of impact cratering in viscous targets, derived from analysis of laboratory experiments (from Greeley *et al.* 1980).

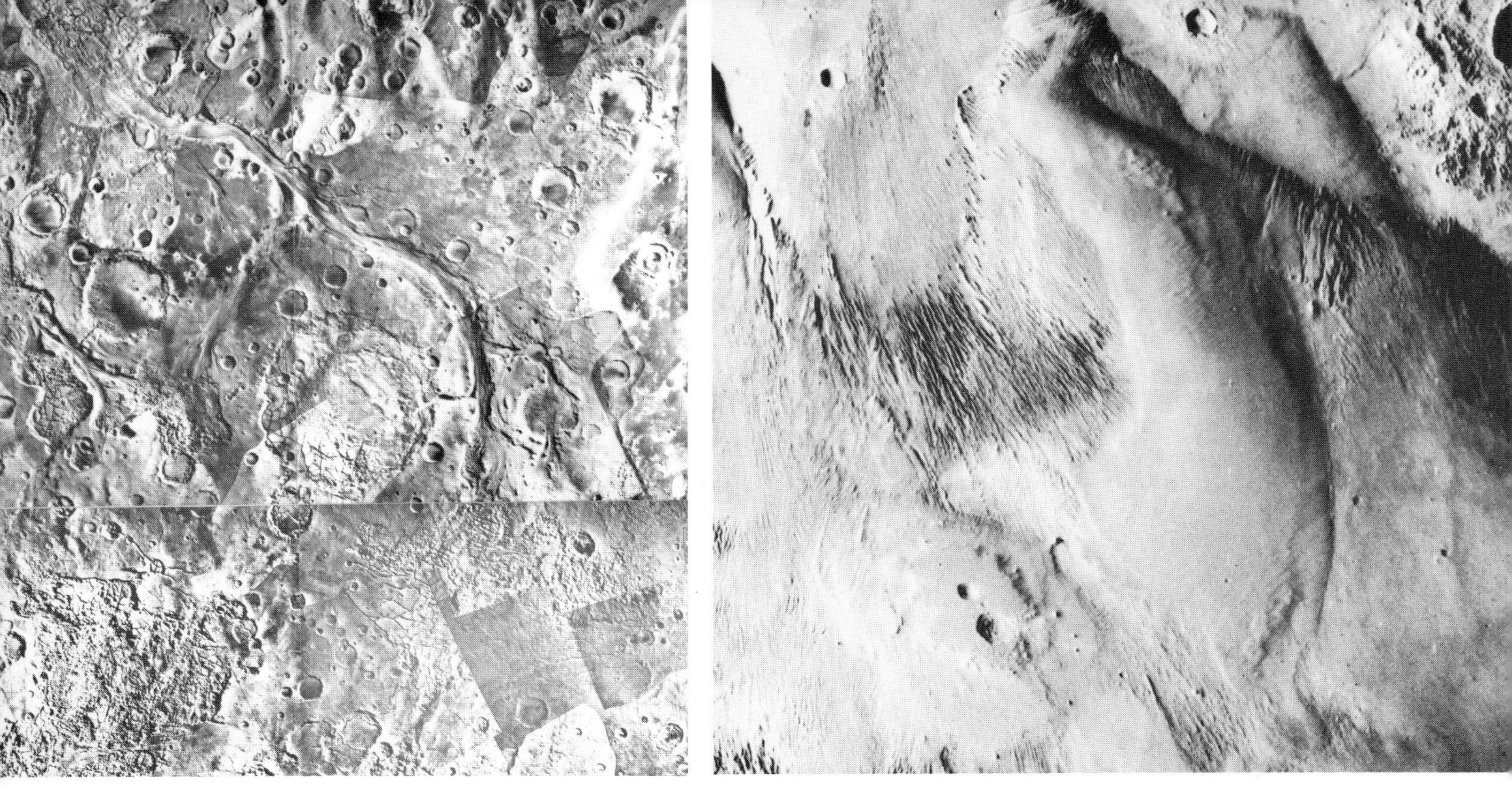

Figure 7.31 (top left) Viking orbiter mosaic showing multi-ring pattern associated with Aram Chaos, centered at 3°N, 22°W. Concentric patterns of unstable chaotic terrain and stable terrains delineate ancient multi-ring basin pattern. Area shown is 1000 by 1000 km (from Schultz *et al.* 1982).

Figure 7.32 (left) Plains in the Cydonia region (41°N, 9°W) showing a sinuous channel that may be analogous to lunar sinuous rilles (Greeley & Spudis 1981). Smooth material flanking the channel and mantling the adjacent terrain could be lava deposits associated with the channel. Area shown is 40 by 40 km (Viking Orbiter 72A11).

Figure 7.33 (above) Postulated ignimbrite deposits that overly ancient cratered terrain in the Memnonia region of Mars, showing possible nonwelded (smooth) and welded (striated) zones (Scott 1982). Area shown is 240 by 290 km (Viking Orbiter 599A53).

Schultz and Lutz-Garihan (1982) suggest that these thick mantle deposits may be related to ancient polar processes.

From photogeologic analysis, Greeley and Spudis (1981) concluded that the highland patera may be similar to ash shields seen on Earth (Fig. 3.18). We speculated that early-stage eruptions through groundwater contained in impact-generated regolith produced phreatomagmatic deposits. With continued eruption, activity evolved into effusion of lava flows, part of which carved channels in the ash deposits leading to the forms exemplified by Tyrrhena Patera (Fig. 7.16).

Phreatic eruptions are also suggested on a smaller scale. In the northern plains, some of the small cones with summit craters have been suggested by Greeley and Theilig (1978) and Frey *et al.* (1979) to be analogous to "pseudocraters" in Iceland. Pseudocraters are small pyroclastic cones (often with summit craters) which form on the surface of lavas that have flowed over water-saturated ground. Local phreatic explosions rupture the crust of the flow and generate piles of tephra.

Except for their size, martian shield volcanoes mimic Hawaiian shields. Careful mapping and analysis of impact crater frequencies by Crumpler and Aubele (1978) show that Arsia Mons, a typical martian shield volcano, developed in the following sequence: (a) construction of the main shield; (b) parasitic eruptions on the northeast and southwest flanks; (c) caldera formation by subsidence of the summit region, perhaps in response to magma withdrawal associated with the parasitic eruptions; and (d) effusive activity through fissures at the summit and on the flanks of the shield.

Olympus Mons (Fig. 7.11) is younger than Arsia Mons and does not appear to have evolved to the same degree, at least so far as the complexity of the caldera is concerned. Its total volume is estimated to be three times that of the entire Hawaiian Emperor chain of volcanoes on Earth. Carr (1973) attributes this to a longer period of activity for Olympus Mons and to differences in tectonic style, noting that the Hawaiian chain of volcanoes are probably produced as the Pacific plate slides over a magma "hot spot". Mars evidently lacks Earth-like plate tectonics. Hot spots in the martian interior may produce lava continually through a "fixed" central vent. Some investigators speculate that, in order for the magma to produce the great heights of the martian volcanoes, the lithosphere must be relatively thick in the Tharsis region.

7.6 Tectonism

Tectonic features, such as faults, provide clues to the style of crustal deformation and can be used to place constraints on models of planetary interiors. When the timing of tectonic deformation can be determined, further knowledge can be gained about the evolution of the interior. Although some tectonic features have been found in association with the Hellas, Argyre and Isidis basins, by far the dominant tectonic province on Mars is the Tharsis region. Much attention has been focused on Tharsis in an attempt to relate the tectonic and volcanic processes to the nature of the interior of Mars and its thermal history.

7.6.1 The Tharsis region

The Tharsis region is a broad plateau (Fig. 7.34) some 4000 km across which stands more than 10 km above the surrounding plains. It is dominated by central volcanoes and volcanic plains which vary greatly in age. The "apex" of the province is the Tharsis bulge, an enormous swelling centered at about 0° latitude, 114°W.

Numerous faults and fractures cut the plains (Fig. 7.35) and some of the volcanoes. In general, these are extensional features that are radial and subradial to the Tharsis bulge. Fractures, faults and grabens, some as long as 3000 km, are particularly well developed in three areas: north of Tharsis around Alba Patera, northeast of Tharsis in the Tempe region, and south of Tharsis in the Claritas region. Extensive mare-type ridges in Lunae Planum, Sinai Planum and Solis Planum are crudely concentric to the bulge and may reflect compressive forces. As discussed earlier, part of the canyonlands (Noctis Labyrinthus) is also associated with the Tharsis bulge. In all, the Tharsis region and its associated tectonic features leave their imprint on more than a third of the martian surface.

By precise tracking of the orbital paths of the Mariner 9 and Viking spacecrafts, gravity data have been derived for parts of Mars. These data show that a large, positive, free-air gravity anomaly is centered over the Tharsis bulge (Sjogren *et al.* 1975), and that local gravity highs occur over the large volcanoes (Sjogren 1979). On Earth, free-air gravity anomalies are typically small because mountains are compensated by low-density "roots". The anomalies on Mars are much larger than on Earth, suggesting a different mode of support, such as mantle convection or a very thick lithosphere.

7.6.2 History and origin of the Tharsis region

Three lines of evidence are used to deduce the history of the Tharsis region (Carr 1981): (a) analysis of the intersections of various sets of fractures; (b) relative age dating of surfaces based on crater counts; and (c) study of the geometric relationships between the fractures and the dated surfaces. Using this process, Don Wise and his co-workers (1979) found that the oldest fractures occur in the cratered terrain of the Tempe region and inferred that uplift of the Tharsis bulge began early in the history of Mars. Carr speculates that the origin of the bulge may be linked to mantle convection associated with the separation of the core. A second, later episode of fracturing occurred in the northern Tharsis area, forming the Ceraunius Fossae and the extensive faults around Alba Patera (Fig. 7.8). Most of the volcanic units in the center of the Tharsis region were emplaced after the main episodes of fracturing. Detailed photogeologic mapping by Scott and Tanaka (1981) led to the derivation of paleostratigraphic maps which show the evolution of the region. They note that the emplacement of volcanic plains and the formation of central volcanoes occurred concurrently throughout the history of Tharsis.

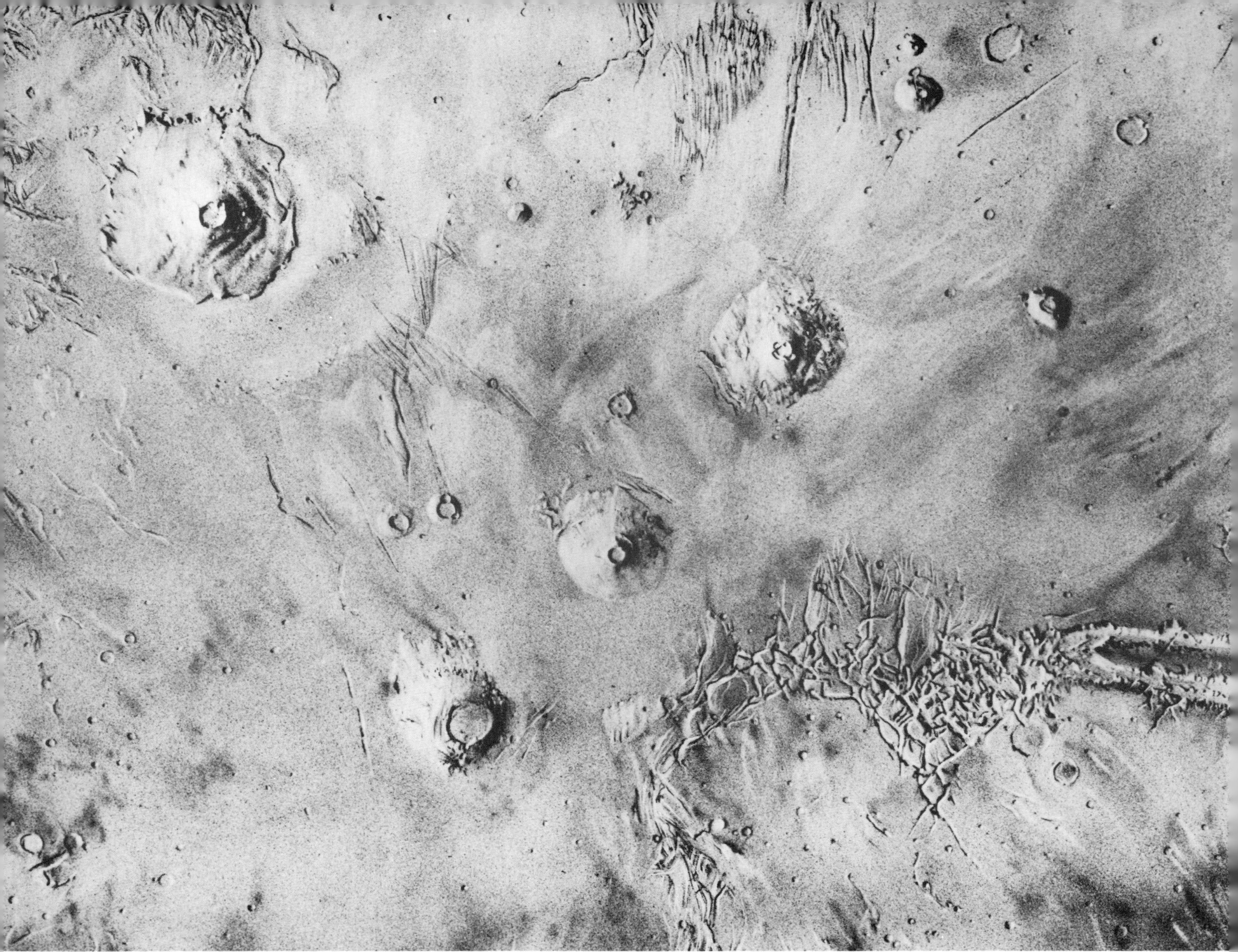

Figure 7.34 Shaded airbrush relief map of the Tharsis region, centered at 7°N, 113°W (courtesy US Geological Survey, Flagstaff).

Sean Solomon (1981) reviewed the various geophysical models that have been proposed to explain the Tharsis bulge and its related surface features. He notes that the "traditional" model involves broad updoming of the martian lithosphere by a thermal or chemical anomaly, leading to extensional fracturing. Calculations of the stresses in the martian lithosphere required by this model, however, raise some serious questions. Phillips and Lambeck (1980) note that the Tharsis bulge is so large in relation to the planetary radius that full spherical calculations must be made in deriving geophysical models. When such calculations are carried out, the stress geometry for the "traditional" model does not fit the fracture patterns seen on the surface. Consequently, two alternative models have been proposed: (a) Sleep and Phillips (1979) suggest an *isostatic model* involving a low-density region of the mantle several hundred kilometers deep which may be partly thermal in origin and (b) a *flexural model* proposed by Solomon and Head (1982a) in which the bulge is largely the result of the accumulation of lava flows supported by a thick lithosphere.

Solomon suggests that some combination of these models may be the best explanation. He notes that older fault systems are best explained by the isostatic model, whereas the younger volcanic terrains fit best with the flexural model. Thus, there may have been a change in the evolution of the interior of Mars from a dynamic mantle to a more static mantle overlain by a relatively thick lithosphere.

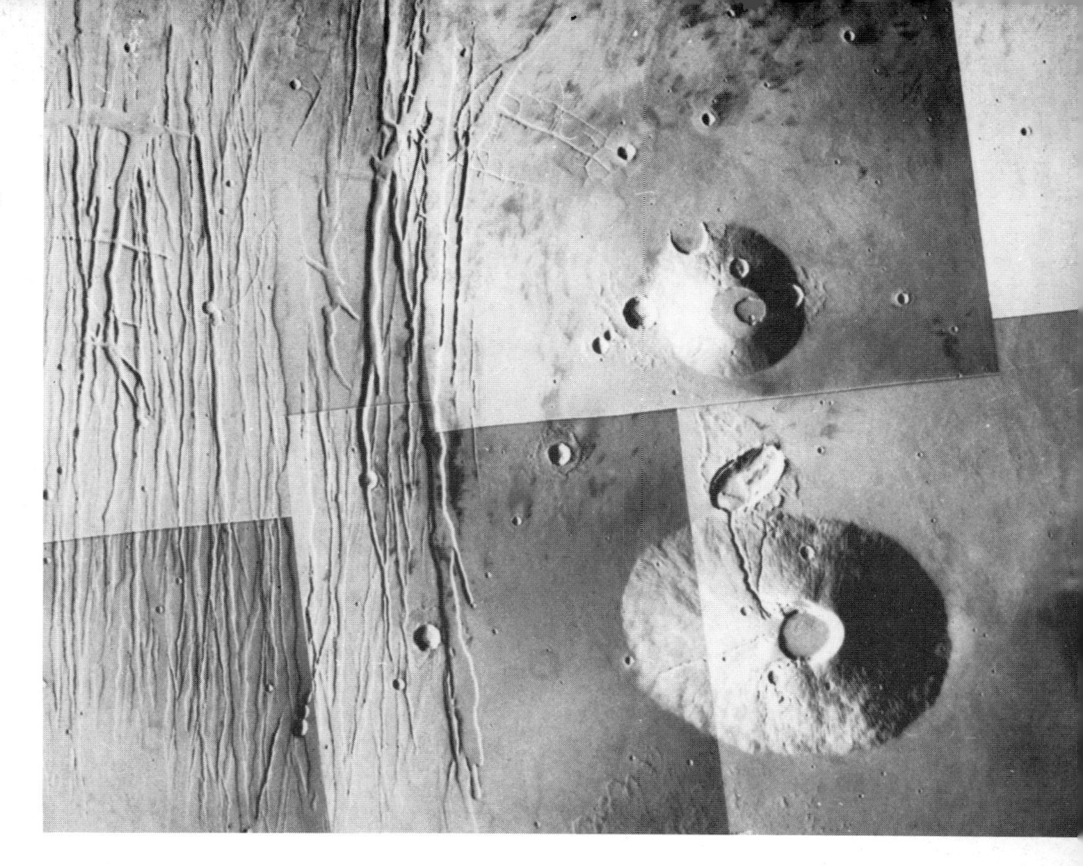

Figure 7.35 Viking orbiter mosaic showing Ceraunius Fossae of the Tharsis region and two of the smaller volcanoes, Uranius Tholus (upper) and Ceraunius Tholus (100 km in diameter; lower) (Viking Orbiter 211-5639 part).

7.7 Gradation

Gradation involves the weathering, erosion, transportation and deposition of materials on planetary surfaces. The present environment on Mars is cold and involves a low-density, CO_2 atmosphere (Table 2.1), and chemical and physical weathering processes probably operate slowly compared to Earth. Nonetheless, views of the martian surface taken from the Viking landers (Figs 1.1 & 1.4) and from orbit show evidence for gradation by aeolian activity, running water, mass wasting and periglacial processes.

7.7.1 Wind

The low-density atmosphere requires high wind speeds for the movement of particles on Mars (Fig. 3.36). Measurements of winds from the Viking landers and observations of dust storms show that threshold velocities are achieved on Mars and that aeolian processes play a key role in the present-day modification of the surface.

Wind erosion and deflation have led to the formation of wind-deflated areas in polar regions (Fig. 7.18), yardangs (wind sculptured hills, Ward 1979) and exhumed terrain (Fig. 7.36). One class of impact crater, termed **pedestal craters** (Fig. 7.7), appears to have formed on plains that were subsequently eroded, perhaps by the wind. In some cases, the cratered plains may have been first mantled and then deflated. In either case, it has been suggested that blocky crater ejecta "armored" the surface to protect it from deflation so that it remained as a high-standing pedestal while the surrounding surface was lowered.

Since their discovery on Mariner 9 images, sand dunes have been found in many areas on Mars. The largest field occurs in the north polar region (Fig. 7.37) and is equal in size to large sand seas on Earth (Tsoar *et al.* 1979). Smaller dune fields are found elsewhere on Mars (Fig. 7.38), often in association with craters (Fig. 7.39). In general, martian dunes exhibit the same morphology as terrestrial dunes, although star dunes and longitudinal dunes appear to be lacking.

Wind streaks are the most common type of aeolian feature on Mars. They can be either bright or dark relative to the surrounding surface and occur in a wide variety of geometries (Figs 7.40 to 7.42). Because some change their size and shape with time, wind streaks are also termed **variable features** (Sagan *et al.* 1972). They are considered to result from wind erosion and deposition and are found in many regions where they can be used as local "wind vanes". For example, Peter Thomas and Joseph Veverka (1979) of Cornell University have mapped wind streaks to determine patterns of atmospheric circulation, following earlier work by Sagan *et al.* (1973).

7.7.2 Water

Depending upon the method of analysis, the total amount of water evolved on Mars would evenly cover the surface from 10 to 160 m. Although liquid water

Figure 7.36 Region southeast of Elysium showing various aeolian features. Cratered terrain in the lower part of picture appears to have been mantled by smooth deposits of possible aeolian origin some of which have subsequently been eroded; half of the crater at (A) has been exhumed. The feature above appears to be a crater buried by the mantling units. Crater at (B) post-dates the mantling deposit, as its ejecta is superimposed on the mantle. Various linear features in the lower part of the picture may be wind-sculptured hills, or yardangs. Pattern in the upper left of picture may be sand dunes composed of weathered particles derived from the mantling deposit. Area shown is approximately 80 by 80 km (Viking Orbiter frame 438S01).

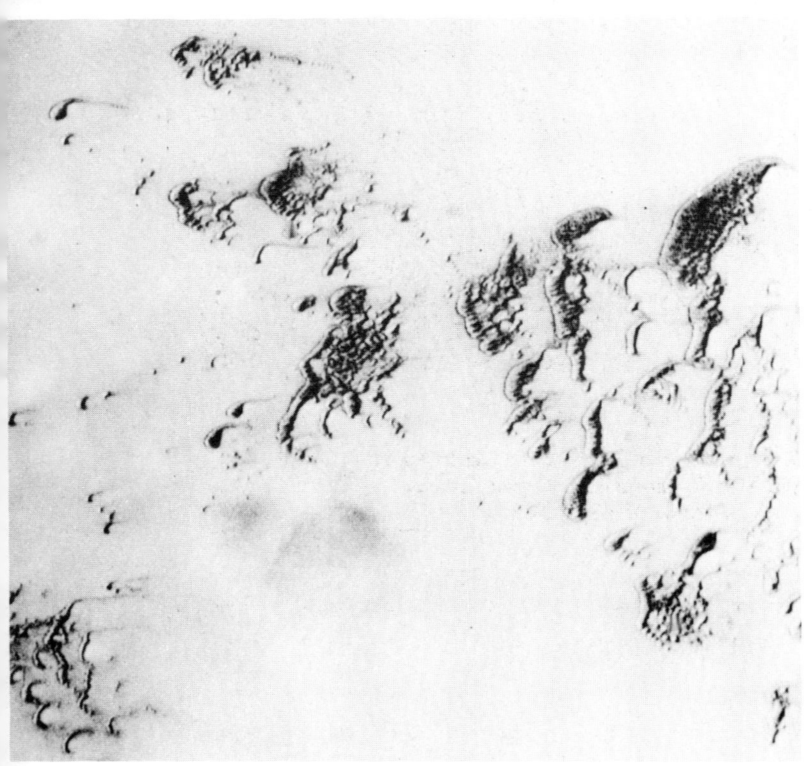

Figure 7.37 Compound barchan dunes in the north polar area. Area shown is 40 by 45 km (Viking Orbiter 544B07).

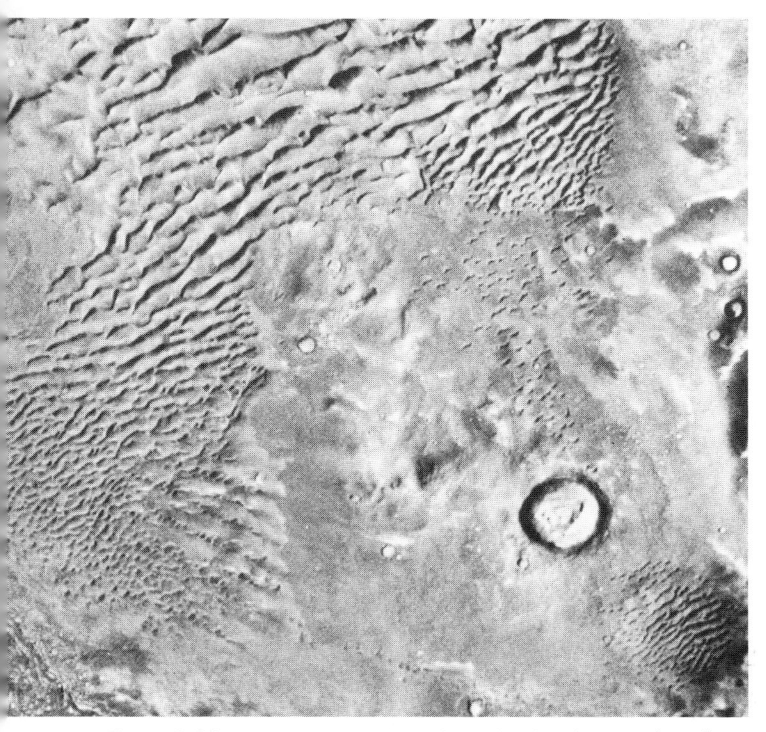

Figure 7.38 Dunes in the cratered terrain showing a series of transverse dunes in the upper left and small barchan dunes in the right half of the picture. Area shown is 50 km across (Viking Orbiter 575B60).

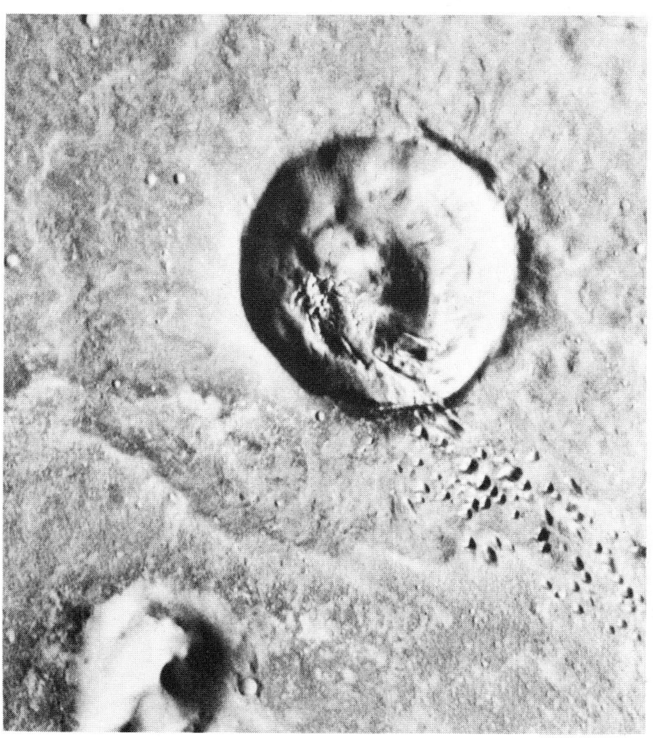

Figure 7.39 "Climbing" dunes which appear to be drifting up and over the rim of a 16 km crater (Viking Orbiter 571B53).

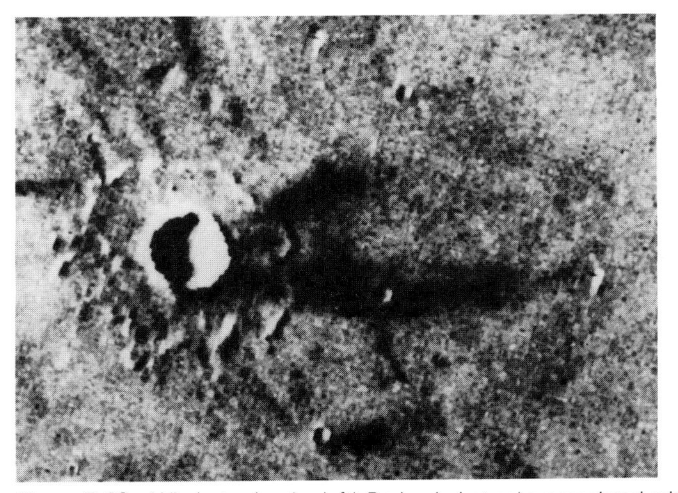

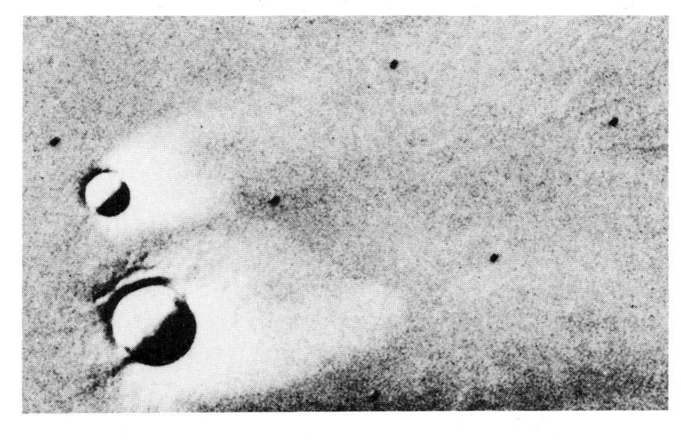

Figure 7.40 Wind streaks. (a, left) Dark wind streaks associated with craters; dark streaks are thought to result from erosional processes. Area is 35 km across (Viking Orbiter 56A20). (b, right) Bright wind streaks associated with craters; most bright streaks are thought to result from depositional processes. Area shown is 35 km across (Viking Orbiter 45B46).

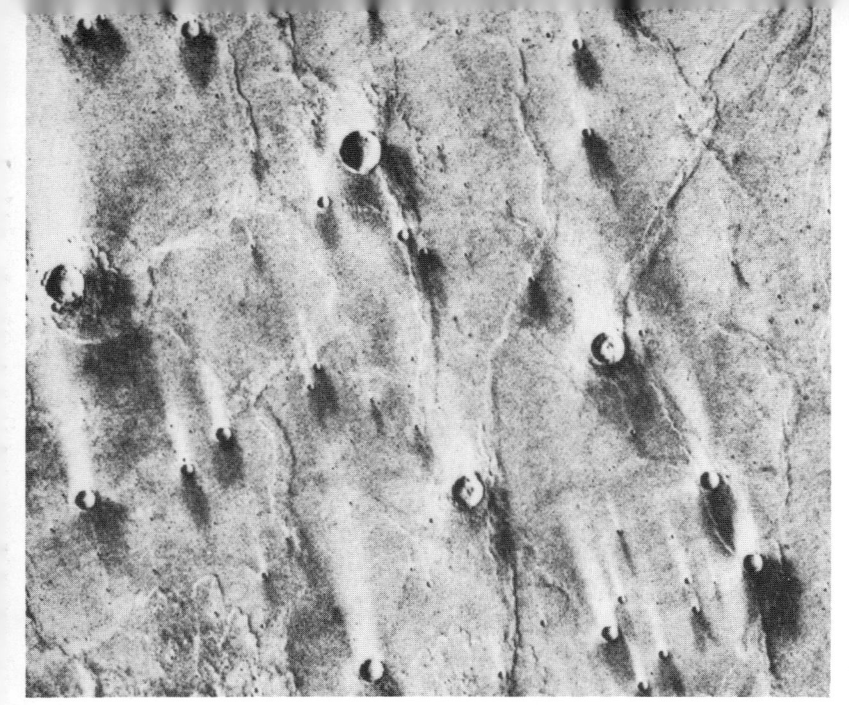

Figure 7.41 Bright and dark streaks associated with craters, suggesting two modes of origin and two winds of opposite directions. Area shown is 200 by 200 km (Viking Orbiter 553A54).

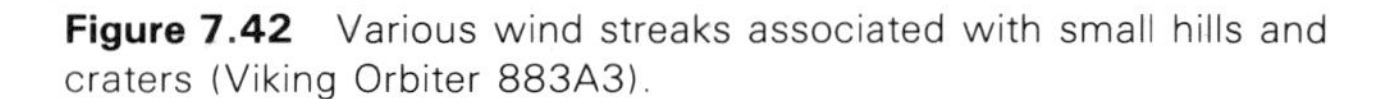

Figure 7.42 Various wind streaks associated with small hills and craters (Viking Orbiter 883A3).

cannot exist on the surface today, various channels and small valleys are generally regarded to be the result of flowing water. Because of the critical implications held by the channels for the history of Mars, NASA formed a working group of scientists to study the topic. Their findings, presented by the Mars Channel Working Group (MCWG 1983), provide an excellent summary of both the outflow channels and the small valleys.

The origin of the outflow channels (Fig. 7.21) remains a controversial topic. A wide range of fluids has been proposed as erosional agents, including running water, ice, lava (Schonfeld 1977), wind (Cutts & Blasius 1981) and mud (Nummedal 1978). Ideas accepted by most students of Mars involve water released by catastrophic floods, an idea first proposed by Hal Masursky (1973) who suggested that the channeled scablands of the Pacific Northwest were a good analog (Fig. 7.43). The scablands formed during the Pleistocene when ice dams broke to release enormous volumes of water that stripped away the basaltic plains. Baker and Nummedal (1978) and Baker (1982) expanded this idea and developed various geomorphologic arguments for the similarities between terrestrial scablands and martian features, although they noted that one to two orders of magnitude greater flow would be required to form the outflow channels on Mars.

Several explanations have been offered for the sudden release of water on Mars. Masursky and colleagues (1977) proposed releases of water as a consequence of ground ice melting, associated with volcanic activity. Scarp retreat was suggested by Soderblom and Wenner (1978) to tap subsurface reservoirs of water. Schultz *et al.* (1982) have noted that many of the outflow channels begin on the rings of heavily degraded impact basins, thereby leading to the suggestion that magma intruded along ancient impact-generated fractures and hydrothermally melted trapped volatiles. Mike Carr (1979) proposed an elegant model for ground collapse and fluid release resulting from outbursts of water from confined aquifers. He estimated the required pore pressures on Mars and concluded that flow could be sustained for long periods, even in today's environment.

Small valleys and valley networks (Fig. 7.6) appear to lack the various features indicative of the types of flow seen in the outflow channels, though the features may simply be too small to be seen on available images. The valleys range in length from a few kilometers to nearly 1000 km, although most are only tens of kilometers long and are typically about 1 km wide. Large, undissected upland surfaces occur between the valleys, suggesting immature development.

The valleys often form networks which superficially

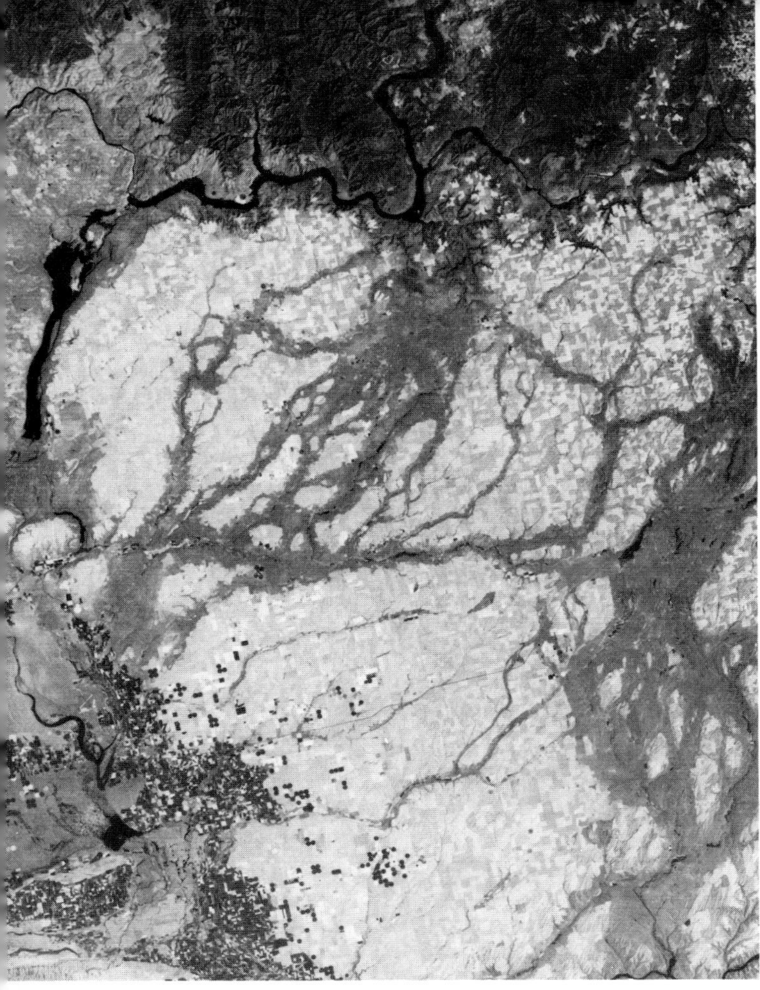

resemble dendritic drainage patterns on Earth. Careful analysis of the junction angles by David Pieri (1979) shows a much narrower range than is typical for Earth, suggesting a different mode of development from the simple runoff of surface water. Furthermore, the valleys often begin and end for no apparent reason (similar to streams in karst regions), although their courses appear to be controlled locally by topography.

Unlike the outflow channels which cut across diverse terrain units, small martian valleys are restricted almost entirely to the ancient cratered terrain. Their origin has been ascribed to both groundwater sapping processes and to runoff of surface water, perhaps associated with rain. Because they are not found in younger terrain, some investigators have suggested that there was a change in the environment which prohibited their formation.

7.7.3 Mass wasting

Mass wasting of several styles has occurred on Mars, as reviewed by Baerbal Lucchitta (1979 & 1981) of the US Geological Survey. Vast landslides have enlarged segments of the canyonlands (Fig. 7.22) and features resembling rock glaciers have been found within the fretted terrain (Fig. 7.44). As discussed by Squyres (1979), many areas, particularly along the boundary between the cratered terrain and the northern lowland plains, exhibit mesas and hills surrounded by debris aprons (Fig. 7.45). The **aureole deposits** (unit g) around Olympus Mons (Fig. 7.26) and similar terrain west of Arsia Mons are ascribed to mass wasting. The precise mode of mass wasting, however, has not been determined; the rate of movement and the amount of water present when the movement occurred are unknown. Although the surface areas of most mass wasting features are too small to date by crater frequency distributions, crater counts on some of the larger landslides suggest that they are relatively ancient.

Figure 7.43 (top left) Landsat image of part of the Columbia Plateau in the Pacific Northwest. Dark areas are enormous channels that resulted from the breakup of an ice dam and flood during the Pleistocene. Flood waters stripped away soil and eroded the underlying basalt flows. Area shown is 150 by 200 km (Landsat frame E-5 854-16504-S).

Figure 7.44 (opposite) View of the Nilosyrtis area at 34°N, 290°W. Ridges and grooves on the material within the valleys may indicate flow of material away from the valley walls and down the valley toward the bottom of the picture. Area shown is about 35 by 40 km (Viking Orbiter frame 884A73).

Figure 7.45 View of debris flows in cratered terrain east of Hellas. Some of the debris aprons extend 20 km from the source. Area shown is 220 by 270 km. North is toward upper left (Viking Orbiter 97A62).

7.7.4 Glacial and periglacial features

Rossbacher and Judson (1981) calculate that the thickness of the **cryosphere** (subsurface zone where water would freeze) on Mars ranges from 3.0 km thick at the north pole to 1.1 km thick at the equator, taking into account the surface temperature and estimates of heat flow. They note that "sinks" for water on Mars include the atmosphere, polar caps and surface frost, exospheric escape and subsurface (or ground) ice. Even using conservative estimates, they conclude that 90 percent of the total volume of water evolved on Mars could be found today as ground ice. Clifford and Hillel (1983), however, argue that ground ice could not exist over martian geologic history; some major recharging mechanism is necessary. In any case, there is opportunity for the formation of various surface features associated with ice.

Several investigators have speculated that glaciers may have existed on Mars. Lucchitta (1982) examined the outflow channels and compared their size and morphology to large, ice-sculptured valleys on Earth. She found that the features characteristic of glaciated terrain (anastomosing valleys, U-shaped valley cross sections, hanging valleys and various linear scour features) can all be identified with many of the martian channels. Lucchitta suggested that at least some of the martian outflow channels could owe their origin and evolution to both fluvial and glacial processes.

As discussed earlier, some small volcanic features may have formed by the interaction of magma and ice on Mars. Hodges and Moore (1979) have carried this possibility to a larger scale and speculated that the scarp around Olympus Mons and its aureole deposits (Fig. 7.26) resulted from subglacial eruptions. They proposed that an ice sheet several kilometers thick may have existed in the Tharsis region. Such an ice sheet may have been related to different locations of the martian polar regions. Peter Schultz (personal communication) argues that polar wandering occurred early in martian history. His evidence includes the existence of thick mantled and layered terrains antipodal to each other and active erosional processes very similar to processes presently acting on and around the present polar deposits. Polar wandering (re-orientation of the spin axis) could account for narrow valley networks occurring within 15° of the present south pole and the distribution of many of the periglacial features found on Mars at a wide range of latitudes.

Numerous other landforms have been suggested to result from ground ice, as reviewed by Lucchitta (1983). Theilig and Greeley (1979) proposed that irregular depressions found in the northern plains (Fig. 7.46) are analogous with terrestrial *alases*, or collapse pits which form from the freezing and thawing of ground ice. Patterned ground of several types also occurs in the northern plains, first described by Carr and Schaber (1977). Polygons (Fig. 7.47) average 5–10 km across and have been proposed to be ice-wedge features similar

Figure 7.46 View of plateau-forming material in Chryse Planitia showing irregular depressions (A) similar in size and form to *alases* which develop on Earth by the collapse of ground containing ice and water. The scalloped edge (B) of the plateau may indicate scarp retreat enhanced by coalescing depressions. Impact craters (C) are clearly distinguished by their raised rims and circular outline. Area shown is 35 by 35 km (Viking Orbiter 8A74).

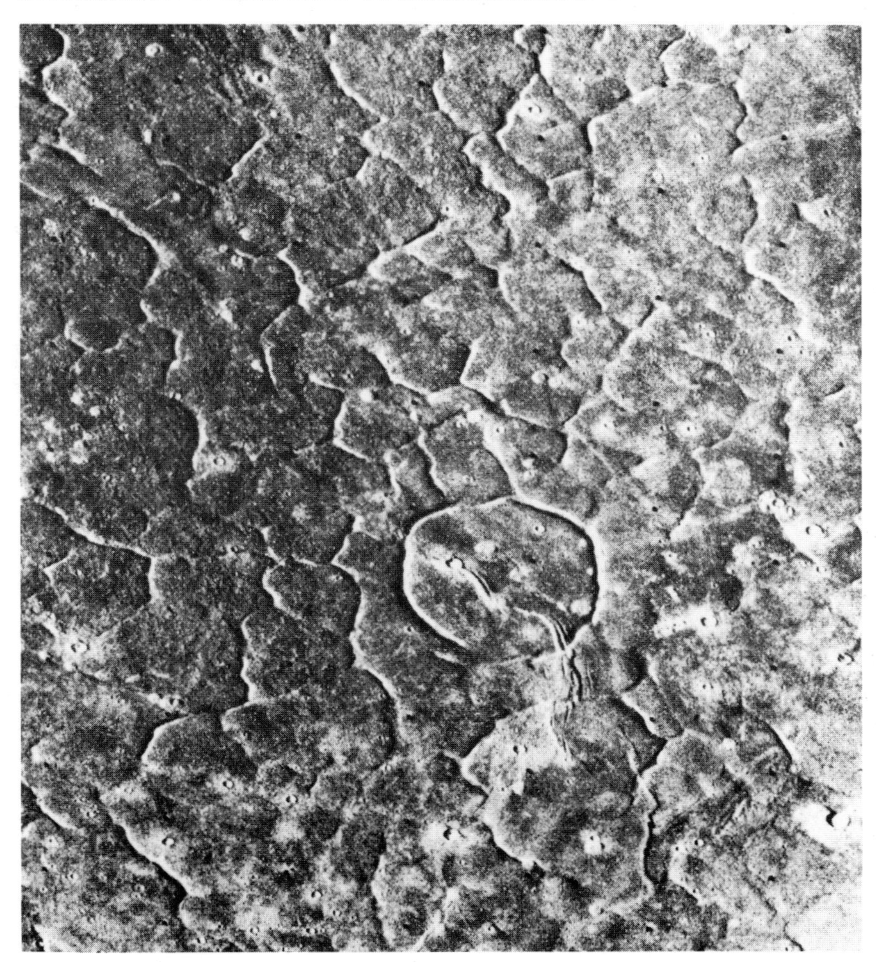

Figure 7.47 Patterned ground in northern plains at 44°N, 18°W; although resembling ice-wedge polygons on Earth, which may be as large as 100 m across, these martian features, 5–10 km across, are orders of magnitude larger (Viking Orbiter 32A18).

to polygonal ground on Earth (Fig. 3.44), tectonic features (Pechmann 1980) or patterns produced by cooling lavas (Morris & Underwood 1978). Equally enigmatic are various curvilinear features (Fig. 7.48) which could be solifluction lobes, linear ice-cored ridges or the result of scarp retreat, as reviewed by Rossbacher and Judson (1981).

7.8 Geologic history

The martian surface displays a rich history of evolution that is only just now beginning to be understood. As demonstrated in Figure 7.49, analysis of landforms and general photogeology enable processes and sequences of formation to be assessed. The early history of Mars is considered to be similar to that of the Moon. Planetary accretion and the period of heavy impact bombardment were accompanied by outgassing and global melting, leading to the formation of a differentiated crust; the earliest discernible history is indicated by remnants of impact basins. Deep-seated structural weaknesses produced by these basins would later control subsequent volcanic activity in a manner much like that on the Moon, but complicated by the presence of ice and the longer-lasting thermal engine of Mars. In the final stages of heavy bombardment, volcanism produced some of the plains in the cratered uplands. By the decline of heavy impact cratering, the terrain dichotomy between the northern and southern hemispheres was established and initial uplift of the Tharsis bulge resulted in crustal fracturing. The reason for the terrain dichotomy remains unsolved in understanding the evolution of Mars.

Outgassing associated with early volcanism and impact bombardment probably generated an atmosphere that was much denser than that found on Mars today. This may have allowed liquid water on the surface and could account for the small valleys which are pervasive throughout the ancient terrains on Mars. This stage may not have lasted very long; the small valleys are not formed in younger terrains and the network patterns are in an immature stage of development.

A change in climate and/or a reduction in the rate of outgassing evidently led to an environment prohibiting surface water. Regolith generated by heavy impact cratering would have been an excellent aquifer and the water responsible for the formation of the valleys may have percolated into the subsurface to form an extensive groundwater system.

Volcanic eruptions, which appear to have operated throughout most of the history of Mars, may have interacted with groundwater to produce volcanic ash plains and patera, the earliest central volcanoes. Volcanism involving lava flows erupted from central vents led to the formation of Alba Patera, the domes and shields in the Tharsis region, and the volcanoes in the Elysium region. Eruptions of very fluid mafic and ultramafic lavas, presumably from fissures, flooded the northern plains.

Catastrophic release of subsurface water appears to have been responsible for the formation of the outflow channels. Although generally younger than the small valleys, these features formed over a wide span of martian history, with the same channels being re-occupied repeatedly.

Tectonic processes continued to deform the Tharsis region, as evidenced by fractures which transect intermediate-age plains. Extension associated with the Tharsis bulge opened the canyonlands, which were subsequently enlarged by various processes including aeolian activity and mass wasting.

In later stages of martian history, volcanism focused in the Tharsis and Elysium regions. Aeolian processes have evidently operated throughout the evolution of the surface and their intensity is probably directly proportional to atmospheric density. These processes and polar-related events are currently active on Mars.

Figure 7.48 Curvilinear features in the northern plains which may be the result of periglacial processes. Area shown is 100 by 100 km (Viking Orbiter 11B01).

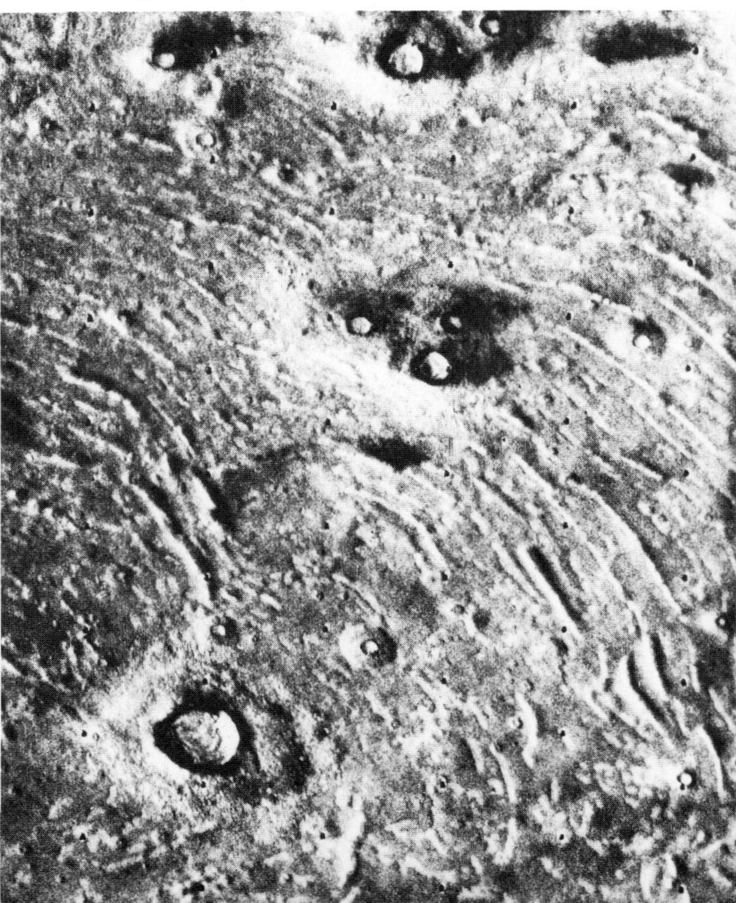

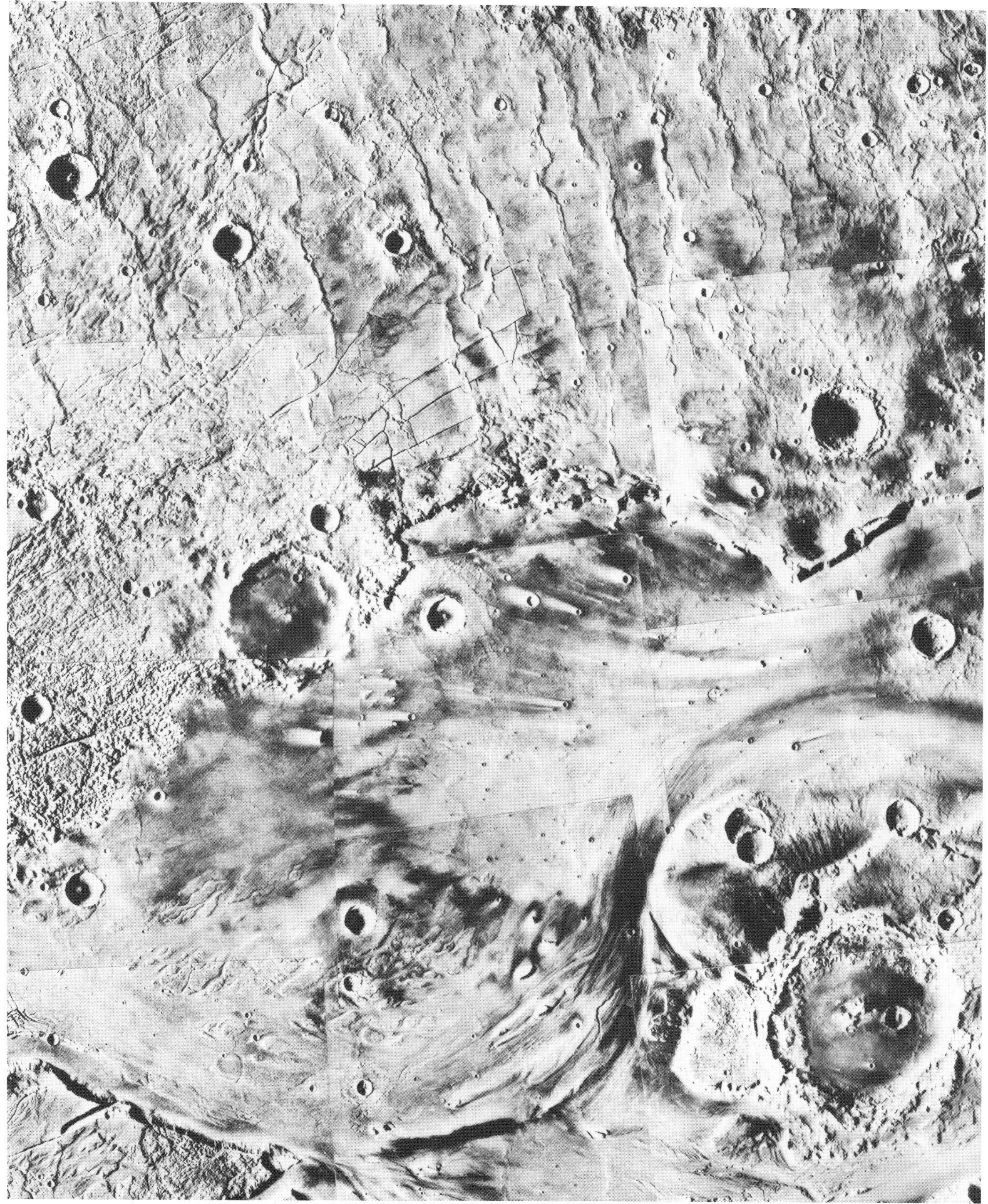

Figure 7.49 Mosaic of Viking orbiter frames showing a northwestern part of the Chryse region which displays the rich variety of geologic processes that have shaped the surface of Mars. Ridged plains in the top center of the picture are similar to lunar mare plains and may represent flood lavas. Impact craters ranging in size from 100 km to the limit of resolution, 180 m, are visible throughout the area. Larger, older craters have been severely degraded. Tectonic processes have caused fracturing of the ridged plains in the upper part of the picture. An enormous channel sweeps through the area in the lower part of the picture and, along with wind streaks, is evidence for processes of gradation. North is toward the top (Viking Orbiters 519A30-36, 520A29-36).

8 The Jupiter system

8.1 Introduction

Voyager 1's dramatic 1979 flyby of Jupiter and its satellites doubled the number of objects available for geologic analysis. Prior to that time, only the surfaces of Earth, Moon, Mercury and Mars were known. Voyager, for the first time, revealed the complexity and diversity of Jupiter's major satellites and provided tantalizing clues to their surface histories. The features seen on the satellites prompted a re-examination of traditional geologic processes. The new data caused planetologists to consider such bizarre processes as volcanism driven by sulfur or sulfur compounds, fracturing of ice-rich planetary crusts, and flooding of surfaces by liquid water.

Summaries of knowledge prior to the Pioneer and Voyager missions are provided in a collection of papers (Burns 1977), based on presentations from an international meeting on planetary satellites held in 1974, and in reviews by Johnson (1978) and Morrison (1982a). Pioneers 10 and 11 arrived at Jupiter in 1973 and 1974 and were the first spacecraft sent to the outer Solar System. Although primarily designed to assess the potential hazards of this "new territory", the Pioneers gathered data on small particles, assessed magnetic fields and obtained low-resolution images of Jupiter and the Galilean satellites.

The flybys of Voyagers 1 and 2 in 1979 and 1980 returned an unparalleled wealth of data on Jupiter and its major satellites (Table 8.1). Among the many discoveries provided by Voyager are the first active volcanoes (other than on Earth) in the Solar System and the ring system around Jupiter. Results are described in various journal special issues (Table 1.2), NASA documents (Table 1.3), a collection of papers prepared for a special meeting (Morrison 1982a) and in reviews by Soderblom (1980), Morrison and Samz (1980) and Morrison (1983).

Pre-space age studies of the Jupiter system began with the discovery of its major satellites (Io, Europa, Ganymede and Callisto) by Galileo Galilei in the early 1600s, using the newly invented telescope. Although he called them the "Medician stars" in honor of his patron, Cosimo de' Medici, they are now generally called the Galilean satellites (although in Italy they are still often referred to as the Medician satellites). The 1950s saw the application of modern photometry, polarimetry and spectrophotometry to assess the colors and albedos of the satellites, while near-infra-red photometry provided clues to the surface compositions. In 1957, Gerard Kuiper suggested that the surfaces of Europa and Ganymede were covered with water ice, a suggestion that was later confirmed by Pilcher and colleagues (1972) who utilized spectroscopic observations.

In the early 1970s, John Lewis (1971) published a theoretical analysis of the interior structure of the satellites. Based on models of Solar System elemental abundances and estimates of the satellite masses, Lewis predicted that Ganymede, Callisto and possibly Europa have rocky cores and mantle/crusts composed predominantly of water ice, while Io is composed mostly of rocky materials (Fig. 8.1). From considerations of the physical properties of ice-rich crusts, Johnson and McGetchin (1973) predicted that topographic features, such as craters, would deform viscously and become "flattened".

Jupiter and its 16 known satellites have often been described as a miniature Solar System. Jupiter emits energy – mimicking a star – and the Galilean satellites appear to be differentiated with respect to their distance from Jupiter. Had it grown more massive, Jupiter quite probably would have become a star. The four innermost moons are very small, apparently rocky bodies. The largest of these four is Amalthea which measures 270 by 165 by 150 km and was imaged by Voyager (Fig. 8.2). As described by Thomas and Veverka (1982), Amalthea is heavily cratered; the largest crater, Pan, is about 90 km across. In addition, there are various grooves and ridges that are tens of kilometers long. The surface of Amalthea is very dark and red, perhaps colored by sulfur emitted from Io, although bright spots also occur in several areas.

The Galilean satellites constitute the next group of four moons and are the primary subject of this chapter. The orbital and rotational geometries of the four, plus Amalthea, is such that the same side of the satellite always points toward Jupiter, leading to the terms **sub-jove** (side facing Jupiter), **anti-jove**, **leading hemisphere** (side facing the direction of orbit) and **trailing hemisphere**. The outer eight satellites range in size from ~10 to 180 km in diameter, but little is known about their surfaces (Cruikshank *et al.* 1982 provide a review).

Table 8.1 Missions to the outer planets.

Spacecraft	Encounter date	Mission	Event
Pioneer 10	3 Dec. 1973	Jupiter flyby	Investigation of the interplanetary medium, the asteroid belt, and the exploration of Jupiter and its environment. Closest approach to Jupiter 130 000 km. Exited Solar System 14 June 1983; still active.
Pioneer 11	2 Dec. 1974	Jupiter flyby	Information beyond the orbit of Mars; investigation of the interplanetary medium; investigation of asteroid belt; exploration of Jupiter and its environment. Closest approach to Jupiter 34 000 km.
Voyager 1	5 Mar. 1979	Jupiter flyby	1408 images of jovian satellites were obtained. Major discoveries were active volcanism on Io and a ring around Jupiter.
Voyager 2	9 July 1979	Jupiter flyby	1364 images of jovian satellites were obtained revealing high resolution of Jupiter's geologically diverse icy surface satellites, Europa and Ganymede.
Pioneer 11	1 Sept. 1979	Saturn flyby	Transmitted low-resolution images and other data about Saturn, discovered additional rings and moons not previously known.
Voyager 1	13 Nov. 1980	Saturn flyby	900 images of the saturnian satellites were obtained. Major discoveries were the unexpected complexity of the rings and the true nature of Titan's atmosphere.
Voyager 2	26 Aug. 1981	Saturn flyby	1150 images of the saturnian satellites were received. Despite a scan platform malfunction, virtually all important objectives were met.

8.2 Io

Earth-based observations have long shown that Io is more red than any other object in the Solar System, including Mars. Possible surface compositions were considered, and it was found that the overall spectrum of Io closely matches that of sulfur (Nelson *et al.* 1982). However, it is also known that the exact color of sulfur is strongly dependent on its temperature and that the presence of even slight quantities of other materials can drastically alter its color. Thus, the interpretation of the surface composition of Io was (and still is) open to question (Sill & Clark 1982).

Io is about the same size and density as Earth's Moon, suggesting a predominantly silicate composition. Prior to the Voyager encounters, it was reasoned that Io ought to be a cratered object similar to Mars. Not only should its surface be capable of preserving craters, but given the proximity to the asteroid belt and the potential gravitational focusing effect of Jupiter, Io should be rather heavily cratered. The early, low-resolution pictures of Io obtained by Voyager 1 as it approached Io (Fig. 8.3) showed numerous dark, circular features which were initially interpreted as impact craters.

However, as the spacecraft neared Io and returned progressively higher-resolution images, the dark features were seen to be quite irregular in form and to lack the features diagnostic of impact craters. Careful and detailed analysis of even the highest-resolution images (~1 km) has failed to reveal structures that could be confidently attributed to an impact origin. Thus, for the first time in Solar System exploration, a planetary object was found which either had not experienced impact cratering – an idea immediately dismissed because of the heavily cratered neighbors of Io (Ganymede and Callisto) – or one whose surface was younger than the rate of formation for impact craters large enough to be seen.

A short time after the discovery of Io's lack of impact craters, another discovery and milestone in Solar System exploration was made – a finding which also explained the apparent absence of impact craters. Analysis of navigation images by Morabito and her colleagues (1979) of the Jet Propulsion Laboratory showed a peculiar bright pattern emanating from the surface of Io (Fig. 8.4). Upon closer inspection, this pattern was found to be an enormous volcanic eruption plume which was showering materials over the surface of Io at a prodigious rate and which was capable of burying not only impact craters but other features as well.

This remarkable discovery was, in fact, predicted. Stan Peale, Pat Cassen and Ray Reynolds (1979) had assessed the tidal stresses generated within Io as a consequence of being "pushed–pulled" between the gravi-

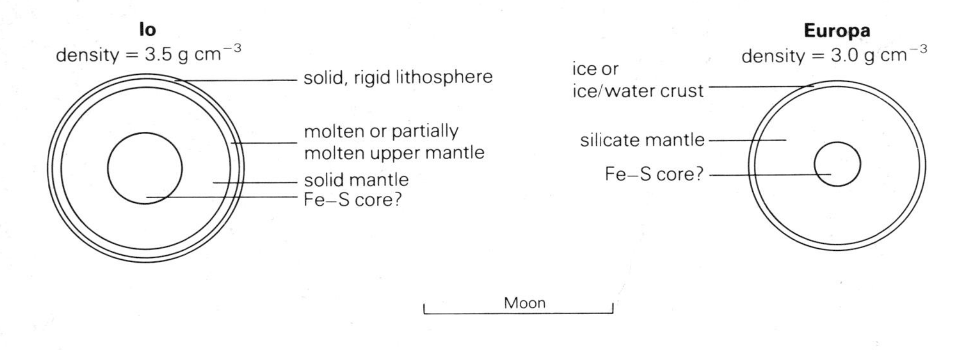

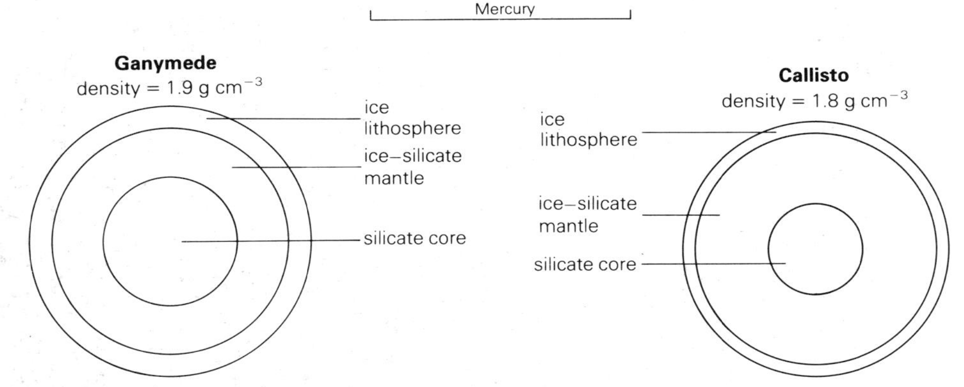

Figure 8.1 Diagram showing the interiors of the Galilean satellites. Io and Europa are both thought to be composed largely of silicate material. Both Ganymede and Callisto may have lithospheres of ice and cores of silicates with mantles composed of undifferentiated materials. Ganymede, being more massive and containing a larger component of silicates, probably began internal differentiation earlier and continued to undergo segregation after the period during which Callisto underwent differentiation. Scale bars indicate the diameters of the Moon and Mercury.

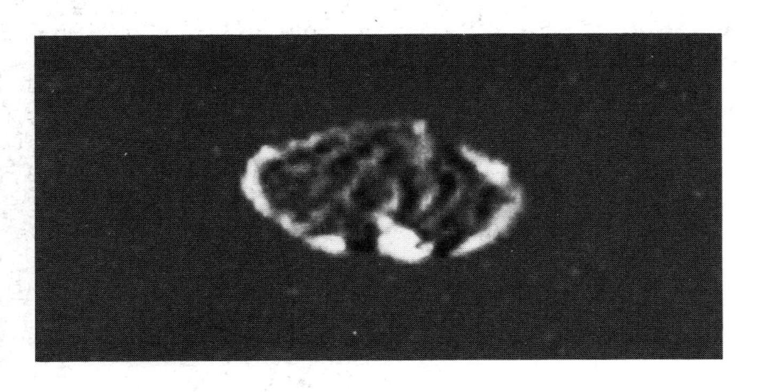

Figure 8.2 Voyager 1 image of Amalthea; this innermost, rocky satellite of the Jupiter system is about 155 by 270 km (Voyager 1 image 1097J1-001).

tational fields of Jupiter and Europa. They calculated the possible heat generated by tidal stresses to be in the order of 10^{13} watts – more than two to three orders of magnitude greater than heat released from normal radioactive decay. This led them to suggest that active volcanoes might be found and they published their suggestions just before the Voyager flyby!

Subsequent to the prediction and the Voyager discovery of active volcanism, Matson and colleagues (1981) studied Earth-based thermal infra-red data and estimated Io's heat flow. They obtained a value about 10 times greater than the amount based solely on tidal heating, suggesting additional heat sources, such as radioactive decay. This value is some 30 times higher than Earth's heat flow. Thus, Io is the most active object in the Solar System explored thus far, an observation borne out by the satellite's myriad volcanic surface features.

8.2.1 Physiography

About 35 percent of the surface of Io was photographed through the combined coverage of Voyagers 1 and 2 at resolutions sufficient to resolve features larger than ~5 km. Most of this coverage is in the equatorial and south polar regions (Fig. 8.5). The highest-resolution images (about 0.5 km per pixel) occur in an equatorial zone between 275°W to 360°W longitude.

Preliminary photogeologic mapping by Schaber (1980 & 1982b) shows that Io can be subdivided into three

Figure 8.3 View taken by Voyager 1 of Io showing the albedo variations and numerous dark spots. The dark spots were initially thought to be impact scars but higher-resolution images revealed them to be volcanic vents. Diffuse, circular zones (arrows) were found to be active volcanic plumes, as shown in Figure 8.4 (Voyager 1 42J1 + 000).

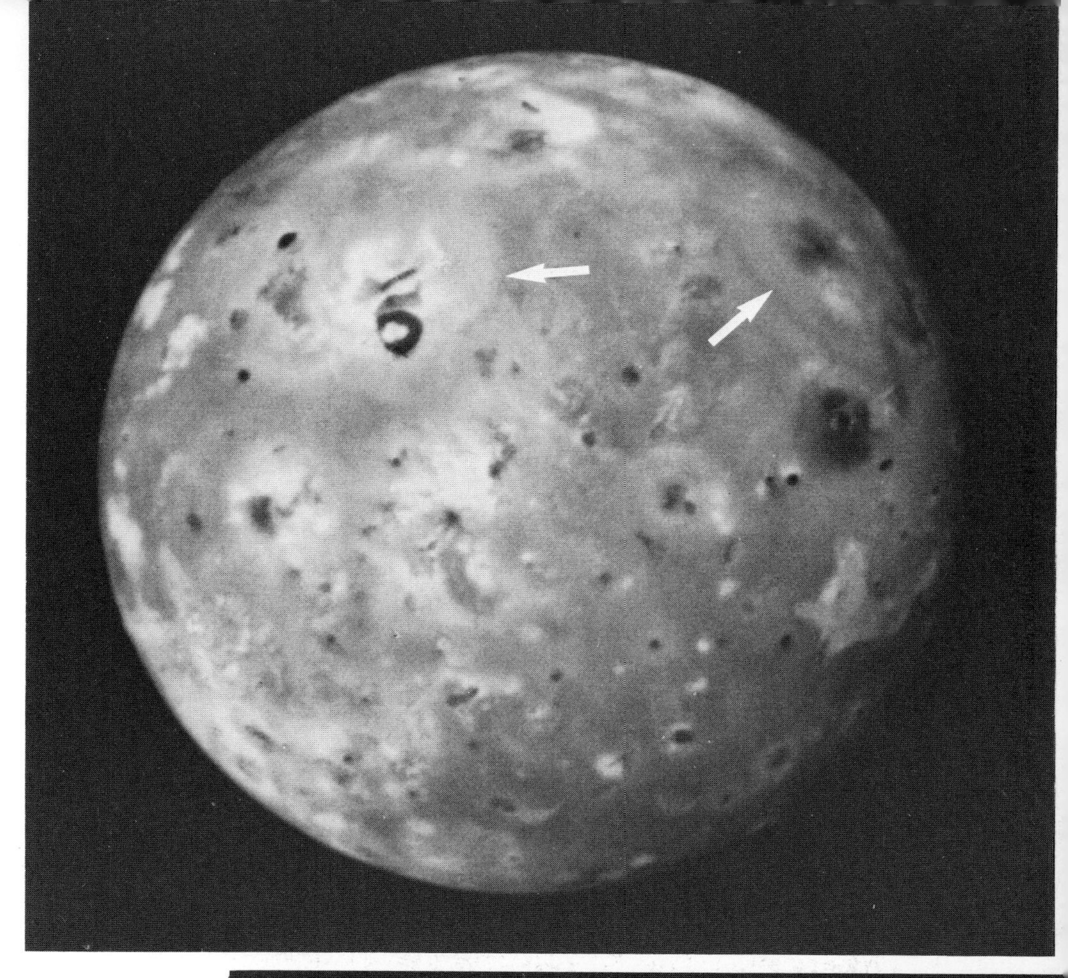

principal units: various plains, vent-related materials and mountain materials. The plains and vent materials are attributed to volcanic processes. The mountains may be crustal materials not necessarily related directly to volcanism. In addition, various possible erosive features and tectonic structures have been identified and mapped. It should be noted that the lack of adequate stereoscopic images and impact craters inhibit the determination of stratigraphic relationships. However, embayment and cross-cutting geometries enable some sequences to be determined, at least locally.

In general, the surface of Io displays a wide variety of colors, including various shades of red, yellow, orange and brown, plus very dark and very light areas. Many of these colors can be attributed to sulfur (Sagan 1979) or anhydrous mixtures of sulfur allotropes plus SO_2 frost and sulfurous salts of sodium and potassium (Fanale *et al.* 1979). The controversial topic of whether these possible compositions are representative of the topographic features (such as flows), or simply represent a thin mantling layer (a few millimeters thick), remains open to speculation.

Figure 8.4 Image showing volcanic plume associated with the vent, Prometheus, as seen in this limb view of Io. The plume extends about 100 km above the surface and forms an umbrella-shaped pattern 300 km wide (Voyager 1 1637748).

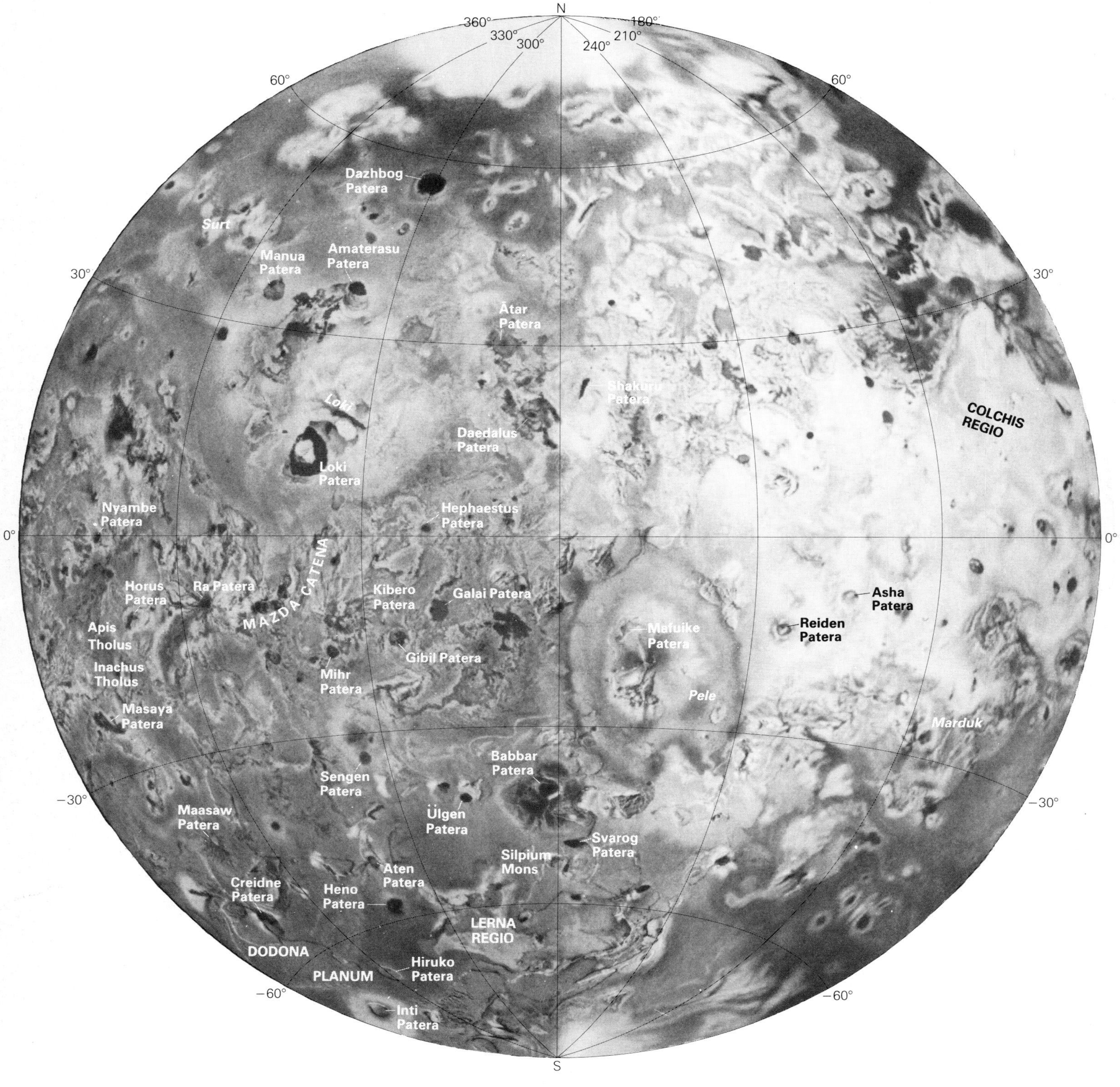

Figure 8.5 Shaded airbrush relief map of Io showing various terrains and selected named features. Features in italics refer to active volcanic plumes (courtesy US Geological Survey).

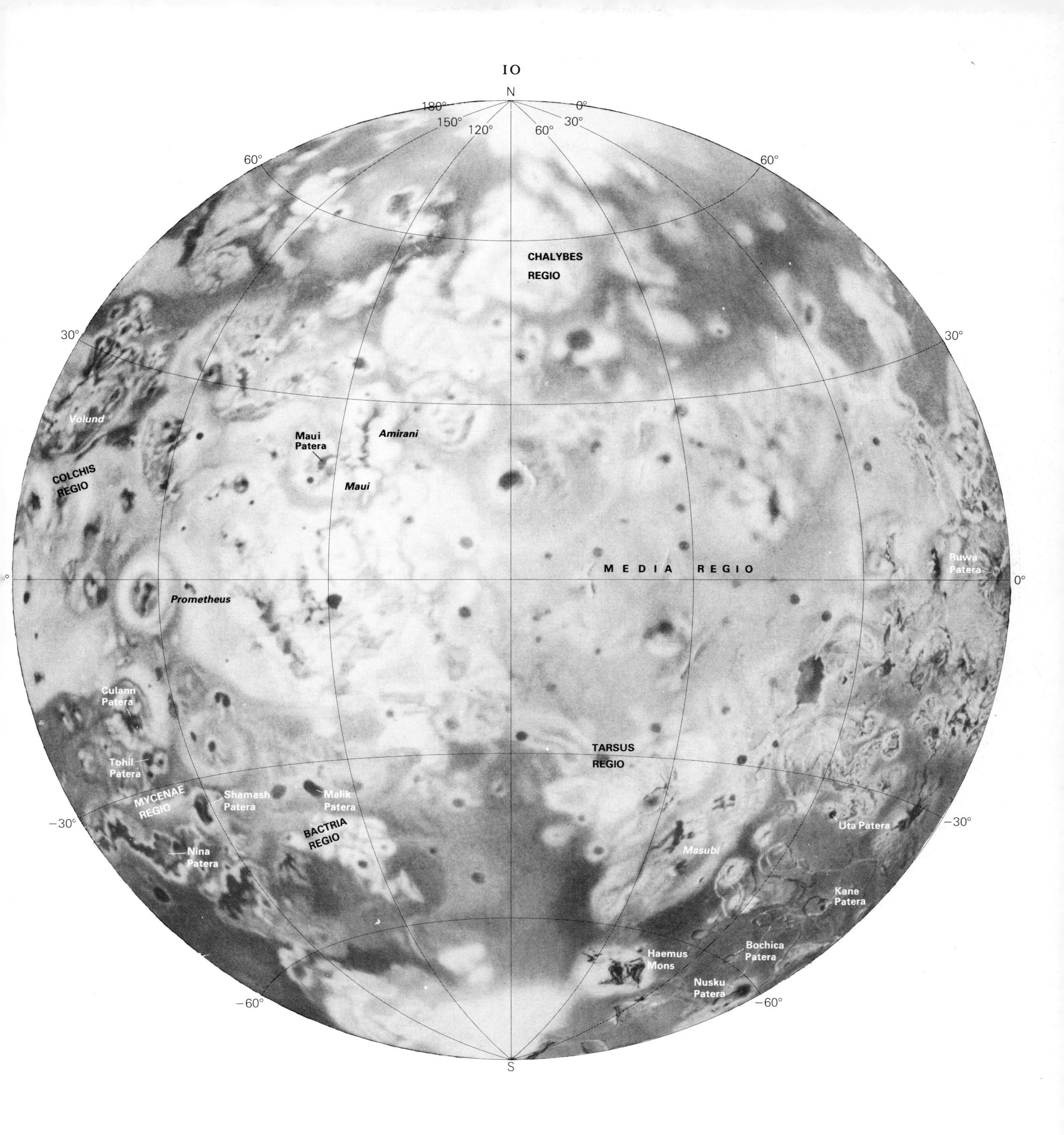
IO
N
180°
0°
150°
120°
60°
30°
60°
60°
CHALYBES
REGIO
30°
30°
Volund
Maui
Patera
Amirani
COLCHIS
REGIO
Maui
MEDIA REGIO
Ruwa
Patera
0°
Prometheus
Culann
Patera
TARSUS
REGIO
Tohil
Patera
MYCENAE
REGIO
Shamash
Patera
Malik
Patera
−30°
BACTRIA
REGIO
Uta Patera
−30°
Nina
Patera
Masubi
Kane
Patera
Haemus
Mons
Bochica
Patera
Nusku
Patera
−60°
−60°
S

Figure 8.6 High-resolution image of mountainous terrain on Io. This massif, Haemus Mons, is about 200 km across at the base and is estimated to stand more than 9 km above the surrounding plains. Bright halo surrounding the base of the mountain may be fumarolic material (Voyager 1 157J1 + 000).

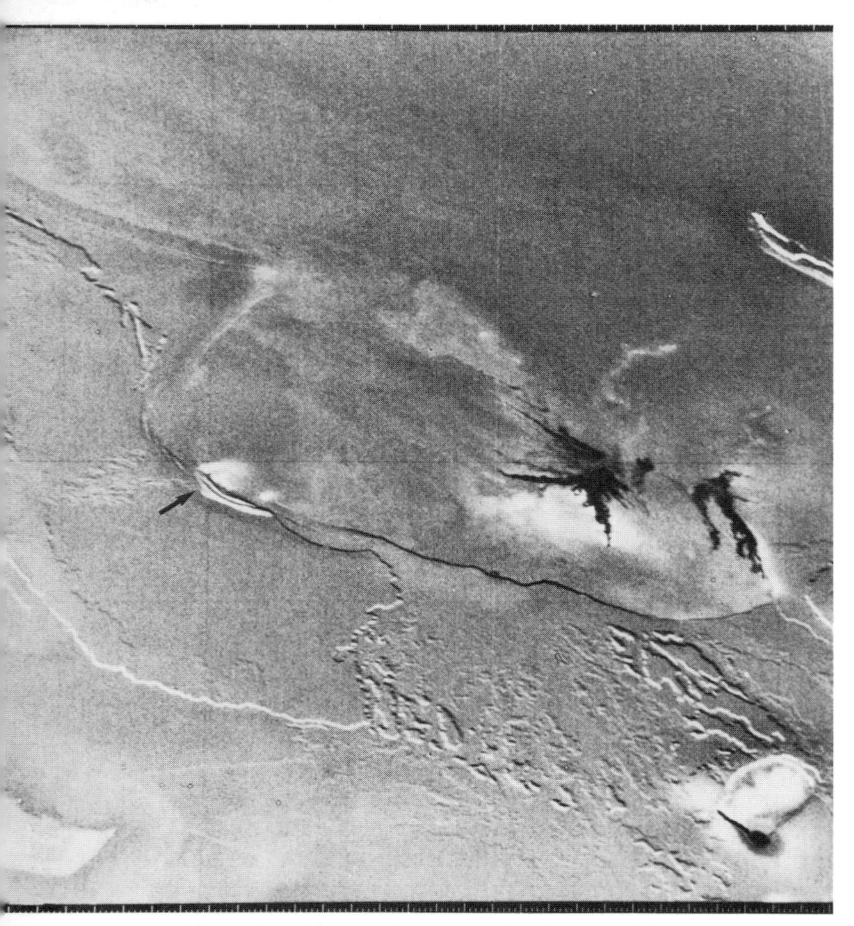

Figure 8.7 Voyager image showing layered plains near Lerna Regio, Io. Layered plains are considered to consist of lava flows with interleaved pyroclastic materials derived from volcanic plumes. Bright, wispy zone at the base of the scarp (arrow) is thought to be sulfur dioxide frost, perhaps generated by geyser-like activity; area shown is 880 km wide (Voyager 1 147J1 + 000).

Mountain material Mountain material forms high-standing blocks of rugged relief. These masses can exceed 100 km across and rise more than 9 km above the surrounding surface (Fig. 8.6). Within the area seen at high resolution, mountain material appears to be evenly distributed in regard to both latitude and longitude. Embayment relationships of surrounding plains suggest that mountains comprise the oldest material on Io. In some cases, though, mountains appear to be associated with recently active vents.

Clow and Carr (1980) have analyzed the physical properties of sulfur and demonstrated that features such as steep scarps that are higher than about 1000 m could not be self-supporting. From this analysis, they conclude that the mountains must be composed of something other than sulfur, most probably silicate materials.

Plains Plains of several types constitute the most widespread units identified on Io. In general, plains range from black and white to various reds and yellows, although plains units in the polar areas tend to be dark brown to black (Masursky *et al.* 1979). *Intervent plains* constitute about 40 percent of the area mapped. These plains are characterized as having smooth surfaces, possible low scarps and intermediate albedos. Intervent plains are interpreted by Schaber (1982b) as being fallout deposits from volcanic plumes, interbedded with local flows and fumarolic materials.

Layered plains (Fig. 8.7) form smooth, flat surfaces and exhibit grabens and scarps up to 1700 m high, again suggesting compositions other than pure sulfur. The most conspicuous layered plains are found in the south polar region, although similar units occur throughout

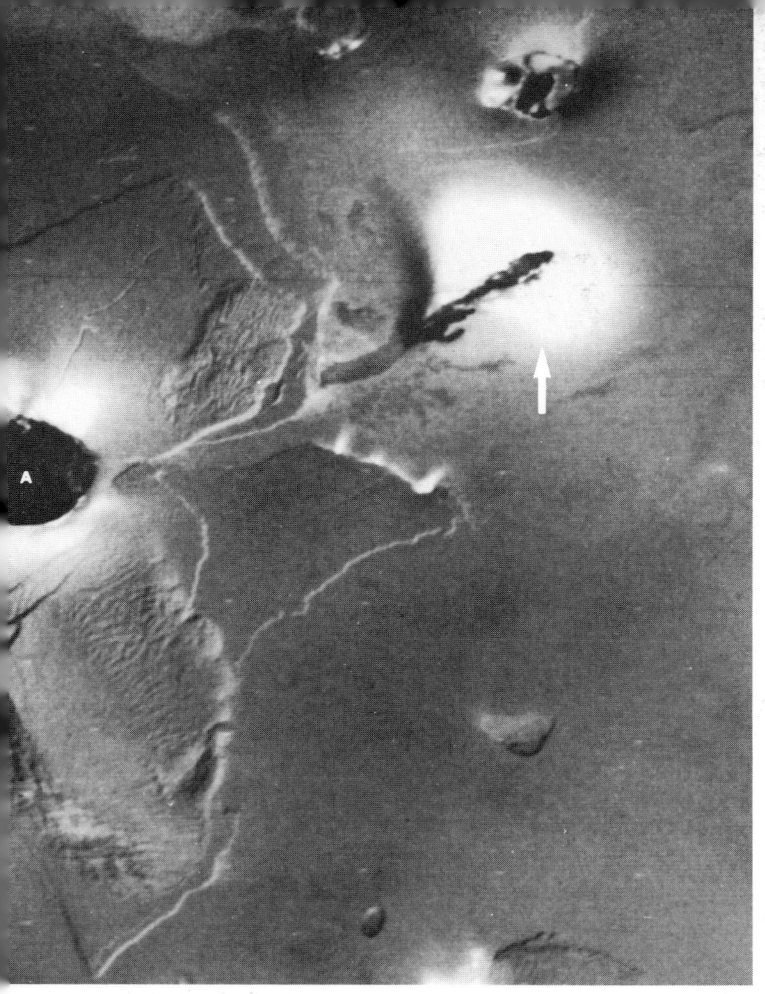

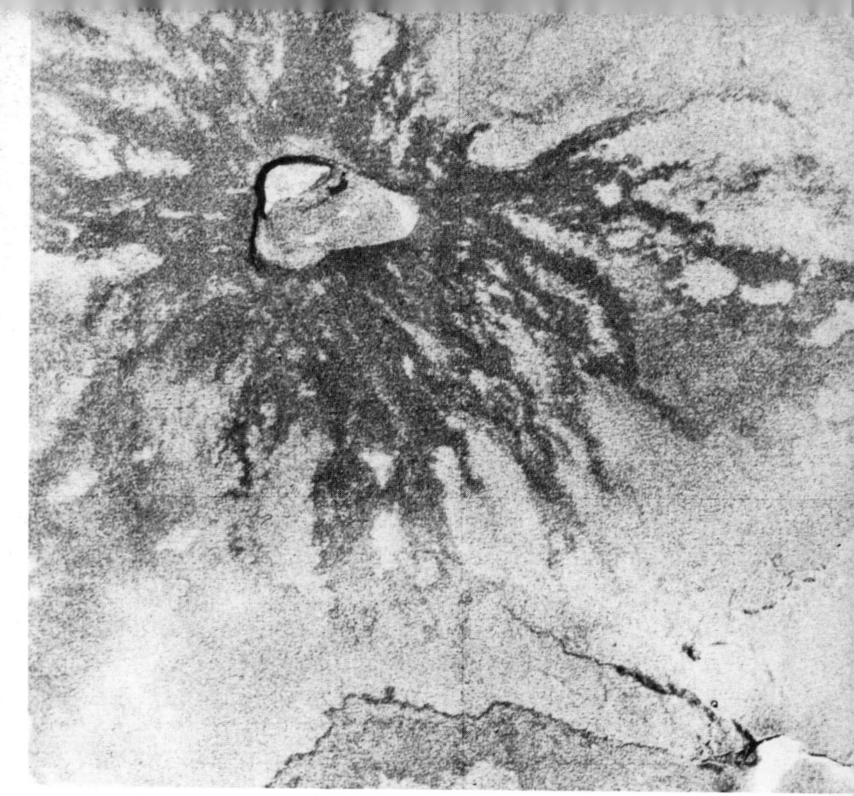

Figure 8.9 High-resolution image of Maasaw Patera showing large, complex caldera measuring 50 by 25 km, plus various flows which form a low-profile shield-like structure (Voyager 1 199J1 + 000).

◁ **Figure 8.8** View of a 100 km diameter caldera (A) near Creidne Patera and various flows. Bright zone around the base of one prominent flow (arrow) may be sulfur dioxide frost, as shown in Figure 8.7 (Voyager 1 75J1 + 000).

the mapped area. Transection by faults, overlap by various plains and possible erosional features suggest complex histories.

McCauley and co-workers (1979) note that mesas, hills and remnants of subdued scarps beneath younger layers attest to multiple cycles of erosion and deposition. These disrupted, "etched" surfaces and irregular "fretted" margins cannot be attributed to fluvial or aeolian processes in the cold, tenuous atmosphere of Io, nor can ionic bombardment from Jupiter account for the features, given their relative youth. Rather, McCauley and his colleagues evoke a sapping mechanism in which liquid SO_2 is the dominant erosional agent. They suggest that a hydrostatic condition could be established if the crust were fractured, as might occur by faulting, driving molten SO_2 toward the surface. At the triple point for SO_2, part of the liquid would begin to crystallize and the system would expand, forming SO_2 vapor. Upon reaching the surface at or near the vent, the solid–fluid mixture would be released energetically at a velocity of $\sim 350 \mathrm{~m~s^{-1}}$. At this velocity, the SO_2 "snow" could be sprayed as far as 70 km from the vent, although most of the material would fall closer to the vent. The numerous bright patches seen on Io (Fig. 8.8) are suggested to have formed by this mechanism. Thus, McCauley *et al.* suggest that newly formed scarps would be rapidly eroded by undercutting and slumping, leading to the formation of irregular margins, and that withdrawal of support from beneath the solid crust could cause collapse, generating irregular surfaces. These processes would continue until the source of SO_2 was depleted locally.

Vent materials Vent materials include all surface units that can be related directly to vents of various types. Vents include calderas (large, complex depressions; Fig. 8.9) and possible fissures. More than 300 vents have been identified, constituting about 5 percent of the total mapped surface area. The calderas average about 40 km across. Calderas on Io tend to be larger (up to 250 km across) and more frequent in equatorial regions than in the south polar region (maximum size is 100 km). They are much more randomly distributed on Io than on Earth, Moon or Mars, where the local and global tectonic framework seems critical to the location of volcanic vents. This suggests that the vents and related "hot spots" on Io are not controlled by strongly patterned convection cells.

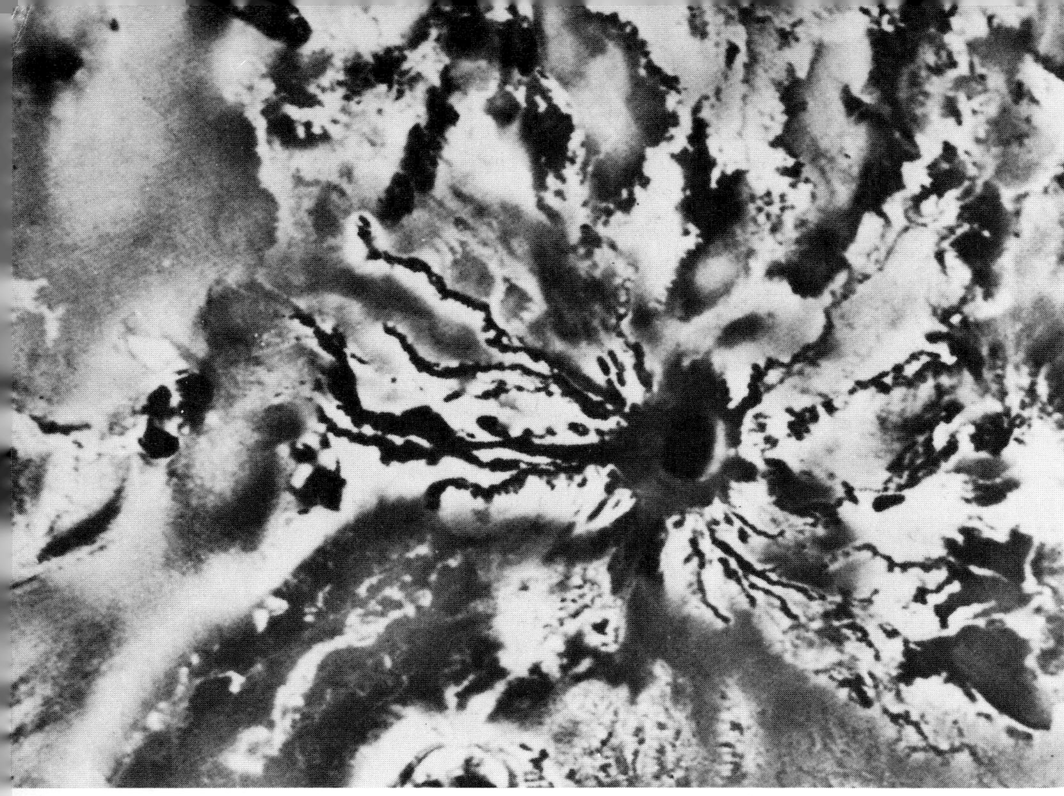

Figure 8.10 Ra Patera, a large shield volcano located at 9°S, 126°W on Io. Measuring more than 600 km across, this is one of the largest volcanic constructs in the Solar System, nearly equal in size to Olympus Mons on Mars (Fig. 7.11).

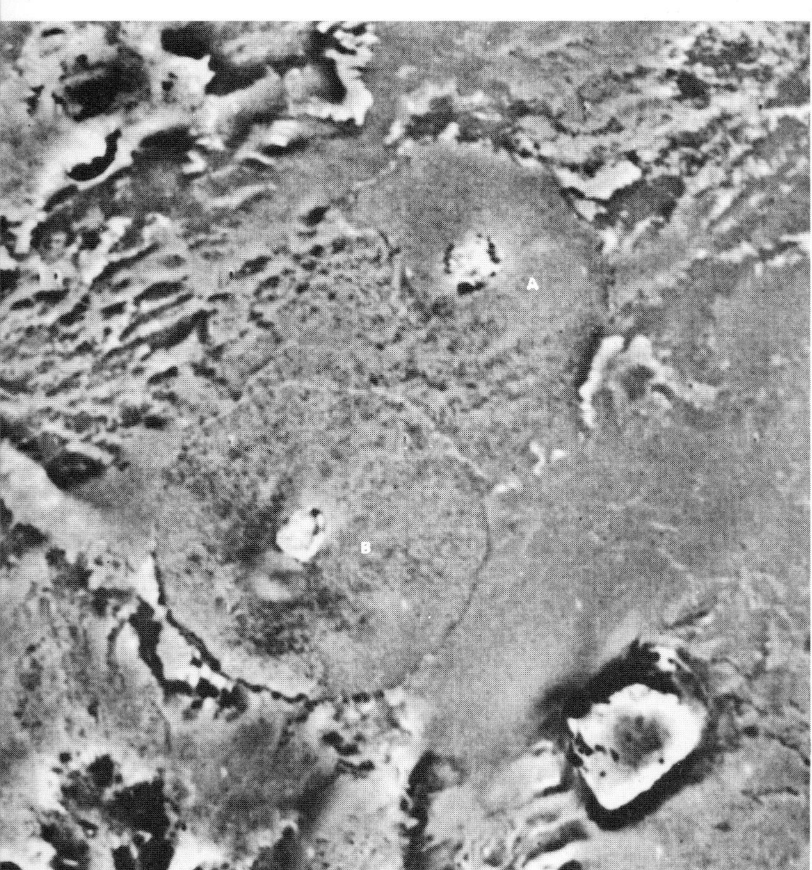

Figure 8.11 Voyager image showing two unusual volcanic constructs, Apis Tholus (A) and Inachus Tholus (B). These disc-shaped volcanoes may have formed by very fluid lavas that spread symmetrically outward from their central vents and then solidified. Area shown is 600 by 800 km, centered at −17° latitude, 350° longitude (Voyager 1 71J1+000).

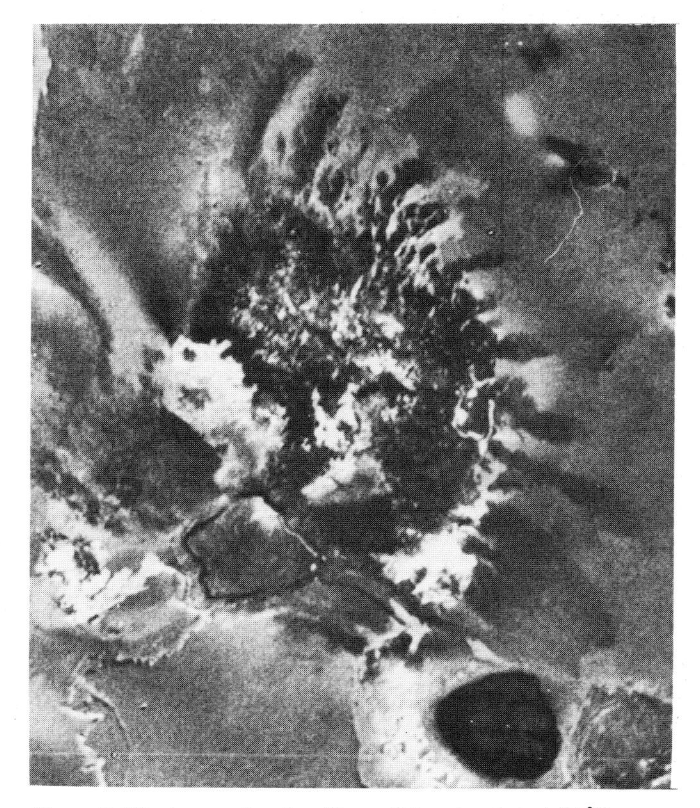

Figure 8.12 Image showing Kibero Patera located at 11°S, 300°W on Io. This complex volcano is composed of multiple flow units which originated from an irregular caldera in the lower left hand corner. Digitate flows extending to the right appear to represent "break-outs" of magmatic materials from the base of the complex flows. This feature measures 180 by 140 km (Voyager 1 105J1+000).

Calderas show complex patterns of multiple eruptions and enlargements by wall-collapse. The depth to the floor of the vents ranges from zero (level with surrounding surface) to more than 2 km. The relatively high, steep walls of some calderas may indicate that they are composed predominantly of silicate materials. Many of the floor units are very dark, suggesting the presence of molten sulfur. Imaging of some calderas by both Voyager 1 and Voyager 2 showed changes that occurred in the 4-month interval; parts of Io's lava lake at Loki Patera may have crusted over, while new vents may have developed in other areas (Terrile *et al.* 1981).

Several different styles of flows have been identified on Io, including massive, coalescing units up to 700 km long (which may indicate high rates of effusion of relatively fluid materials) and narrow, sinuous flows up to 300 km in length which radiate from central vents. The flows may be composed of sulfur which melts at a low temperature ($\sim 115\ ^\circ C$) and is very fluid. Furthermore, analyses of its thermal properties (Fink *et al.* 1983) suggest that, once mobilized, sulfur could flow long distances. Molten sulfur shows a variety of colors as it cools. Pieri *et al.* (1982 & 1984) have analyzed Voyager images and, based on an assumption of sulfur composition, concluded that the observed colors of individual flows correlate with their distance from the vent.

Flows have accumulated to form several different types of volcanoes (Carr *et al.* 1979) including shields (Fig. 8.10), "discoid" volcanoes (Fig. 8.11), and more complex volcanoes that have multiple, digitate flows emanating from a more massive central construct (Fig. 8.12). Lack of topographic information, however, prohibits assessment of the exact morphology. Schaber (1982b) notes that shield-like volcanoes tend to be limited to the area between 45°S and 30°N, coinciding approximately with the equatorial band where active volcanic plumes are found (Strom & Schneider 1982).

8.2.2 Volcanic plumes

Nine active volcanic plumes (Fig. 8.4) were observed during the Voyager 1 encounter (Strom *et al.* 1979). As described by McEwen and Soderblom (1983), these plumes may involve more than one style of eruption. The most energetic sprayed sulfurous gases and solids as high as 300 km above the surface, at velocities up to $1\ km\ s^{-1}$. Pyroclastics rained down on the surface up to 600 km from the vent.

Of the nine active plumes observed during the Voyager 1 flyby, one cut off four months later when Voyager 2 imaged Io, and two new ones were observed. The volcanic plumes are driven by internal heating, possibly in a geyser-like mechanism involving SO_2 (Kieffer 1982). Johnson *et al.* (1979) estimate that upwards of 10^{10} tons of material are erupted from Io each year. Averaged over the entire area, the surface of Io is being buried at the rate of 100 m per million years. This figure closely correlates with the rates of surface renewal necessary to account for an absence of impact craters.

Combined with the various young flows observed on the surface, volcanic plumes indicate a highly active planetary object – one that has been "cooked" and differentiated through geologic time. Based on current activity, it is estimated that the entire mass of Io may have been recycled in its lifetime. Volatiles such as water and carbon dioxide have long been lost (Johnson & Soderblom 1982), while most heavier materials have sunk to form a core. Sulfur and various sulfur compounds, aided perhaps by bodies of silicate magmas, are constantly recycled, forming the complex surface observed today.

8.3 Europa

Of the Galilean satellites, Voyager imaging is poorest for Europa. The best resolution is only about 2 km per pixel, and the total coverage is rather small (Fig. 8.13). Despite these limitations, the surface of Europa displays features that pique the imagination.

Europa has an overall density of $3.0\ g\ cm^{-3}$ and is considered to be differentiated into a silicate interior surrounded by a shell ~ 100 km thick which is composed of soft ice or water overlain by a veneer of ice. Depending upon the amount of heat generated by radioactive sources and by tidal stresses, like Io, Europa may have a partly liquid mantle. Its surface has the highest albedo of the Galilean satellites and is considered to be composed principally of water ice. However, the surface reflectivity and spectral characteristics are not uniform, suggesting that the ice has various impurities. These impurities could be derived from interior sources (such as silicates erupted onto the surface) or from exterior sources, such as sulfur emitted by Io and infall of meteoritic material.

8.3.1 Physiography

Surface features and terrain units have been mapped by Lucchitta and Soderblom (1982) of the US Geological Survey, using Voyager images. In general, Europa displays little topographic relief, but has a variety of smooth and textured terrains of different colors, along with networks of grooves, ridges and other linear features. The part of the satellite imaged by Voyager is divided into two principal units, mottled terrains and various plains (Fig. 8.14). Mottled terrain is further

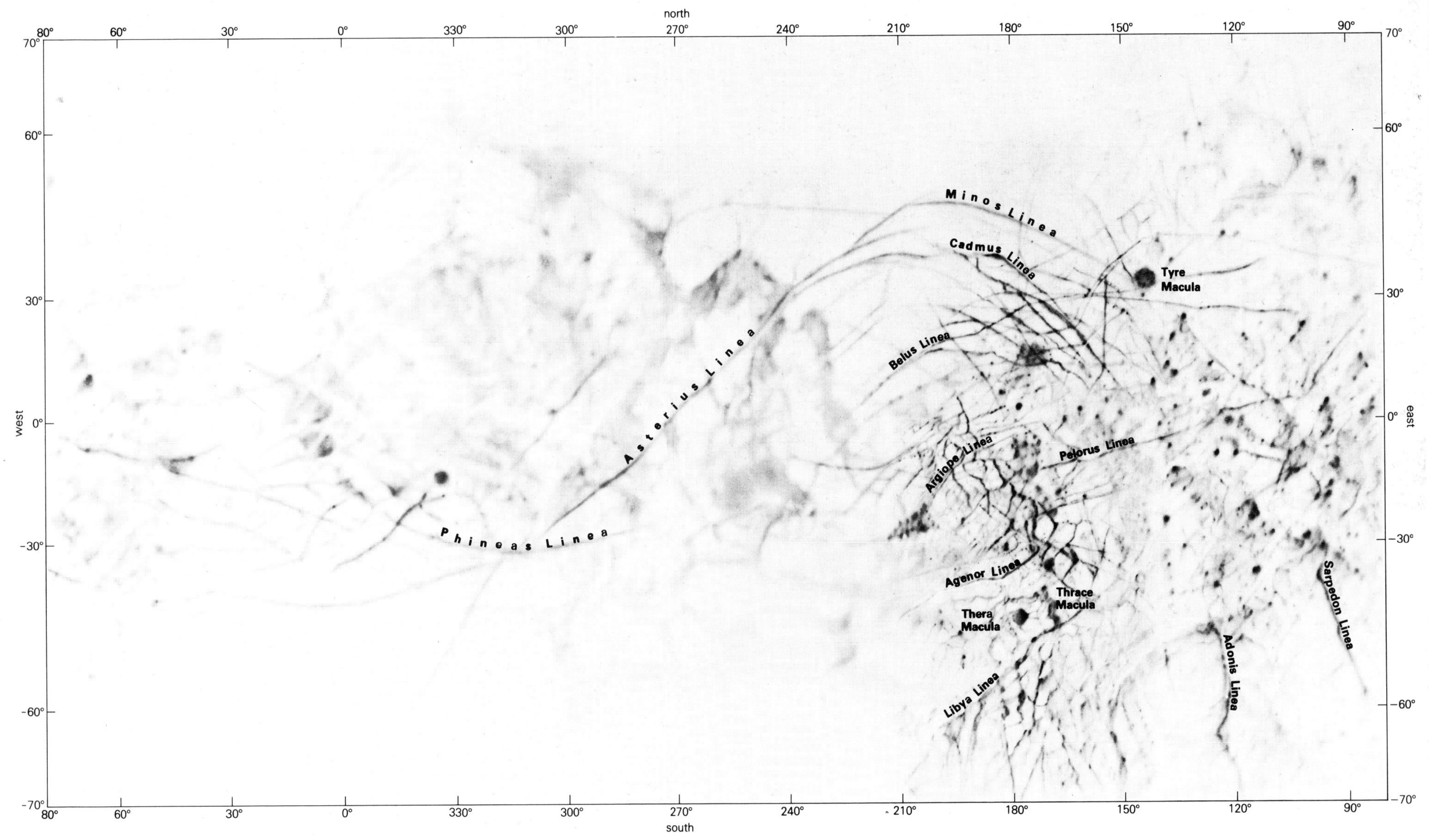

Figure 8.13 Shaded airbrush relief map of Europa showing prominent terrains and named features (courtesy US Geological Survey).

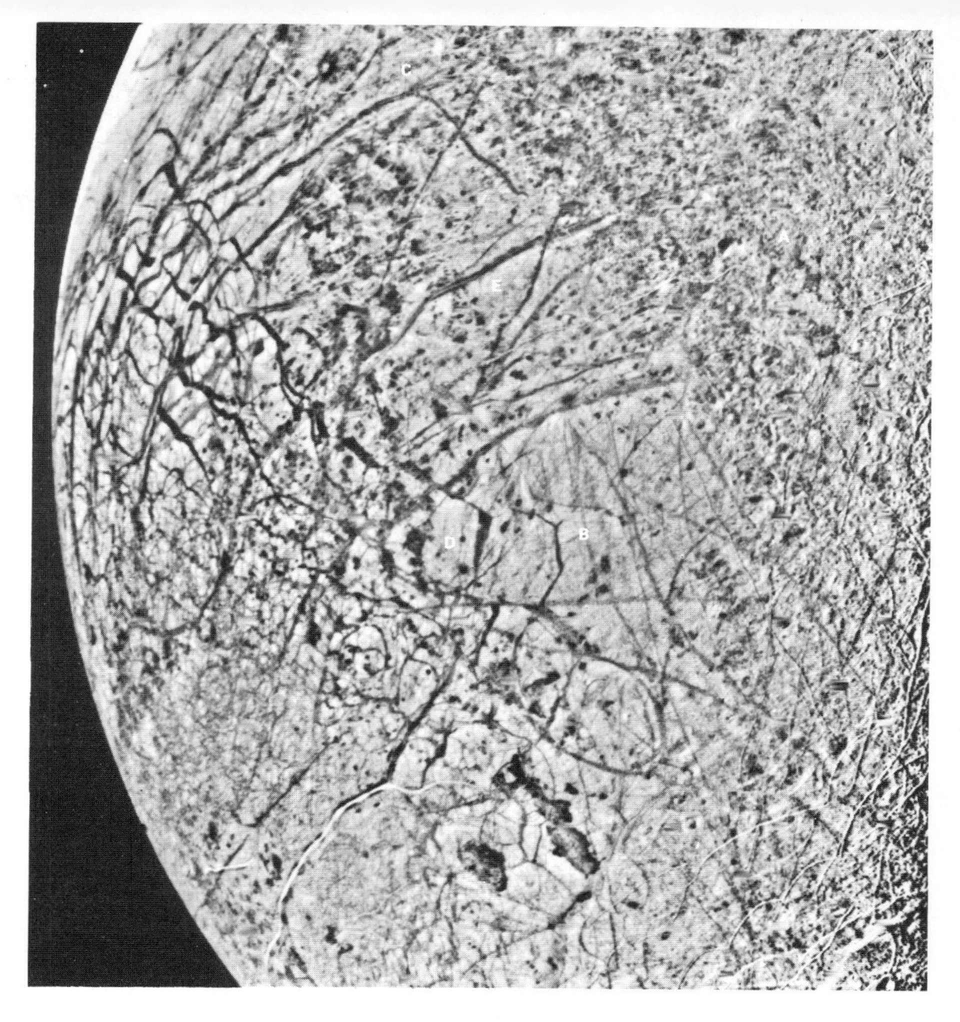

Figure 8.14 Voyager 2 image showing mottled terrain and various plains units (Lucchitta & Soderblom 1982): brown-mottled terrain (A), plains (B), gray-mottled terrain (C), and fractured plains (on the left hand side); also shown are wedge-shaped bands (D) and triple bands (E). Area shown is 2680 by 3160 km, centered at 10°S, 160°W (Voyager 2 1255J2-001).

divided into *brown mottled terrain*, which is moderately textured by small hills and forms sharp contacts with other units, suggesting that it is younger, and *gray mottled terrain*, which is smooth and has diffuse contacts. In general, the mottled terrain on the leading hemisphere is relatively bright, whereas the mottled terrain in the trailing hemisphere is darker. Lanzerotti *et al.* (1978) suggested that this difference results from the bombardment of ions originating from Jupiter's magnetosphere. Plasma within the oribital path of Europa travels faster than the satellite and would impact the trailing hemisphere, implanting ions in the ice and darkening the surface.

Plains units are subdivided into four types based on albedo and patterns of lineations. *Undifferentiated plains* are smooth surfaces cut by numerous linear features. This unit tends to be gradational with other units. *Bright plains* occur as polar deposits and have criss-crossing lineations that range from broad bands to faint streaks. Bright plains are the most highly reflective unit observed on Europa. *Dark plains* have a lower albedo than other plains units and it has been suggested that ice has been removed to leave a lag deposit of silicates. Alternatively, the low albedo could result from ions that are swept up by Europa, as described above. In either case, one would expect surfaces on Europa to darken with age. *Fractured plains* are found in the southwest part of the area mapped and are characterized by short, curved gray streaks and numerous brown spots.

In addition to the units described above, small spots and patches of brown and gray material occur on Europa. For example, Figure 8.15 shows Thrace Macula and Thera Macula, two brown patches in the southern hemisphere. Brown material is also found in small spots, along presumed fractures, and associated with some craters. Most investigators consider brown material to be of internal origin.

8.3.2 Craters

Lucchitta and Soderblom (1982) recognized two types of possible impact features on Europa. Class 1 craters are a few to a few tens of kilometers across and have raised rims, central peaks and ejecta deposits. Class 2 craters are large, flat, brown, circular areas 100 km or larger across. Craterforms of both types are very rare, at least in the areas photographed. Figure 8.16 shows various class 1 craters and includes a bowl-shaped crater, a central peak crater, a possible multi-ringed crater and a dark-halo crater. Figure 8.17 shows Tyre Macula, a class 2 feature. It consists of a circular area

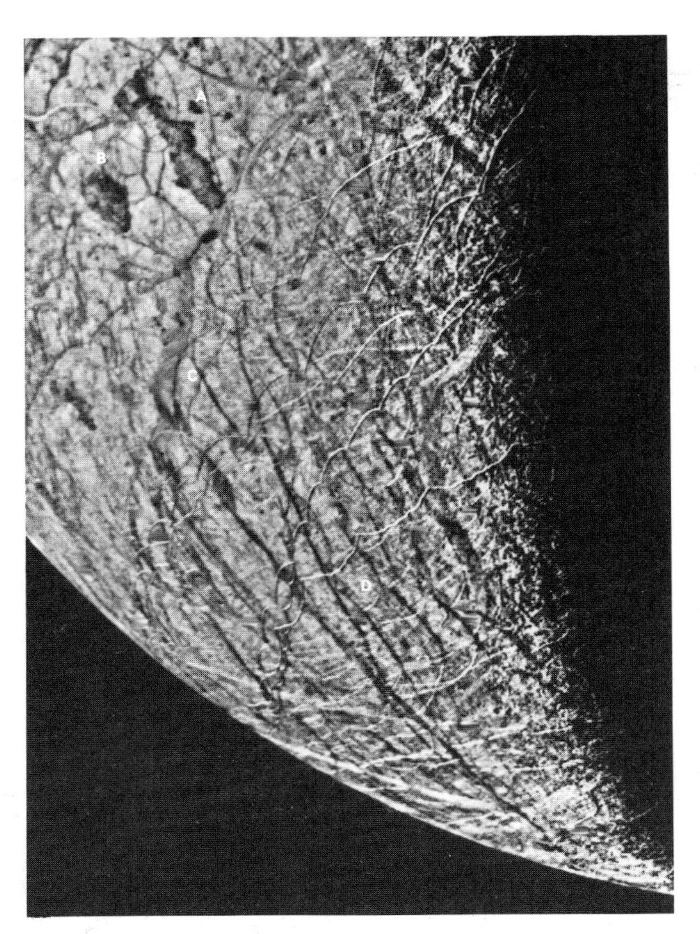

Figure 8.15 Image showing Thrace Macula (A), Thera Macula (B), gray bands (C) and cycloidal ridges (D). Thrace Macula is about 170 km long (Voyager 2 1372J2-001).

~100 km across that has faint, concentric rings, but no apparent relief. Smith *et al.* (1979a) have described similar features on Ganymede as being impact craters that formed in ice-rich crusts and which have "relaxed" with time by viscous flow (see Section 8.4.2) or, alternatively, impacted into a material with a different rheology than silicate material, such as thin ice over a water substrate (Greeley *et al.* 1982).

In all, fewer than a dozen features that resemble impact structures have been found on Europa. The surface may be relatively young, with older craters having been destroyed or buried. Like Io, Europa experiences tidal stresses which may generate heat leading to volcanism. Furthermore, various lineaments may represent fractures which could have released liquids to the surface to flood and bury craters. However, calculations by Cassen *et al.* (1980) show that the tidal heating may be insufficient to melt internal ice, though interior water may still be liquid because it has never frozen. Instead, they suggest that the smooth and relatively uncratered surface may reflect the very earliest history in which liquid water was generated during initial differentiation and then persisted beyond the period of heavy bombardment.

8.3.3 Tectonic features

Global views of Europa (Fig. 8.18) display numerous linear features, some of which radiate from central spots. These form patterns which bear a remarkable similarity to maps of Mars showing fanciful networks of "canali". Pieri (1980) classified several different linear and curved markings based on size, color and pattern. Along with various ridges, these were described by Lucchitta *et al.* (1981) and mapped by Lucchitta and Soderblom (1982), and include *dark wedge-shaped bands*, *triple bands* and *gray bands*.

Dark wedge-shaped bands occur south of the equator where they form a NW–SE trending zone. Individual bands can be 25 km across at their widest end and up to ~300 km long. They cut across other bands and hence are younger. Most investigators interpret these bands as representing a form of incipient plate tectonics. Helfenstein and Parmentier (1980) suggested that tidal stresses could rupture the surface, although Finnerty *et al.* (1980) consider the stresses to be inadequate for crustal failure. Ransford *et al.* (1980) offered an alternative model involving mantle upwelling and crustal extension.

Triple bands are among the most conspicuous features on Europa (Fig. 8.19). They consist of a pair of dark bands separated by a narrow, bright stripe or ridge. Triple bands range up to 18 km wide and >1000 km long and trend northwest in the northern hemisphere and southwest south of the equator. Locally, some triple bands merge with single brown streaks. Others emanate from dark circular spots (due to impacts?). Most triple bands disappear where they enter brown mottled terrain. Helfenstein and Parmentier (1980) analyzed triple band patterns and concluded that they could have been generated by crustal stresses resulting from tidal deformation related to orbital eccentricities. Finnerty *et al.* (1980) proposed that the bands resulted from fracturing as a consequence of global expansion and that the fractures were filled with aqueous fluids and silicates to form breccia dikes. They suggest that relatively pure water was segregated and expanded upon freezing to form the central bright ridge. Gray bands (Fig. 8.15) are found in the south polar area. They are cut by all other lineations, including ridges, and are considered to be relatively old features.

Some Voyager images, taken along the terminator, enable subtle topography to be defined. A series of complex ridges was found, with the most extensive ridge occurring south of the equator. Some ridges are relatively straight, others (especially in the south polar area)

a

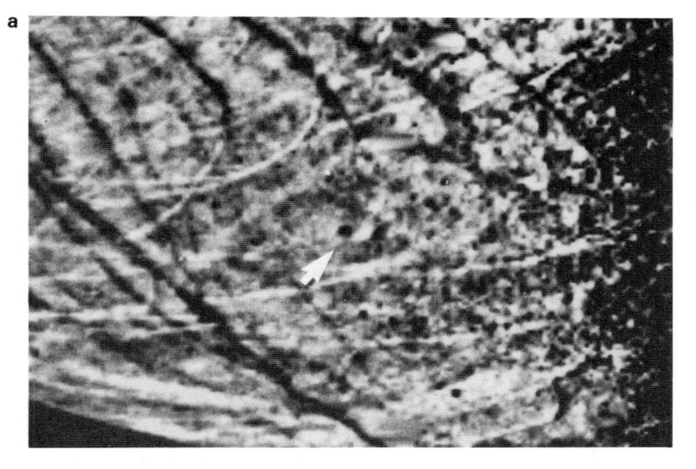

d

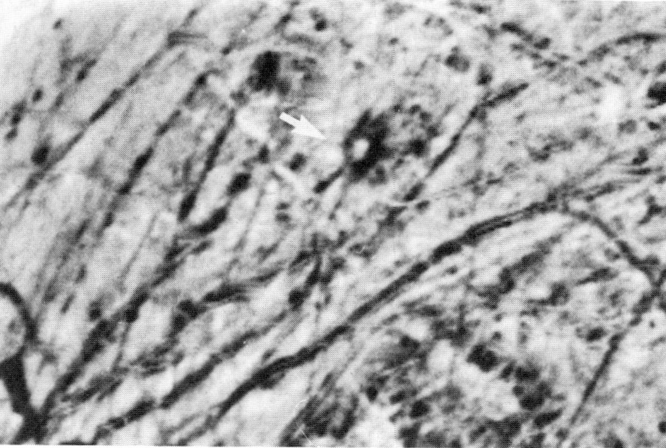

Figure 8.16 Views of possible Class 1 impact craters on Europa (Lucchitta & Soderblom 1982). (a) Bowl-shaped crater 15 km in diameter (Voyager 2 1234J2-001). (b) 15 km crater with central peak and possible bright ejecta (Voyager 2 1368J2-001). (c) Multi-ringed crater (arrow), inner ring 30 km in diameter (Voyager 2 1219J2-001). (d) Dark-halo crater (arrow), 50 km in diameter (Voyager 2 1255J2-001).

b

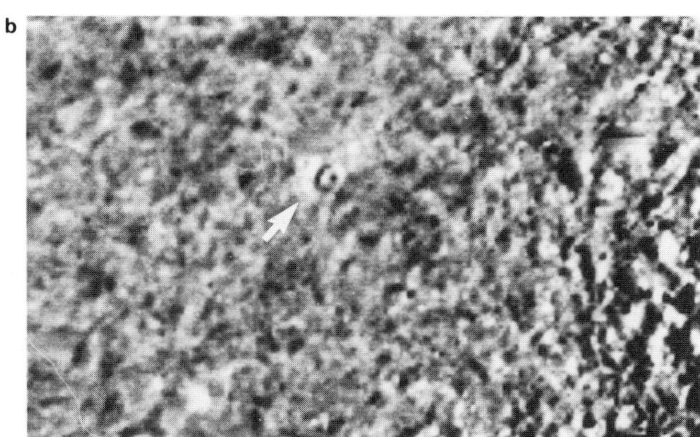

c

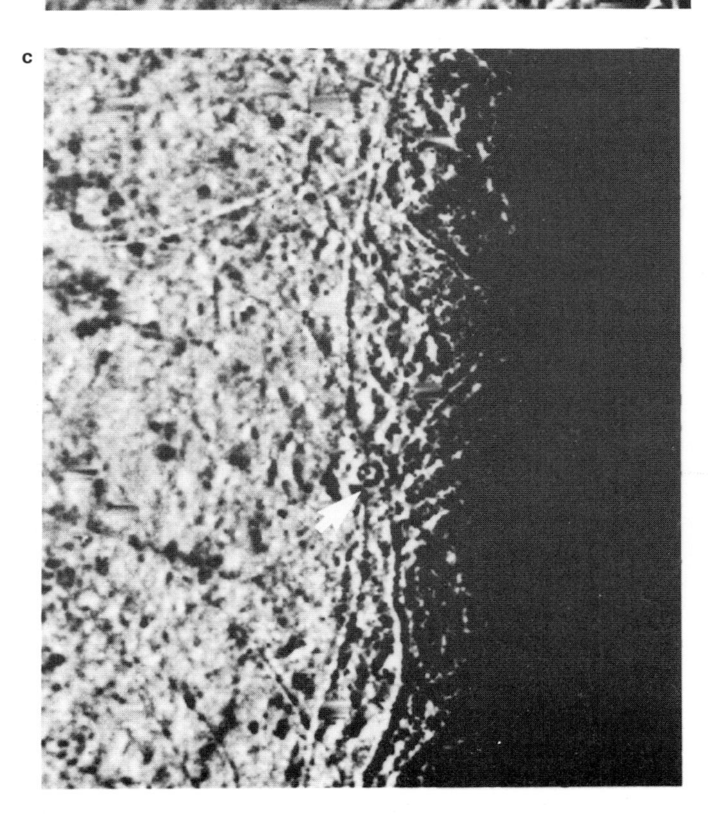

Figure 8.17 Voyager 2 image of Tyre Macula (Class 2 crater); this 100 km diameter feature (arrow) seen in Figure 8.18 displays faint concentric rings and may be a form of impact crater (Voyager 2 1557J2-002).

are cycloidal (Fig. 8.15). From cross-cutting relations, the ridges appear to be the youngest features on Europa. Mapping by Dave Pieri (1980) shows that the ridges form small circles centered approximately on the anti-jovian point, which suggests that they may be related to tidal deformation.

8.3.4 Geologic history

Based on mapping and models of thermal evolution, Lucchitta and Soderblom (1982) have derived a tentative geologic history for Europa that includes four general stages (from oldest to youngest):

(a) Formation of bright plains material, probably by the freezing and solidification of an ice crust. Subsequent units appear to have formed at the expense of bright plains.
(b) Formation of the dark plains, the fractured plains and possibly the gray mottled terrain. All involve the development of units which may result from intrusion of materials from subsurface sources or alteration or ageing of other material.
(c) Emplacement of brown material predominantly through fractures, but also localized as patches which could result either from eruptions or by impact excavation. Near the end of the emplacement, the brown mottled terrain was disrupted to produce complex topography.
(d) Formation of ridges, possibly from dike-like intrusions.

Lucchitta and Soderblom emphasize that Europa displays a surface that is more the result of tectonism than of Earth-like "layercake" stratigraphy. Europa can be summarized (Morrison 1983) as a body having a rocky, lunar-like interior overlain by an icy crust that has been repeatedly fractured by tectonism and intrusion by water-rich materials.

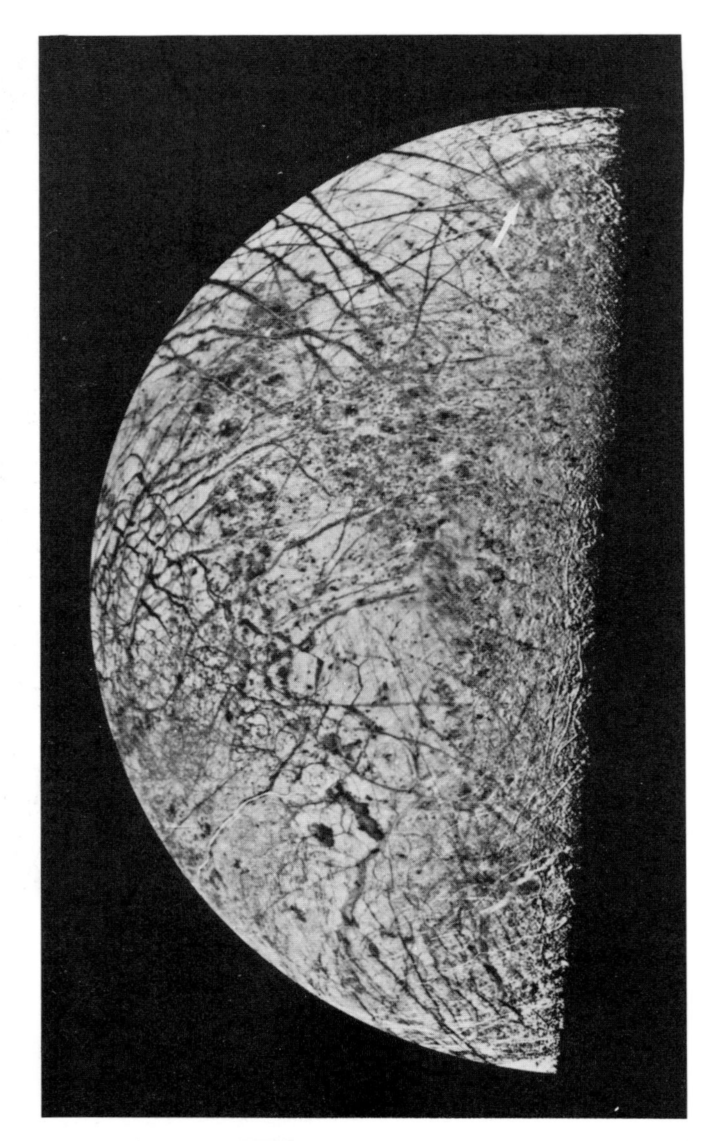

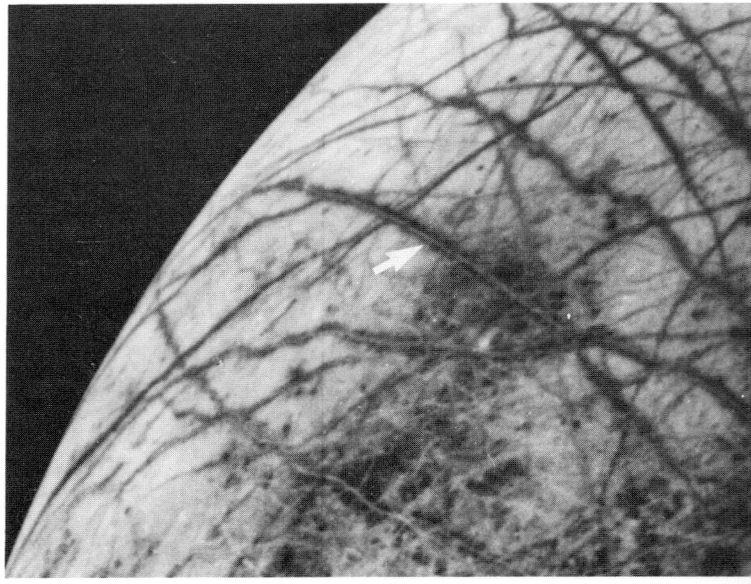

Figure 8.18 (top right) Mosaic showing various linear features which criss-cross Europa; note Tyre Macula's detailed structure (arrow).

Figure 8.19 (opposite) View of "triple bands", ranging up to 18 km wide, which consist of two parallel dark zones separated by a central bright zone (arrow), which, in some cases, are ridges. Triple bands may be associated with global lineation patterns that form great circles on Europa (Voyager 2 1186J2-001).

Figure 8.20 Low-resolution view of Ganymede showing dark, cratered terrain, light grooved terrain, and bright halo craters. Circular dark zone in the upper right is Galileo Regio. Light, circular patches within Galileo Regio are palimpsests (Voyager 2 528J2-002).

8.4 Ganymede

Voyager views of Ganymede show a planetary crust that has been severely deformed and which may be the closest analog to Earth's plate tectonics. Voyagers 1 and 2 imaged approximately one-half of Ganymede's surface (Fig. 8.21) and returned photographs with resolutions as good as 500 m/pixel for some areas. Ganymede is larger than the planet Mercury and is the largest satellite in the Solar System. It has a density estimated to be slightly less than 2.0 g cm^{-3}, suggesting that it is about half water.

In addition to tectonic deformation, the surface of Ganymede displays terrain units of various albedos (Fig. 8.20). As discussed for Europa, lower albedo areas could result from continued infall of meteoritic material, or from the removal of ice to leave a lag deposit of dark materials. In any event, the greater impact crater frequencies observed on dark terrains indicate greater age.

8.4.1 Physiography

Preliminary mapping by Shoemaker *et al.* (1982) shows that Ganymede consists of two primary terrains (dark, cratered terrain and light, grooved terrain), two less extensive terrains (smooth plains and reticulate terrain) and basin terrains, plus various tectonic features and craters. The terrain units are subdivided primarily on the basis of albedo and crater frequency.

Dark, cratered terrain consists of low albedo, heavily cratered areas (Fig. 8.22) considered to be remnants of the ancient crust. This unit occurs as large, rounded areas or as small polygonal patches. The largest expanses of this unit include Galileo Regio on the leading hemisphere north of the equator, Nicholson Regio on the trailing hemisphere south of the equator, and Marius Regio which straddles the equator on the leading hemisphere.

Grooved terrain constitutes about 60 percent of the surface photographed. Most of both polar areas are composed of grooved terrain, at least of the areas imaged. Grooved terrain is characterized by subparallel grooves and ridges that are $\geqslant$100 km in length, separated by 3–10 km. Local relief from groove-to-ridge averages 300–400 m, but may be as great as 700 m. The grooves and ridges occur as "bundles" a few tens of kilometers wide by hundreds of kilometers long. Bundles and sets of bundles display complex cross-cutting relations which can be used locally to derive sequences of formation.

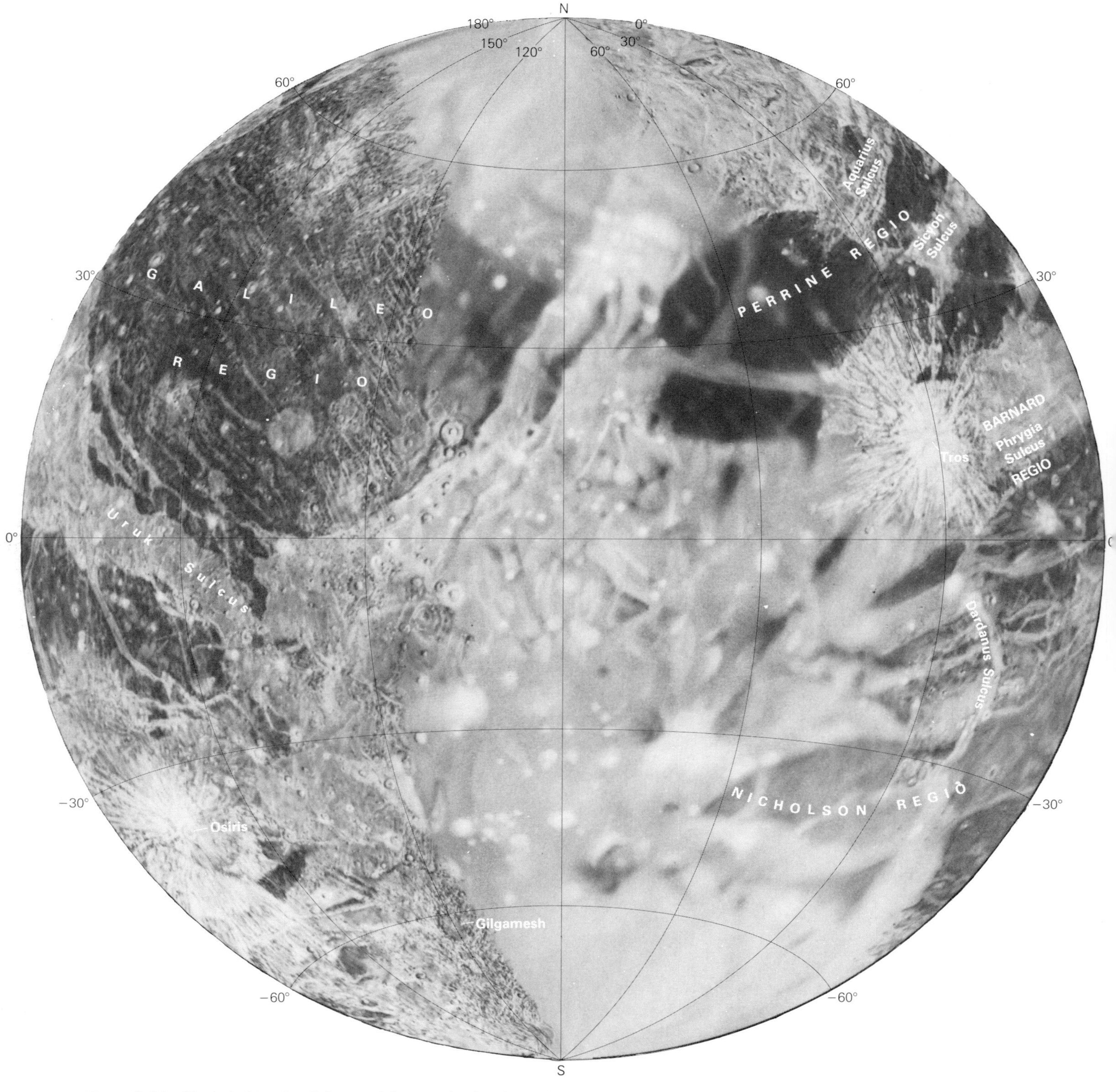

Figure 8.21 Shaded airbrush relief map of Ganymede showing various terrains and selected named features (courtesy US Geological Survey).

GANYMEDE

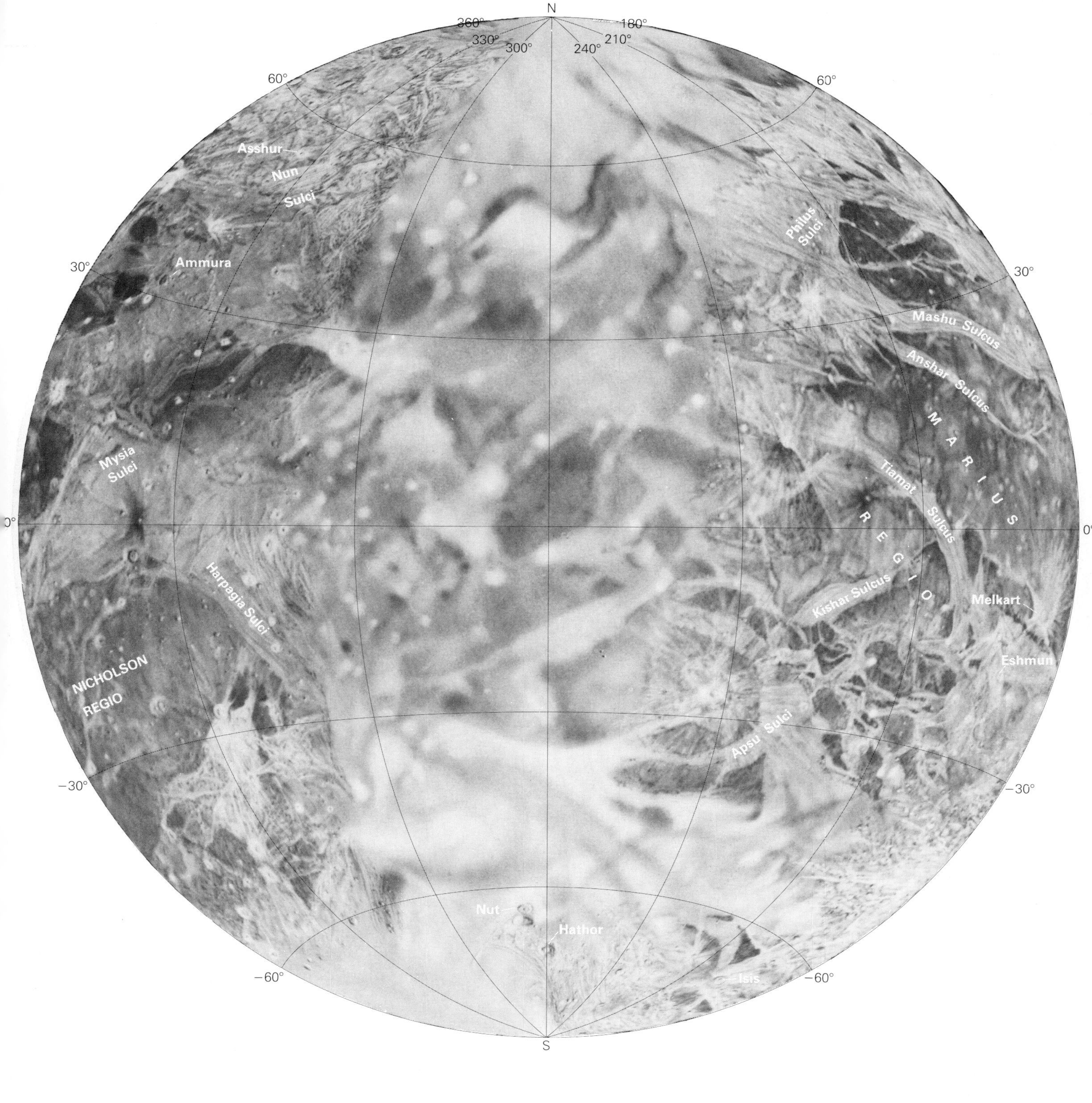

N
360°
330°
300°
240°
210°
180°
60°
60°
Asshur
Nun
Sulci
Ammura
Philus Sulci
30°
30°
Mashu Sulcus
Anshar Sulcus
MARIUS
Mysia Sulci
Tiamat Sulcus
REGIO
0°
0°
Harpagia Sulci
Kishar Sulcus
Melkart
Eshmun
NICHOLSON
REGIO
Apsu Sulci
−30°
−30°
Nut
Hathor
−60°
Isis
−60°
S

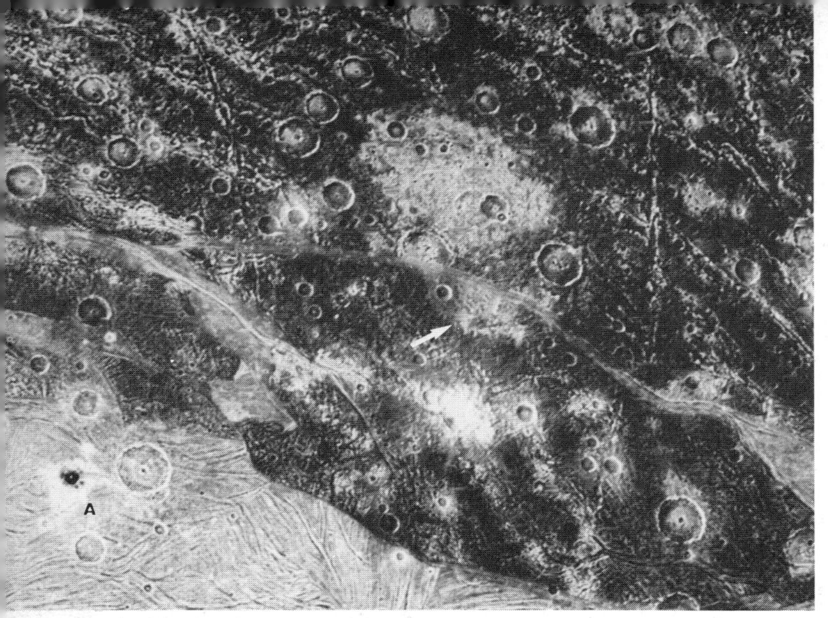

Figure 8.22 View showing dark, cratered terrain in southern Galileo Regio and part of Uruk Sulcus (grooved terrain). Palimpsests (light, oval areas) are considered to be impact scars. One palimpsest (arrow) has been cut by a narrow band of grooved terrain. Also shown are furrows which cut the dark terrain. Note the dark material excavated by the crater on the lower left (A). Area shown is 1700 km across, centered at 9°N, 145°W (Voyager 2 464J2-001).

In general, grooved terrain has formed at the expense of dark, cratered terrain, as shown in Figure 8.22, where bundles of grooves and ridges are seen to cut into darker, more heavily cratered terrain. Most investigators consider the processes of formation to involve rupture of ancient crust, perhaps accompanied by the release of liquid water or icy slush as a flood over the surface. Compression, extension, rotational shear or some combination of deformation have been proposed to explain the grooves.

Smooth plains occur as small patches and appear to be facies of grooved terrain (Shoemaker *et al.* 1982). Smooth plains have about the same albedo as grooved terrain, but lack ridges and grooves (Fig. 8.23a). There is little topographic relief on the smooth plains, and they appear to be superimposed on adjacent units, suggesting that they may be relatively young flows.

Reticulate terrain appears to be a transitional unit between dark, cratered terrain and grooved terrain. The distinctive pattern (Fig. 8.23a) may result from two or more sets of grooves that are superimposed. In addi-

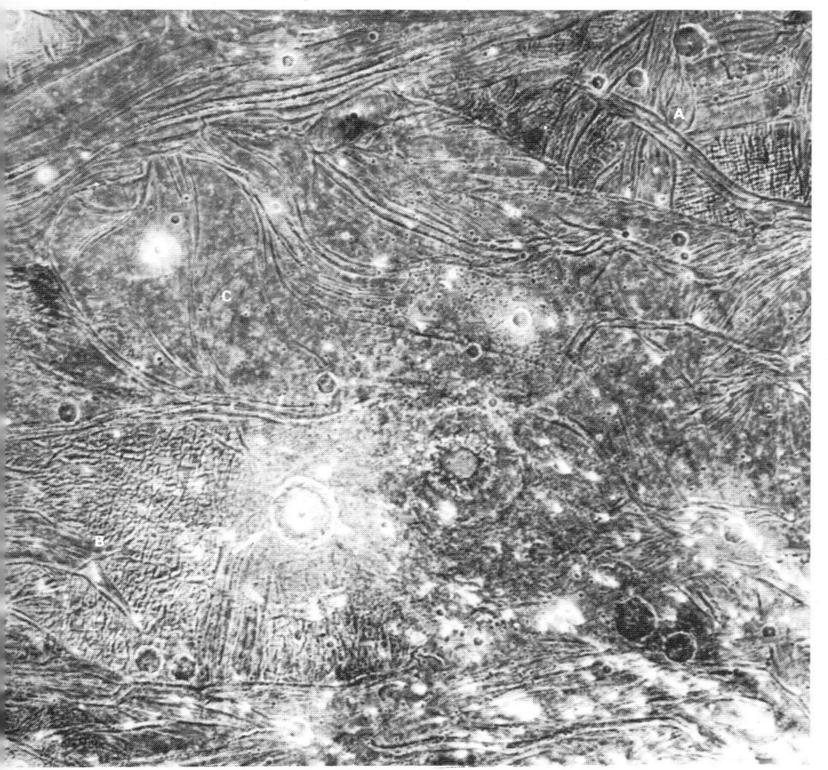

Figure 8.23a View showing two types of reticulate terrain near Galileo Regio. Reticulate terrain at (A) exhibits orthogonally intersecting grooves, while that at (B) is more irregular and hummocky. Various sets of grooved terrain crosscut dark terrain, older sets of grooved terrain, smooth plains (C), and the reticulate terrain. Arrow indicates a pit crater with a high-albedo central dome. Area shown is about 1500 km across, centered at 25°S, 176°W (Voyager 2 494J2-001).

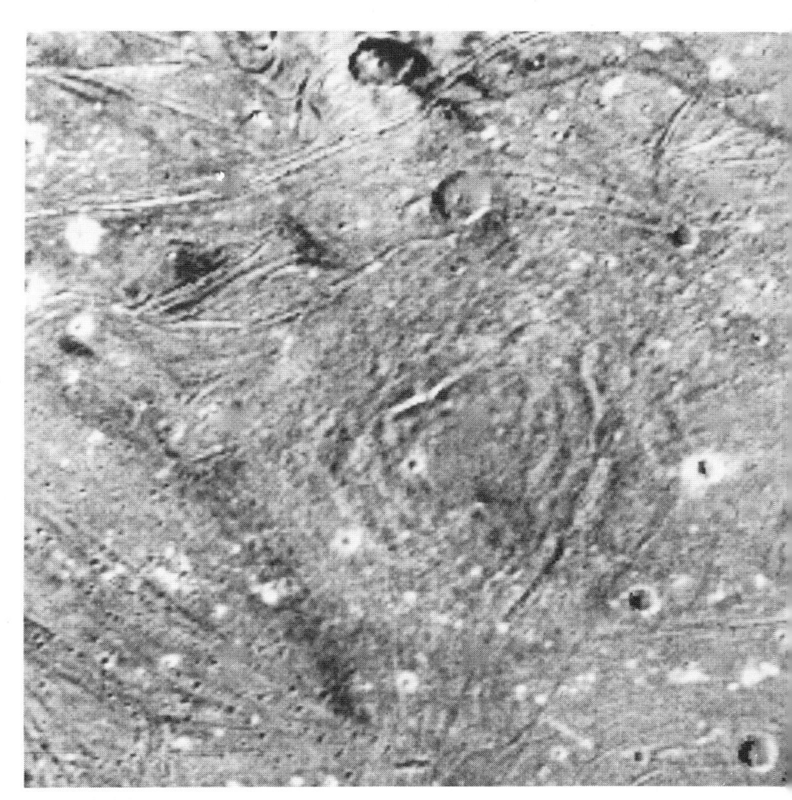

Figure 8.23b Image of a domed palimpsest in the region of Nun Sulci, Ganymede. A swarm of secondary craters on the surrounding grooved terrain attests to the original presence of an impact structure near the center of the dome. The dome, more than 350 km across and ~2 km high at its center, may have formed by the combined processes of ice-volcanism and isostatic uplift. Image centered at 35°N, 328°W (Voyager 1 859J1+000).

tion, the unit often includes numerous small hills which produce a hummocky texture.

Basins of probable impact origin also occur on Ganymede. Thus far, two such structures have been found, the largest of which is the Gilgamesh basin (Fig. 8.24). Although poorly defined, it has features suggestive of multi-ringed basins on terrestrial planets. Gilgamesh consists of a 150 km wide central depression filled with relatively smooth, featureless plains that are surrounded by rugged, mountainous ejecta deposits which decrease in relief radially outward. Secondary craters, many of which occur in chains, can be traced outward as far as 1000 km. A prominent ring occurs at a radius of 275 km; Shoemaker *et al.* (1982) consider this to represent the boundary of excavation for the basin. From stratigraphic relations and crater frequency distributions, Gilgamesh appears to have formed shortly after the development of the adjacent grooved terrain.

In addition to the Gilgamesh basin, an unnamed circular feature (Fig. 8.25) centered at ~7°S, 115° has been suggested by Shoemaker *et al.* (1982) as a possible impact feature. Superimposed on grooved terrain, its rim defines a diameter of about 200 km. Hummocky, ejecta-like deposits and secondary craters extend outward 150–450 km from the center. The floor of this structure is convex and has about the same radius of curvature as Ganymede, suggesting that hydrostatic equilibrium has been achieved.

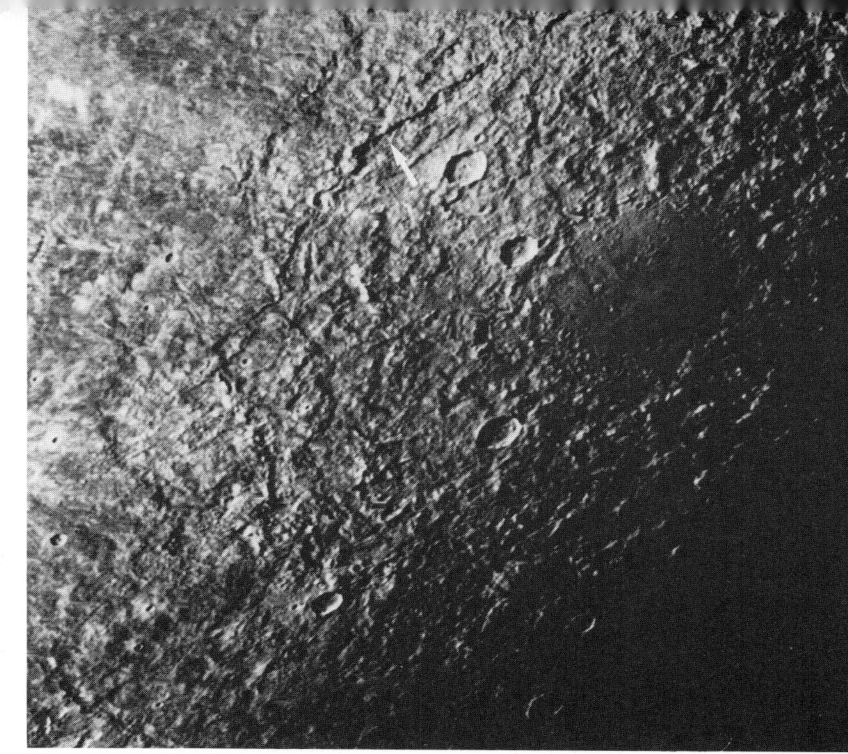

Figure 8.24 The Gilgamesh basin is defined by a smooth floor, 150 km across, in the center of the basin, surrounded by blocky terrain of 1.5–2 km relief. Prominent scarp (arrow) is considered by Shoemaker *et al.* (1982) to be the outer boundary of excavation. Area shown is centered at 61°S, 130°W (Voyager 2 527J2-001).

8.4.2 Craters

One of the more striking discoveries of the Voyager mission is the unusual appearance of the impact craters on Ganymede and Callisto in comparison with craters on the terrestrial planets. With increasing diameter, the craters on the icy satellites become progressively flatter. A possible process responsible for this "flatness" was predicted by Johnson and McGetchin (1973) well before the Voyager missions. They reasoned that plastic or viscous flow of icy crusts would reduce the relief of topographic features. Given the properties of ice and estimation of temperatures as a function of depth, Johnson, McGetchin and subsequent investigators (Parmentier & Head 1981) suggested that, following the impact excavation stage, crater floors first flattened, then bowed upward to a convex form as the crater rims subsided. As discussed by Passey and Shoemaker (1982), under these conditions, long wavelength features would be more affected by this **viscous relaxation** than short wavelength features. Thus, larger and deeper craters would deform faster and to a proportionately greater extent than smaller craters. For example, examination of craters on Ganymede shows that, while large primary craters are considerably "flattened", many of their secondary craters are bowl-shaped depressions.

Craters on Ganymede have been described by Passey and Shoemaker (1982) as including bowl-shaped, smooth-floored, central peak and central pit craters, plus an unusual craterform termed **palimpsests**. *Bowl-shaped craters* range in size from the limit in resolution to ~20 km in diameter (Fig. 8.26). Similar to small craters on terrestrial planets, they typically have depth-to-diameter ratios of 1:6 to 1:12 . Many of the craters in this size range occur in chains and appear to be secondary craters. *Smooth-floored craters* have flat to slightly convex floors and range in diameter from ~20 to ~40 km. They appear to be transitional in morphology between smaller, bowl-shaped craters and large craters which have complex interiors. Floor convexity, or "up-bowing", seems to be more pronounced in craters formed in the ancient cratered terrains than elsewhere. *Central peak craters* are ~5 to ~35 km in diameter. Peaks have basal diameters of 1–5 km and are as high as 700 m. In contrast to craters on the Moon and Mercury, craters >35 km on Ganymede lack central peaks. *Central pit craters* are commonly 16–120 km in diameter and essentially all craters on Ganymede larger than 40 km have pits. Generally, the diameter of the pit increases with crater size. Some pits have high albedo domes situated at their centers (Fig. 8.23a) which might be the result of ice diaperism (Malin 1980).

Figure 8.25 This unnamed western equatorial basin in one of the largest craterforms on Ganymede. It is 250 km across and highly "flattened" compared to the more recently formed Gilgamesh basin (see Fig. 8.24). Beyond the hummocky rim deposit to the upper left and lower left are swarms of secondary craters. Image centered at 4°S, 118°W (Voyager 2 552J2-001).

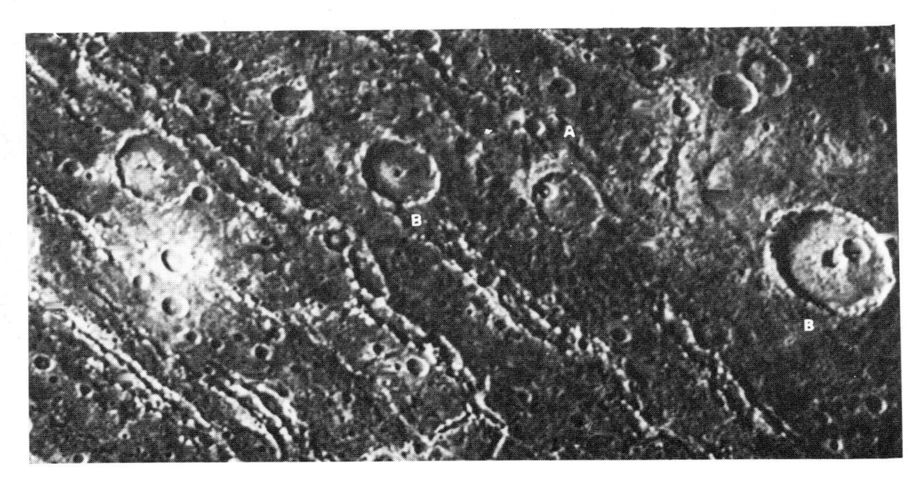

Figure 8.26 View of dark, cratered terrain showing small bowl-shaped crater (A) and central-pit craters (B). Area shown is in southeast Galileo Regio and is 600 km across (Voyager 2 546J2-001).

The term **palimpsest** refers to writing materials, such as parchment, that were reused in ancient times by removing the writing, although the original text was still faintly visible. In a similar vein, Smith and his fellow Voyager team members (1979b) applied the term to circular features they found on Ganymede that appeared to be the imprint of impacts (Figs 8.22, 8.26 & 8.27). Palimpsests range in size from 50 to 400 km in diameter and occur as bright circular-to-oval patches with little topographic relief. The total area of palimpsests equals about 25 percent of the surface imaged. Most of them are zoned with an inner brighter area – thought to represent the excavation zone – and an outer rugged area representing part of the ejecta field. Some palimpsests form broad topographic domes (Fig. 8.23b) which might be the result of the combined processes of ice volcanism and isostatic upwelling (Squyres 1980).

Shoemaker and co-workers (1982) suggest that the high albedo interior zone of palimpsests represents relatively "clean" ice excavated from the subsurface. From the size relations and the inference of the excavation zone, they suggest that when most palimpsests formed, the lithosphere was about 10 km thick and was underlain by a water or slush mantle. They note that craters <50 km in diameter do not have bright rims but exhibit bright floors, suggesting that the excavation bowl failed to penetrate through the lithosphere into the water mantle. Greeley *et al.* (1982) and Croft (1983) suggest that palimpsests and some shallow craters may be relatively pristine forms that result from impacts into ice which melts and deforms as part of the cratering processes. These craterforms are thought to represent the final stages of the period of heavy bombardment as the crust of Ganymede was solidifying.

Ejecta rays (both bright-ray and dark-ray) are prominent on Ganymede for some craters (Fig. 8.27). In general, rays are brightest when formed on grooved terrain and are darker for craters formed in dark, cratered terrain. As noted by Shoemaker *et al.* (1982), some rays exhibit abrupt changes in albedo along their length, some craters may have bright rays extending from one side and dark rays on the opposite side, while still other craters may have bright rays, but dark floors. Most of these variations can be attributed to differences in layers of materials being excavated.

Horner and Greeley (1982) found that many craters <90 km in diameter on Ganymede exhibit ejecta deposits that end abruptly in a scarp (Fig. 8.28), similar to some craters on Mars. We suggested that these could be a type of ejecta-flow crater in which ice-rich target material was melted during impact and at least partly emplaced as a fluidized mass.

Crater frequency distributions have been obtained for several areas on Ganymede (Greeley *et al.* 1982, Passey & Shoemaker 1982, Woronow *et al.* 1982). Although the

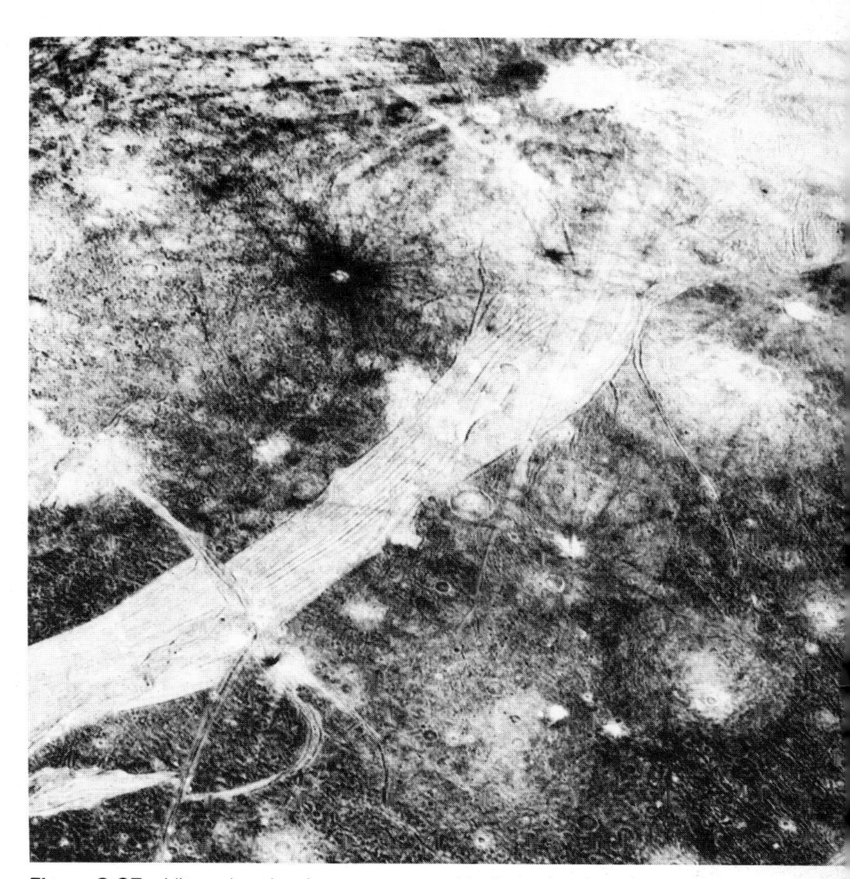

Figure 8.27 View showing impact crater with dark ejecta and bands of grooved terrain that have been cross-cut by younger grooved terrain (lower left corner); area shown is 2400 by 2400 km (Voyager 2 378J2-001).

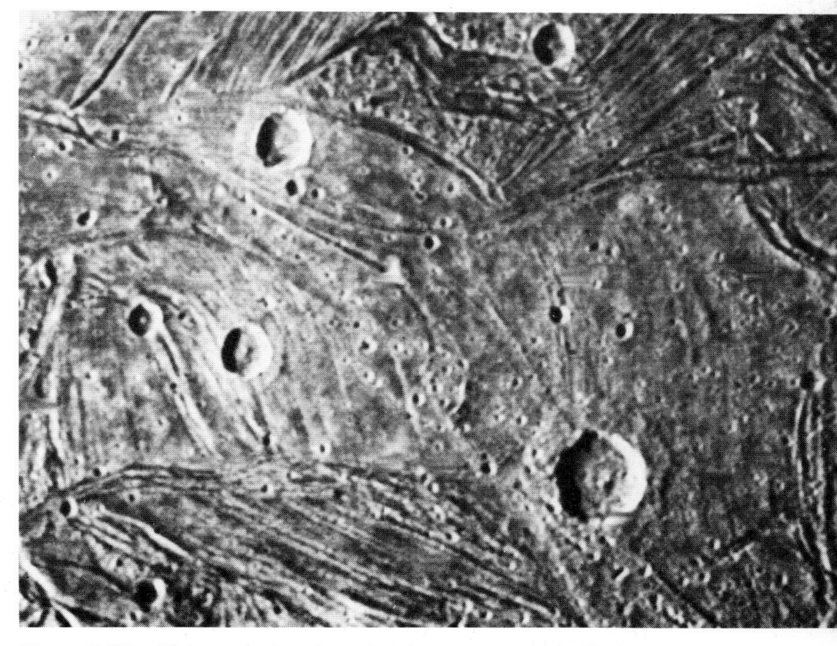

Figure 8.28 High-resolution view showing craters which display ejecta deposits that terminate in a scarp similar to ejecta-flow craters on Mars. Area shown is 100 by 100 km, centered at 38°S, 125°W (Voyager 2 572J2-001).

dark, cratered terrain (the oldest unit on the satellite) has about ten times the number of craters than lunar maria, it has fewer craters than the lunar highlands or the heavily cratered terrain of Mars. Crater counts on the grooved terrain are highly variable, with some south polar areas having about the same frequency as parts of the dark, cratered terrain.

8.4.3 Tectonism and volcanism

The prominent tectonic features on Ganymede include various grooves, fractures and grabens. Furrows and troughs cut the dark, cratered terrain, as shown in Figure 8.26. In Galileo Regio, these features can extend several hundred kilometers and are broadly arcuate, with the radius of curvature focused approximately on the anti-jovian point. McKinnon and Melosh (1980) have analyzed these features and suggested that they are regional grabens that formed in response to global stresses.

Grooves are typically flat-floored and occur in bundles, or packets, or subparallel scarps a few hundred kilometers long, although individual grooves may extend 1000 km. A few groove-sets exhibit some strike-slip displacement (Fig. 8.27), but for the most part, grooves appear to represent vertical displacement and are probably block faults.

Squyres (1980) and Parmentier *et al.* (1982) analyzed the global tectonic pattern exhibited on Ganymede and concluded that phase changes of ice could account for the fracturing and faulting. Ice occurs in several different crystal forms, called **polymorphs**, which have different densities and other physical properties as a function of temperature and pressure. These authors note that as Ganymede evolved, transitions from one ice phase to another would cause differences in global density and volume which could lead to a 7 percent increase in surface area. Expansion of this magnitude could have caused widespread normal faulting of the style exhibited in the grooved terrain. A subsequent analysis by Golombek (1982) shows that much less surface expansion could also produce the observed faulting in the grooved terrain. He estimates that the minimum thickness of the lithosphere was 5–9 km at the time of furrow formation and was about 4 km at the time of groove formation.

Crustal extension and normal faulting may also have released subsurface liquids. As discussed for Europa, this flooding can be considered a style of volcanism in which the "magma" consists of water and ice slush. The higher albedo of the grooved terrain and the presence of the smooth terrain support this interpretation.

8.4.4 Geologic history

Based on physiographic mapping, interpretation of surface-modifying processes and inferences of the interior properties and evolution, Shoemaker *et al.* (1982) have proposed the following history for Ganymede. The earliest record is characterized by the dark, cratered terrain which represents the early-formed lithosphere. During the early heavy bombardment (Shoemaker & Wolfe 1982), the lithosphere was too thin to preserve craters, but with time the crust thickened. Because of the physical properties of ice-rich materials, small craters would be preserved earlier than large ones. Palimpsests are thought to represent extreme cases of "relaxed" large craters or structures resulting from impacts into a plastic-like, deformable slurry underlying a thin, rigid crust and probably represent the final stages of the period of heavy bombardment. From preservation of rim structures, it is estimated that the lithosphere had thickened considerably by the time of formation of Gilgamesh, a multi-ring basin >300 km in diameter.

Expansion of the core related to phase changes of the icy interior led to the development of grooved terrain through tectonic deformation and release of liquids to the surface. Formation of grooved terrain may also have resulted from upwelling convection cells within the mantle. As the lithosphere thickened, the formation of grooved terrain ceased. Local eruptions producing small patches of smooth terrain and impact cratering – principally by comets – mark the last major phases of Ganymede history. Surface modifications by infrequent impacts and ion bombardment continue today.

8.5 Callisto

Callisto is the darkest of the Galilean satellites, yet is still twice as bright as the Moon. Its density is only about 1.8 g cm^{-3}, making it the least dense of the Galilean satellites and suggesting that it has the greatest proportion of water. Voyager images show that the surface of Callisto is heavily cratered (Fig. 8.29) but has relatively little relief. Except for provinces related to large impact events, the surface viewed by the Voyagers is essentially uniform, consisting of relatively dark, heavily cratered terrain (Fig. 8.31).

Crater counts for Ganymede and Callisto (Fig. 8.30) show that Callisto is more heavily cratered than even the oldest terrain on Ganymede, despite the suggestion that the gravitational focusing effect of Jupiter should lead to more large craters on Ganymede. Thus, it is inferred that the surface of Callisto records a longer history than that of Ganymede.

The most prominent individual surface feature on Callisto is a multi-ring structure named Valhalla. As

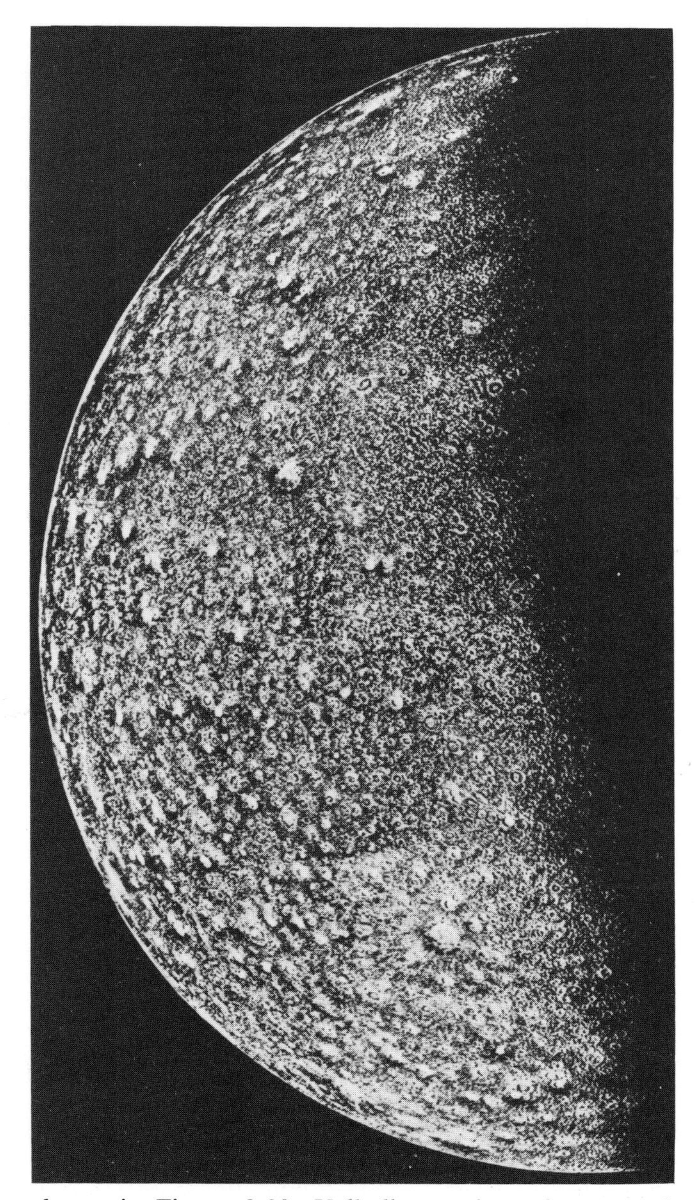

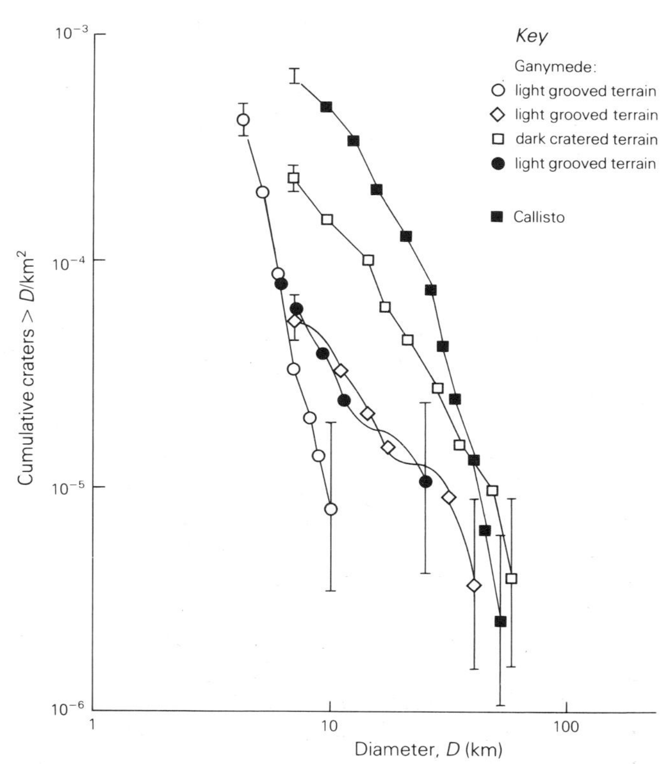

Figure 8.30 Crater frequency distributions comparing various terrains on Ganymede with the substantially more heavily cratered surface of Callisto (from Smith *et al.* 1979b).

Figure 8.29 Mosaic showing the heavily cratered terrain of Callisto as viewed from a distance of 215 000 km (Voyager mosaic 260–586).

shown in Figure 8.32, Valhalla consists of a central bright zone about 600 km across surrounded by numerous concentric rings extending outward for nearly 2000 km from the center of the structure. The rings are spaced 20–100 km apart and ring spacing tends to increase with radial distance from the center. Mapping by Smith *et al.* (1979b) shows that there are relatively few superposed craters in the center of Valhalla and that the density of craters increases outward (Fig. 8.33a) until it reaches about the same frequency as the average of the whole satellite.

First recognized in a study by Hale *et al.* (1980), the rings of Valhalla may be divided into three zones based on morphology (Fig. 8.33b). The rings in the innermost zone consist of fairly continuous, narrow bands separated by wider bands of intermediate-albedo terrain. These bands are sinuous and scalloped in plan view and may be ridges, as are the innermost rings of the Asgard basin seen elsewhere on Callisto (Fig. 8.34). The width of this inner zone varies from ~200 km in the southeast to ~300 km in the northeast. The middle zone is characterized by weak to discontinuous ring development and varies in width from ~400 km in the southeast to ~200 km in the northeast. The outermost zone displays a wide variation in width and ring morphology. In the south and east the zone is >500 km wide and consists of narrow, bright-floored, sinuous troughs or furrows separated by wider, low-albedo terrain.

The resemblance of troughs or furrows to the furrows seen on the Galileo Regio of Ganymede (Figs 8.22 & 8.26) led to the argument that the furrows on Galileo Regio were also formed by a large impact whose center has been destroyed by grooved terrain formation

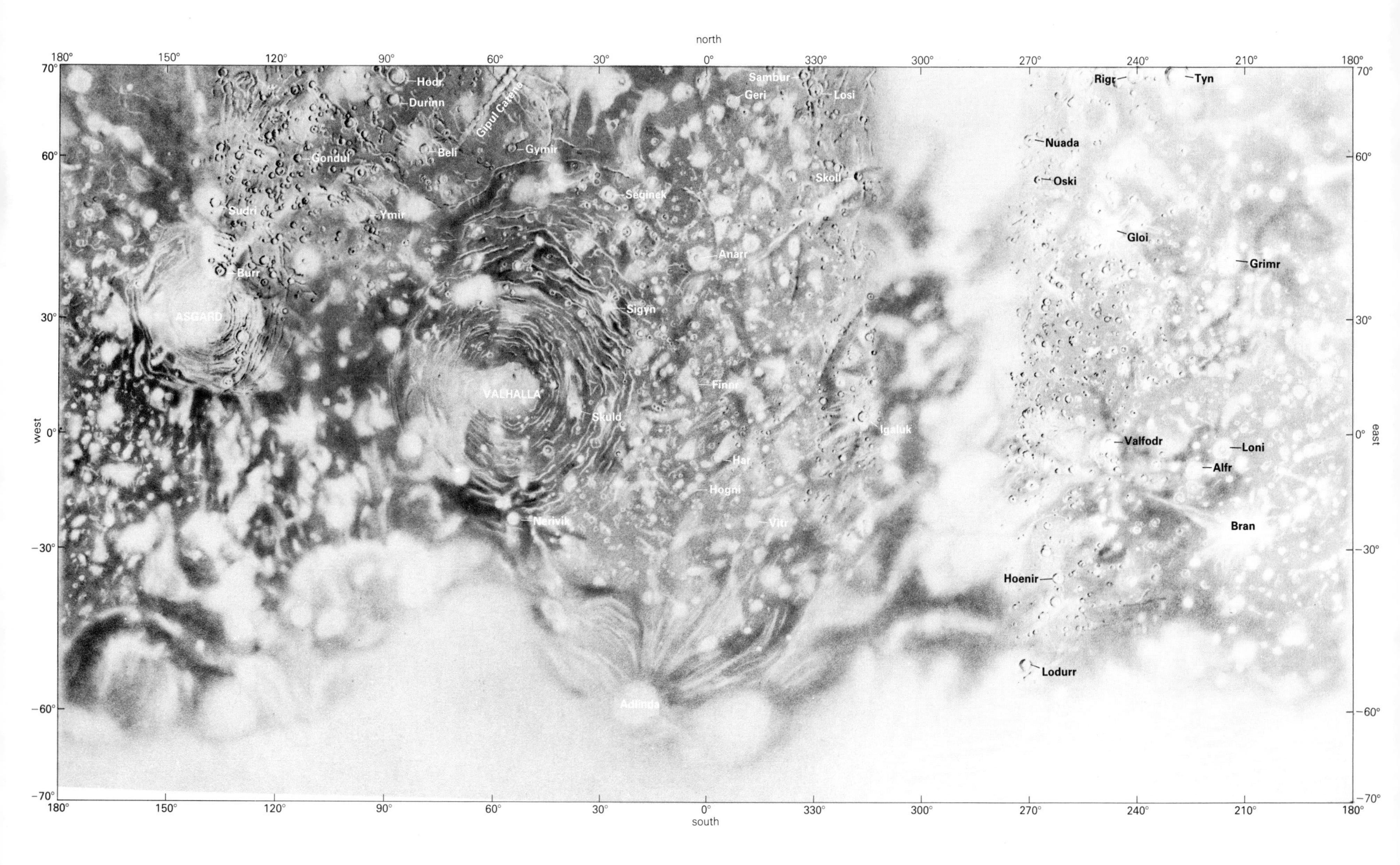
north
south
west
east
180°
150°
120°
90°
60°
30°
0°
330°
300°
270°
240°
210°
70°
-30°
-60°
-70°
Hodr
Durinn
Gipul Catena
Gymir
Beli
Gondul
Sudri
Ymir
Burr
ASGARD
Seginek
Sigyn
VALHALLA
Skuld
Nerivik
Adlinda
Sambur
Geri
Losi
Skoll
Anarr
Finnr
Har
Hogni
Vitr
Igaluk
Rigr
Tyn
Nuada
Oski
Gloi
Grimr
Valfodr
Loni
Alfr
Bran
Hoenir
Lodurr

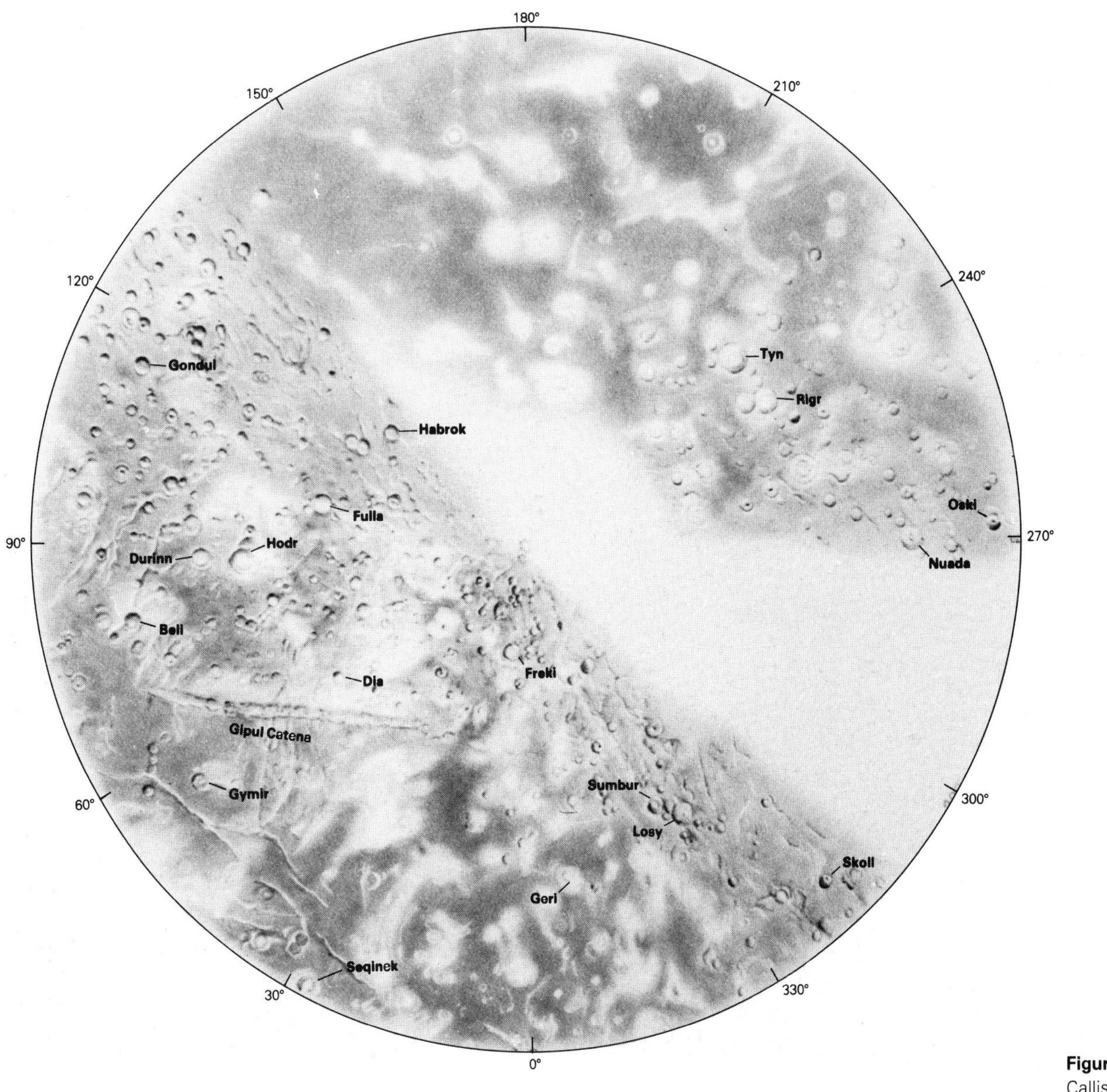

Figure 8.31 Shaded airbrush relief map of Callisto and selected place names (courtesy US Geological Survey).

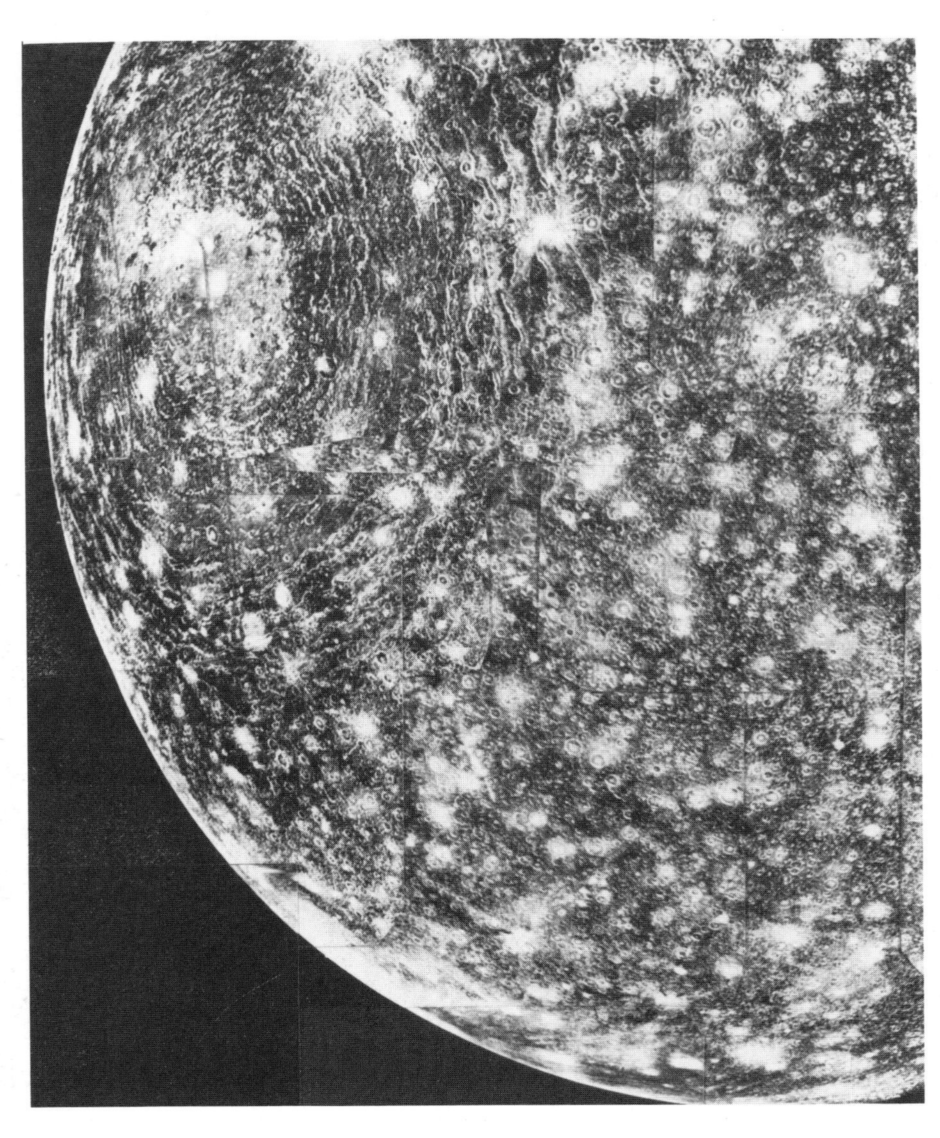

Figure 8.32 Voyager mosaic showing the Valhalla basin, defined by a central bright spot about 600 km in diameter surrounded by numerous concentric rings (Voyager mosaic P-21282).

(McKinnon & Melosh 1980). To the northwest the outermost Valhalla ring zone increases greatly in width to 1400 km; the rings here consist of outward-facing scarps with dark, heavily cratered backslopes. Discrete light-toned bands which occurred below the scarps are thought to consist of material extruded from the scarp bases. In several places this light, extrusive material floods craters.

Jay Melosh (1982) developed a model which may explain the variation in ring morphology by determining the effect of the inward flow of the underlying asthenosphere toward the crater cavity as the initial impact cavity collapsed, producing a characteristic pattern of faults in the disrupted lithosphere as a function of distance from the point of impact (Fig. 8.35).

At least seven other multi-ring features have been recognized on Callisto (Passey & Shoemaker 1982). Asgard (Fig. 8.34), located at 25°N, 145°W has a central bright zone ~230 km across and is surrounded by concentric rings out to 800 km. Other, smaller features are also composed of central bright zones and concentric rings. The bright zones may be palimpsestlike features and the concentric ridges may have formed in the icy lithosphere as impact-generated adjustments.

Cassen *et al.* (1980), Passey and Shoemaker (1982) and others have considered the comparative evolution of Ganymede and Callisto. They note that in the early history both objects formed from materials leading to a water-silicate body. Heating from various sources, including impact cratering, radionuclides and tidal stresses led to differentiation and the formation of a "mud" core. In comparison to Ganymede, the smaller size of Callisto (and hence its lesser amount of radionuclides), the much smaller tidal effects and an inferred lower impact energy resulting from the reduced gravitational focusing of Jupiter because of the greater distance, all equate to heat being substantially lower on Callisto. Thus, its lithosphere probably thickened faster than Ganymede's, preserving the cratering record over a longer period and preventing tectonic deformation of the style which produced grooved terrain on Ganymede.

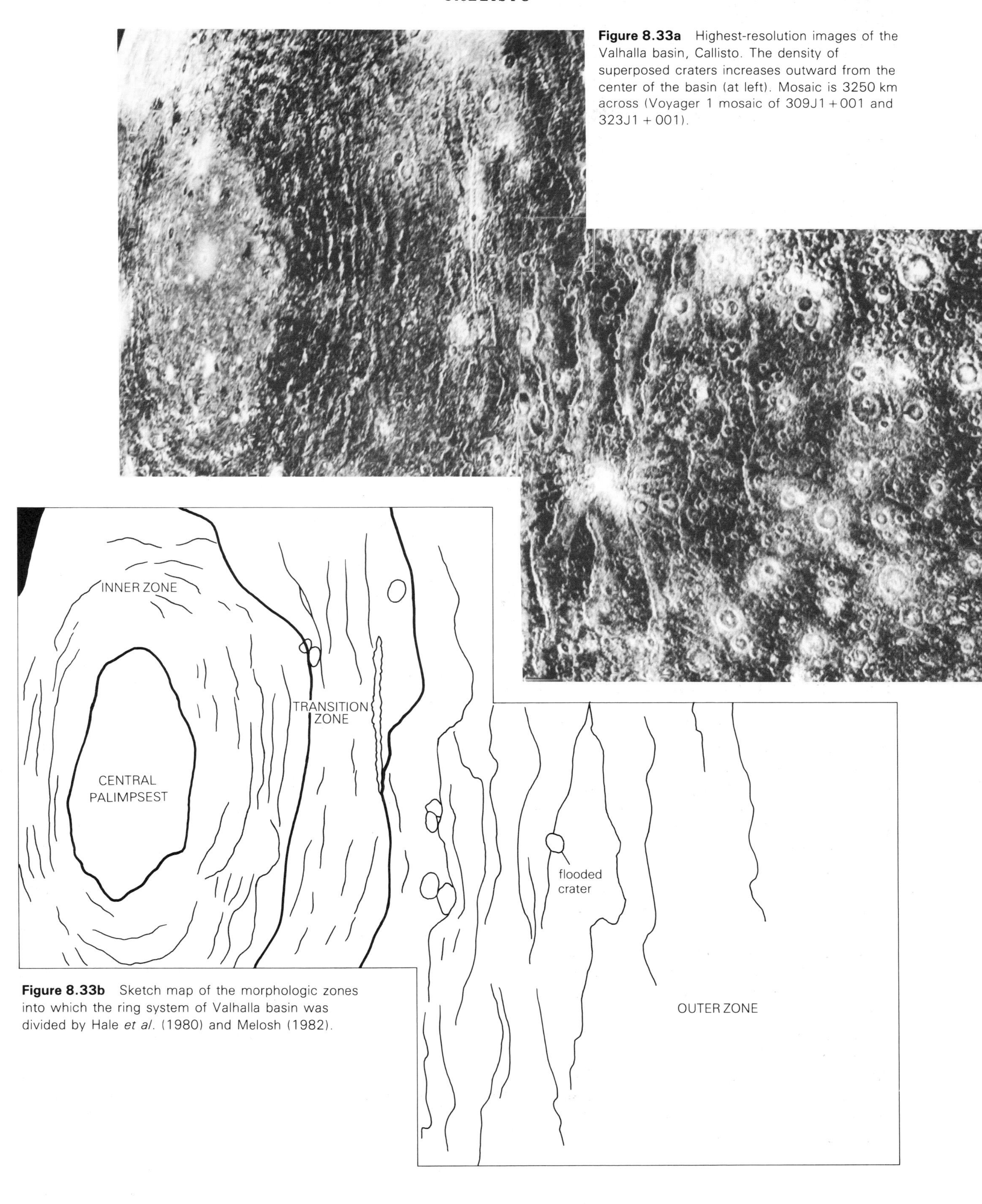

Figure 8.33a Highest-resolution images of the Valhalla basin, Callisto. The density of superposed craters increases outward from the center of the basin (at left). Mosaic is 3250 km across (Voyager 1 mosaic of 309J1 + 001 and 323J1 + 001).

Figure 8.33b Sketch map of the morphologic zones into which the ring system of Valhalla basin was divided by Hale *et al*. (1980) and Melosh (1982).

8.6 The Galileo mission

Many of the questions raised by the Voyager mission for the Jupiter system will be addressed by a mission scheduled for the late 1980s. Named the Galileo mission, a dual spacecraft will be launched from the shuttle in 1986 and will complete the journey of nearly 10^9 km in just under two years. Prior to arrival, an entry probe will separate from the main spacecraft and will be sent on a trajectory to carry it into the atmosphere of Jupiter, measuring gas composition, pressure, temperature, wind velocities and other characteristics of the atmosphere. When the probe penetrates the upper atmosphere, it will travel $\sim 48 \text{ km s}^{-1}$, and it is estimated that frictional heat will ablate ~ 100 kg of the protective shield as the probe descends. The probe will carry six instruments and is designed to function to a maximum pressure of about 30 bar, after which data return is extremely reduced.

The orbiter carries an imaging system, a near-infrared mapping spectrometer, and seven other instruments. With a nominal lifetime of 20 months in orbit about the Jupiter system, the spacecraft will acquire at least 70 000 images of Jupiter and its satellites. During 12 orbits, numerous "encounters" or flybys will be made of the satellites. Although only one near flyby will be made of Io, images of ~ 15 m resolution will be obtained for a small part of the planet. More distant passes will provide ~ 1 km resolution images that will enable monitoring of volcanoes. Several flybys will be made of Europa. Galileo coverage for this satellite will be especially interesting because Voyager images are relatively low resolution, yet show many intriguing landforms. Ganymede and Callisto will also be extensively imaged with some regions observed at < 100 m resolution.

In general, Galileo should provide images with about an order of magnitude higher resolution than Voyager. Some encounters, such as the Io flyby, will yield even higher-resolution pictures. In addition, Galileo will image areas not seen by Voyager, expanding our knowledge of the global characteristics of all the Galilean satellites.

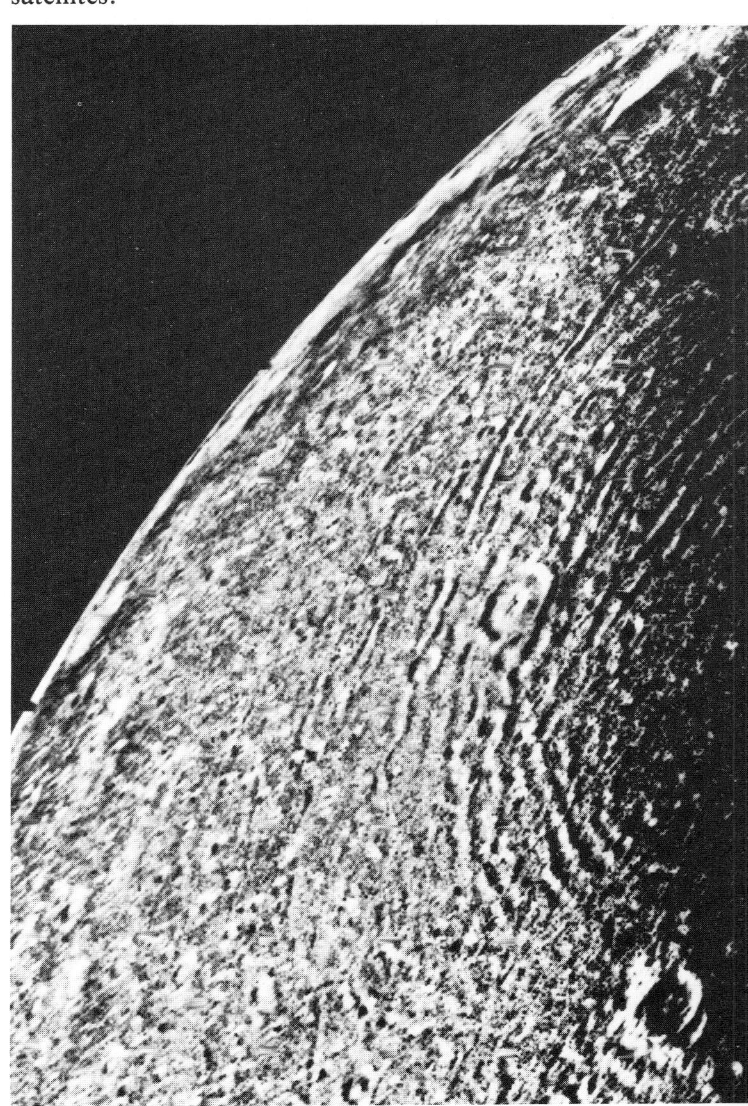

Figure 8.34 Voyager image of Asgard basin, Callisto. This multi-ring system is very similar to Valhalla but is only about 1500 km in diameter. Low sun angle reveals that the inner rings are ridges which appear to be flat-topped. Image is about 900 km across, centered at 14°N, 126°W (Voyager 1 526J1 + 001).

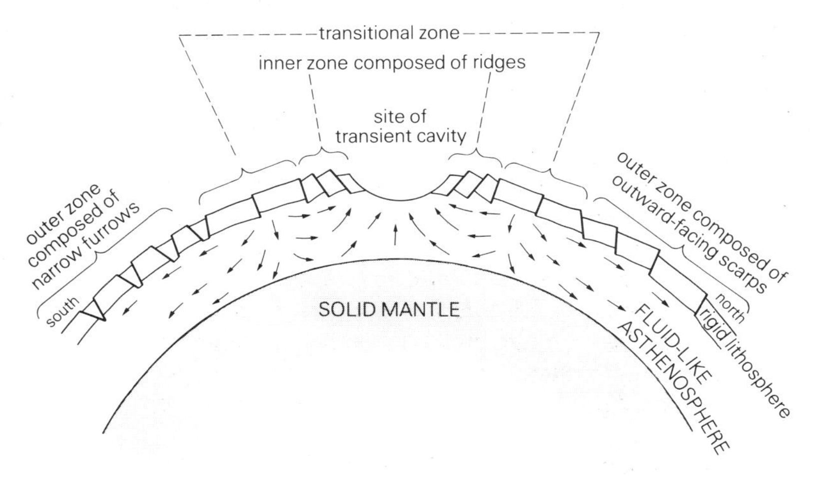

Figure 8.35 Lithosphere–asthenosphere response to the impact which formed the Valhalla basin, Callisto. Note that flow of the asthenosphere, associated with the infilling of the transient cavity, placed the lower lithosphere in motion having different directions of stress with radial distance from the center of impact. This results in different physiographic zones that are concentrically arranged around Valhalla (from Melosh 1982).

9 The Saturn system

9.1 Introduction

Prior to the Voyager flybys in 1980 and 1981 (Table 8.1), the satellites of Saturn appeared as little more than tiny spots of light, even when seen through the most powerful Earth-based telescopes. The Voyager mission not only transformed these spots into objects available for geologic study, but also enabled discovery of at least eight additional satellites, bringing the total known number to 17 (Fig. 9.1). These represent a remarkable diversity of objects, including various icy moons (many of which show signs of repeated resurfacing), the only known satellite in the Solar System to have an appreciable atmosphere (Titan), and a host of "small bodies" (a few tens of kilometers in size) which can be compared with the moons of Mars (Phobos and Deimos) and the innermost moon of Jupiter (Amalthea).

Although Pioneer 11 preceded the first Voyager to Saturn by a little over a year, most of the instruments carried by the Pioneers were designed to make measurements, such as the presence and strength of magnetic fields. However, a simple imaging system also enabled the discovery of a new satellite, as well as a new ring of Saturn – the F Ring – which lies a few thousand kilometers beyond the A Ring. Pioneer also discovered a second satellite, but not by imaging: the spacecraft nearly collided with an object estimated at the time to be ~200 km in diameter. Analysis of data related to magnetic fields showed that the spacecraft passed through the magnetic "wake" of the unknown satellite which was imaged later by Voyager.

Nearly all the geologic data on the saturnian satellites come from the Voyager missions. The paths of both Voyagers were planned to use the gravity of Jupiter to "sling-shot" them on their way to Saturn. Voyager 1 flew closer to Jupiter and received a greater boost in speed that allowed it to arrive at Saturn in November 1980, nine months earlier than Voyager 2. Aimed to make a close pass by Titan, Voyager 1 also viewed the dark (shaded) side of the rings and obtained high-resolution views of Mimas, Dione and Rhea. However, programming the flight path for a close flyby of Titan meant that the trajectory would carry Voyager 1 out of the ecliptic plane of the Solar System and prohibit it from journeying onward to the other outer planets.

The path of Voyager 2 was planned so that eventually it will encounter Uranus and Neptune. Although this requirement placed some constraints on its passage through the Saturn system, the trajectory nicely complemented that of Voyager 1. Voyager 2 made a pass of the illuminated side of the rings and provided close-up views of the satellites Iapetus, Hyperion, Enceladus and Tethys. Unfortunately, during the close approach to Enceladus and Tethys, the scan platform which carries the camera and other instruments became stuck and was unable to move in an azimuth (back and forth) direction. Consequently, the highest-resolution images of these satellites, as well as stereoscopic views of the F Ring were lost because the cameras were pointing in the wrong direction. At first blamed on the possible impact of particles associated with the ring system, analysis showed that the jamming of the scan platform resulted from leakage of lubricant. Subsequent study by spaceflight engineers shows that the platform can be moved, if done very slowly, and that imaging can be accomplished when Voyager 2 arrives at Uranus and Neptune.

Despite the disappointment caused by loss of the highest-resolution images, the Voyagers yielded a wealth of information on Saturn, its rings and the satellites, as described by Smith *et al.* (1981 & 1982) and reviewed by Morrison (1982b) and Soderblom and Johnson (1983).

Like Jupiter, in many respects Saturn also is a miniature Solar System – but with some important differences. For example, in both the Jupiter system and the Solar System in general, the densities of the orbiting bodies decrease with radial distance from the primary object. Analysis of the densities of saturnian satellites shows no such relationship (Table 2.1); rather, water-rich and rocky objects seem to be randomly distributed throughout the system.

Estimates of the satellite densities indicate that they are mixtures of rock and water ice (up to 70 percent ice by mass). Titan, a Mercury-sized object, is the most dense at 1.9 g cm^{-3} and is thought to be about half rock and half ice. Analysis of reflectance spectra shows that the surfaces of all the satellites are dominated by water frost. In fact, Enceladus is the brightest object in the Solar System, being over ten times more reflective than Earth's Moon.

In this chapter, each of the major satellites is briefly described and illustrated. A section then outlines the processes that appear to have modified their surfaces.

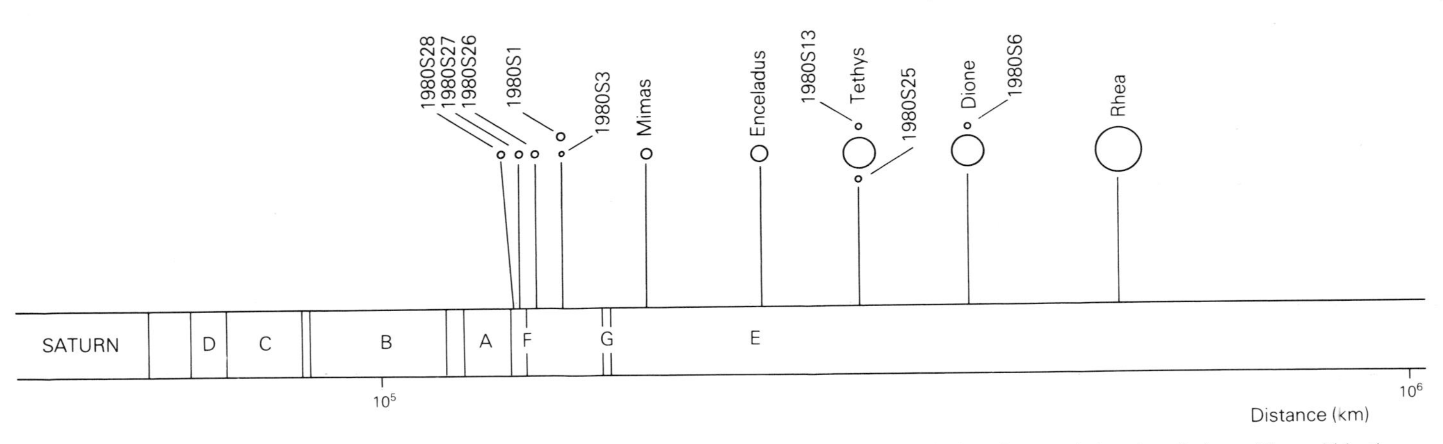

Figure 9.1 Diagram showing the known satellites of Saturn drawn to their correct relative sizes and showing their position within the saturnian system. The distance scale is logarithmic, and the distances are from the center of Saturn. Also shown are the locations of the major components of the ring system.

9.2 Geomorphology of the satellites

The nine largest satellites were discovered telescopically, some as early as 1655. Most of their names were proposed by Sir John Herschel in the 1700s. In Greek mythology, Dione, Rhea, Tethys, Mimas, Enceladus, Titan and Phoebe were giants, while Hyperion and Iapetus were brothers of Saturn.

All of the satellites but two (Hyperion and Phoebe) are in synchronous rotation, i.e. they keep the same side facing toward the primary, as do Earth's Moon and the Galilean satellites. All but two travel in circular prograde orbits in the equatorial plane, which is also the plane of Saturn's rings (Fig. 9.2). The exceptions, Iapetus and Phoebe, travel in paths inclined to the equatorial plane, and Phoebe travels in a retrograde direction. The innermost satellites have orbits interspaced between some of the prominent rings (Fig. 9.1).

Image resolution ranges from very poor to a couple of kilometers for most of Saturn's major satellites. Thus, terrain mapping is difficult for many of the satellites and caution must be exercised in comparing geologic processes and histories because of the uneven image quality for the different objects.

9.2.1 Mimas

Discovered in 1789 by Herschel, Mimas is a ~390 km size object lying between the G Ring and the E Ring of Saturn. During the flyby of Voyager 1, it was imaged at a resolution of about 1 km and shows a cratered surface (Fig. 9.3). Although the distribution of craters is not uniform, most of the surface of this icy object appears to be in equilibrium for craters <30 km; that is, craters up to this size are destroyed at the same rate as they are formed. Most of the craters are deep, bowl-shaped depressions (Fig. 9.4), and many larger than about 20 km have central peaks. Well defined ejecta deposits and bright rays are apparently lacking, although the high albedo of the surface would make rays difficult to see, even if they do exist.

The most striking aspect of Mimas is an impact crater, named Herschel, found on the leading hemisphere (Fig. 9.5). With a diameter of 130 km, it is about one-third the diameter of Mimas. This crater is nearly 10 km deep, with a central peak that rises some 6 km from the crater floor. Larger than Tycho and Copernicus on the Moon (Fig. 4.31), Herschel must be near the maximum size impact that the satellite could survive without shattering apart.

In addition to craters, Mimas displays grooves up to 90 km long, 10 km wide, and 1–2 km deep (Fig. 9.4). The small size of Mimas may preclude internal differentiation (supposedly there would be insufficient radionuclides to melt the ice) and related tectonic deformation, and these grooves could be related to the formation of the crater Herschel or some other impact event. The trailing hemisphere of Mimas shows local clusters of hills 5–10 km across and ~1 km high which may be ejecta deposits.

9.2.2 Enceladus

Like Mimas, Enceladus was discovered by Herschel in 1789. Voyager 1 images were of relatively low resolution and showed this small moon to be a bright, smooth sphere. Voyager 2 obtained images of ~1 km resolution and revealed complex terrains (Figs 9.6 & 9.7) which, in many respects, are similar to those of the Galilean moon, Ganymede.

Given the similar sizes of Mimas and Enceladus and their positions relative to Saturn, before the Voyager mission it was generally thought that the surfaces of these two satellites would be similar. Yet, they represent two extremes. While Mimas preserves a cratering record back to the time of its formation and shows no sign of resurfacing, Enceladus displays a complex geologic history.

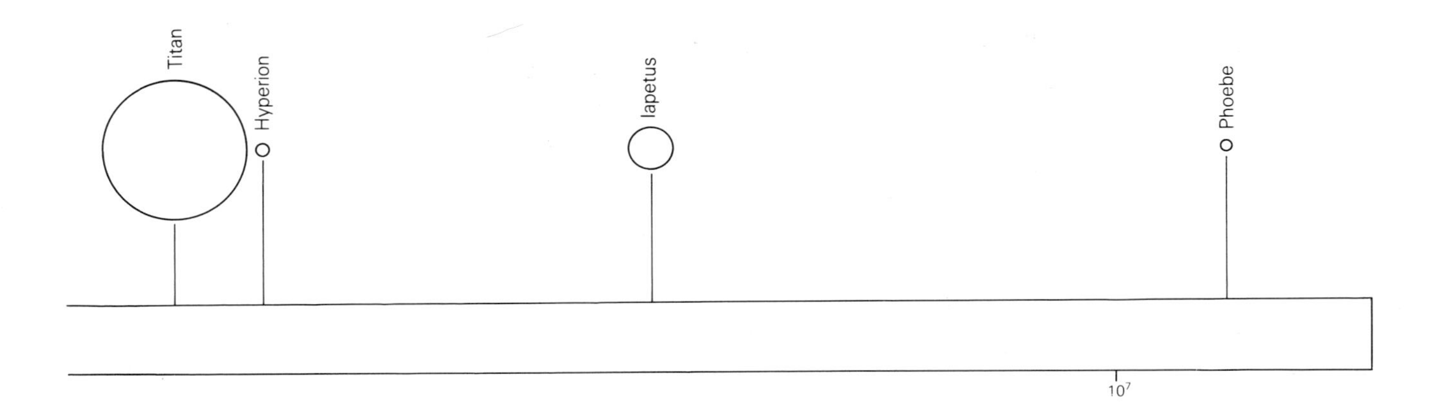

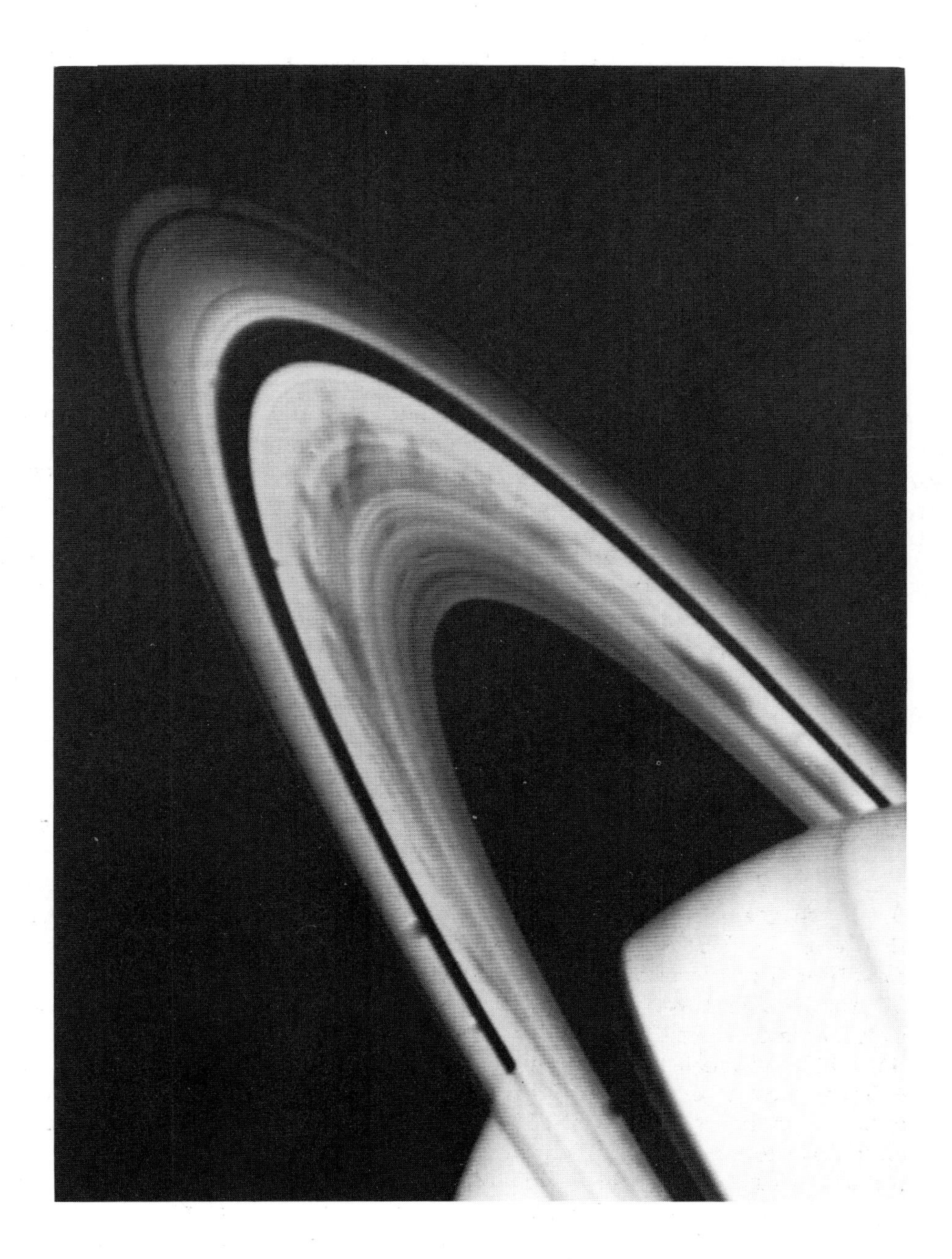

Figure 9.2 Voyager image showing prominent dark "spokes" in the outer half of Saturn's B ring (Voyager 2 JPL P-23881).

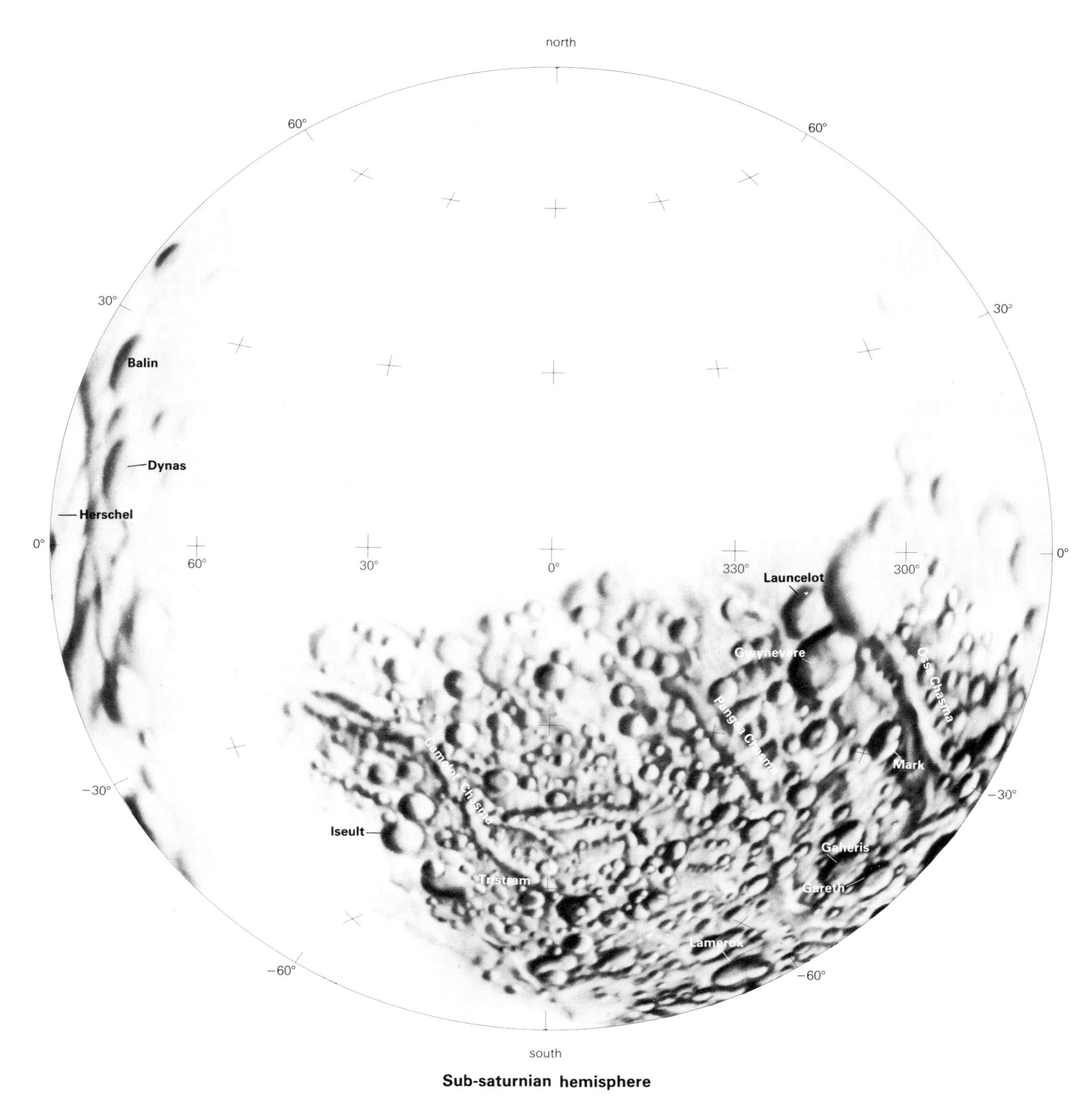

Sub-saturnian hemisphere

Figure 9.3 Shaded airbrush relief map of Mimas (courtesy US Geological Survey).

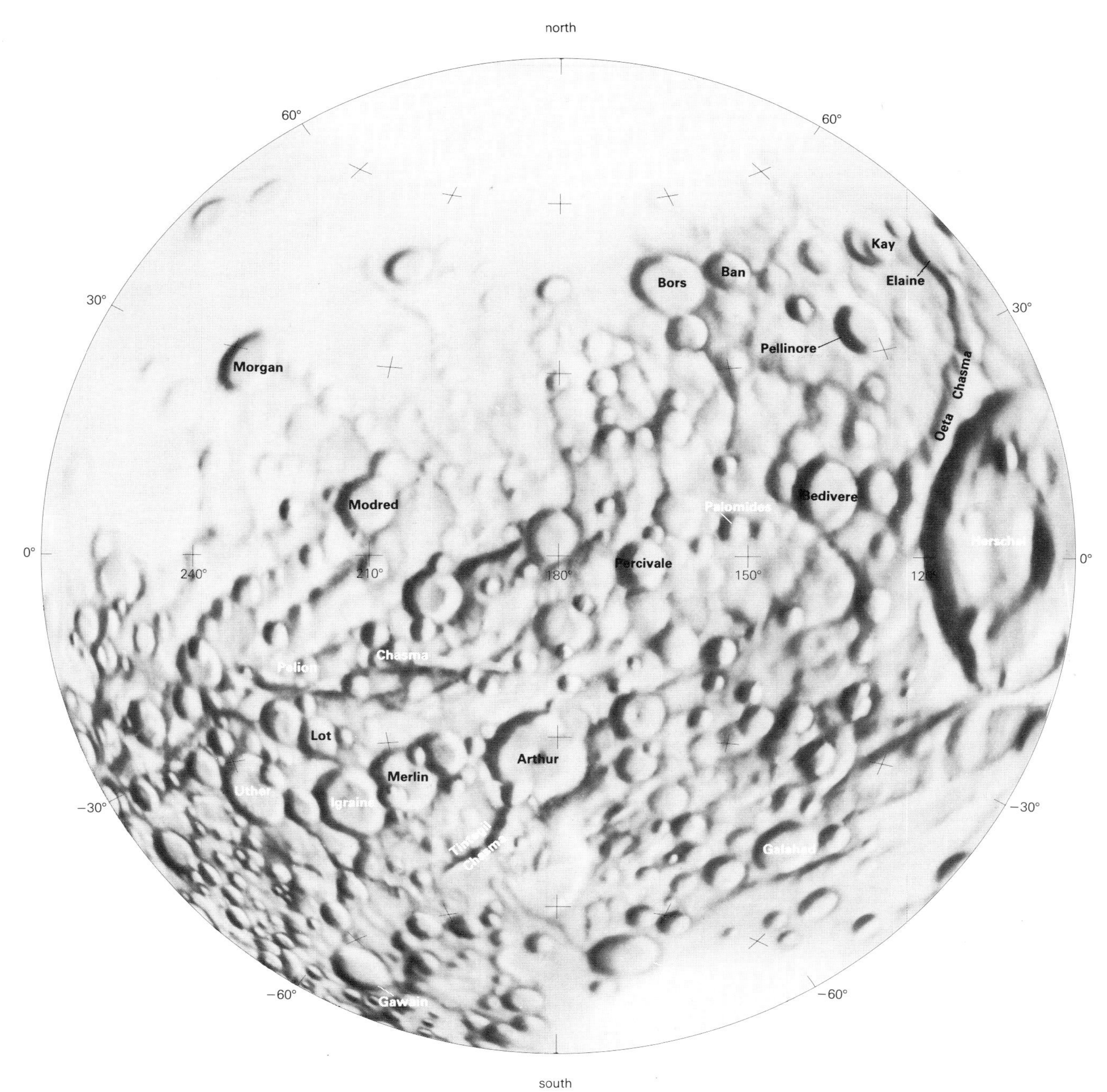

Anti-saturnian hemisphere

Figure 9.4 High-resolution image of Mimas showing the south polar region and prominent grooves (Voyager 1 25S1 + 000).

Figure 9.5 Voyager image of Mimas showing impact crater, Herschel (Voyager 1 JPL P-23210).

Preliminary terrain mapping by Smith *et al.* (1982) and Passey (1983) shows various smooth plains, ridged plains and cratered terrains (Fig. 9.9). *Cratered terrain* is characterized by abundant craters 10–20 km in diameter and is subdivided into two units – in one, the large craters are shallow (perhaps as a result of viscous deformation), while the second is younger terrain in which the craters are not flattened and which may reflect a change in the properties of the lithosphere to a more brittle material. Passey (1983) has carefully analyzed crater profiles to infer properties of the lithosphere of Enceladus and to model the evolution of the interior. He concluded that Enceladus's lithosphere must have a mixture of ammonia ice and water ice, in contrast to the relatively pure water ice of the Galilean satellites.

Cratered plains, centered at 30°N, 345°W, are characterized by bowl-shaped craters 5–10 km in diameter; *smooth plains* are lightly cratered and display a rectilinear pattern of grooves in some areas; *ridged plains* are characterized by complex, subparallel ridges up to 1 km high with interspersed smooth plains. Much of the ridged plains terrain occurs on the trailing hemisphere and appears to have formed at the expense of older cratered terrain.

The landforms seen on Enceladus exhibit tectonic and resurfacing processes unexpected for such a small object. The various episodes of resurfacing probably involved extrusion of water or water ice slurries. The grooves are considered to be grabens resulting from extension and brittle failure of the lithosphere. The ridges, which are similar to those on Ganymede, could result either from compression, from upwelling by convection from the interior, or from expansion of freezing water which intruded fractures.

What could have generated the tectonic deformation and heat to release water? Steve Squyres and colleagues (1983) have analyzed the problem, taking into account the potential sources of heating. Because of Enceladus's small size, neither primordial heat generated by accretionary impact nor radionuclide heating could account for its activity – Enceladus would have frozen solid very early in its history. It is unlikely that there are sufficient radioactive elements unless the satellite's composition is extremely unusual in comparison to known Solar System elemental abundances. The most likely source is tidal heating, similar to that which affects Io. Enceladus has an orbital eccentricity forced by a resonance with Dione. This eccentricity would "push–pull" the satellite and would, in turn, generate tidal heating. Although Squyres *et al.* (1983) conclude that greater eccentricities than present values are required to melt an initially frozen body, especially if pure water ice is involved, they note that the inclusion of hydrates of ammonia or a methane clathrate could lower the melting temperature substantially (by 100 °C or more).

It has even been suggested that Enceladus may experience active volcanism. Although there is no direct evidence of eruptions, it has been noted that the densest part of Saturn's E Ring is coincident with the orbit of Enceladus (Fig. 9.1), suggesting that the satellite could be the source of the material which makes up the ring (Smith *et al.* 1982).

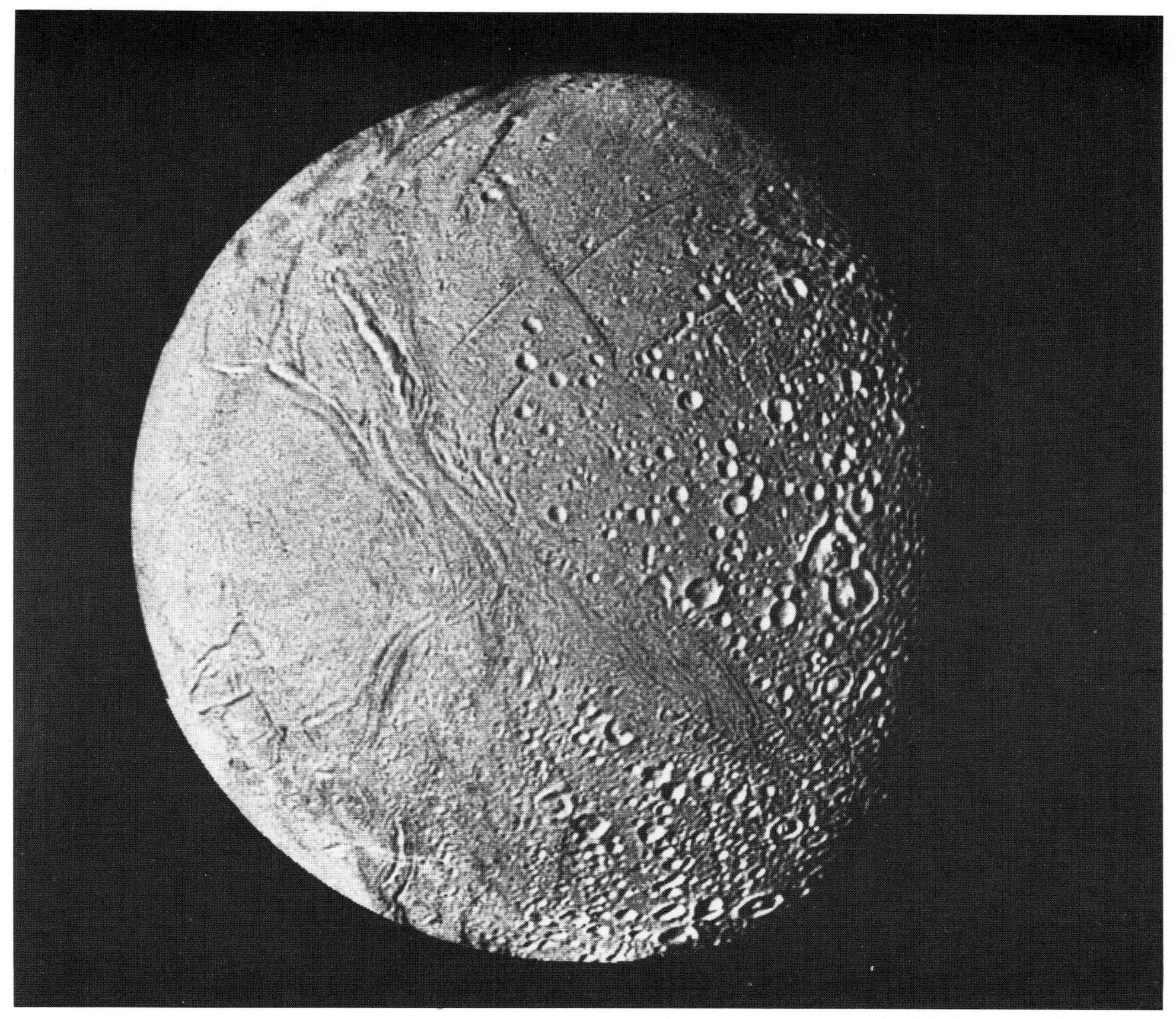

Figure 9.6 Image of Enceladus showing cratered terrain, grooved terrain, and smooth terrain; note the lateral offset along a long fracture. The smooth terrain at the subsolar point (left side) may be an illusion due to the effects of illumination (Voyager 2 JPL P-23956).

9.2.3 Tethys

Tethys is more than twice the diameter of Enceladus and displays landforms indicative of tectonic deformation and resurfacing, although these features are not as extensive as on Enceladus. The two most striking surface features of Tethys are *Odysseus*, an impact crater 400 km in diameter (about 40 percent of the diameter of the satellite), and an enormous canyonland, *Ithaca Chasma*, which wraps three-quarters of the way around the globe (Fig. 9.8). The general geology of Tethys has been described by Moore and Ahern (1983) who mapped the major terrains (Fig. 9.10). *Hilly cratered terrain* is the oldest recognized unit and is characterized by rugged topography and a high frequency of craters >20 km in diameter, most of which are degraded. *Plains terrain* is centered at -10° latitude, 310° longitude, coincident with the trailing hemisphere. It consists of less rugged, more sparsely cratered surfaces in comparison with the hilly cratered terrain. In addition, plains terrain displays faint lineations and variations in its albedo.

The crater Odysseus is equal in size to Mimas and occurs on the leading hemisphere. It is a relatively shallow feature with a ring-like central peak complex and a floor that matches the planetary curvature (Fig. 9.11). Smith *et al.* (1982) note that the low relief of Odysseus may be a consequence of viscous deformation of an icy lithosphere. Soderblom and Johnson (1983) propose that the impact occurred when the interior of Tethys was partly liquid or still "soft" (plastic-like) ice. Moore and Ahern (1983) note that only about 1–2 per-

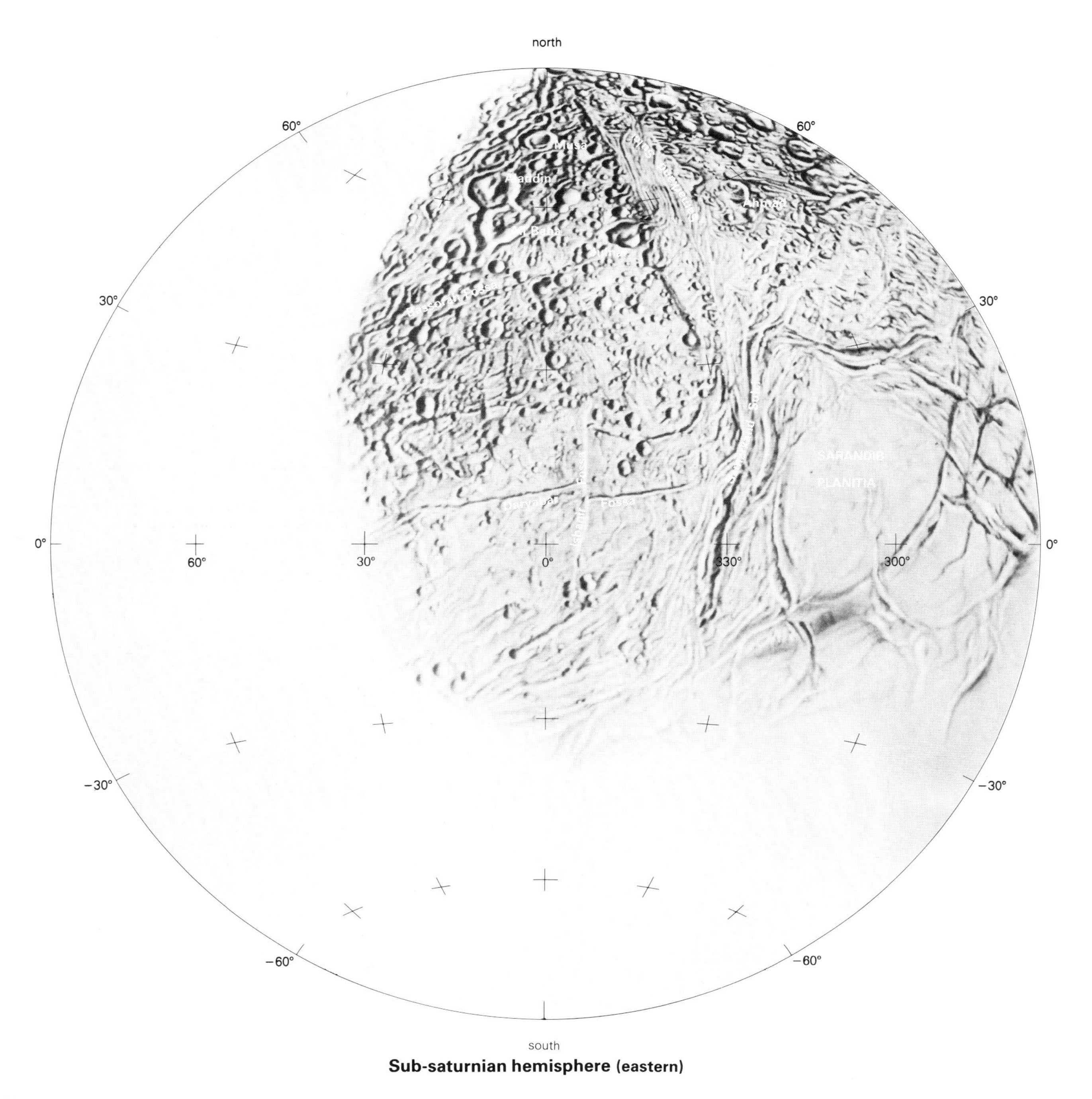

Figure 9.7 Shaded airbrush relief map of Enceladus (courtesy US Geological Survey).

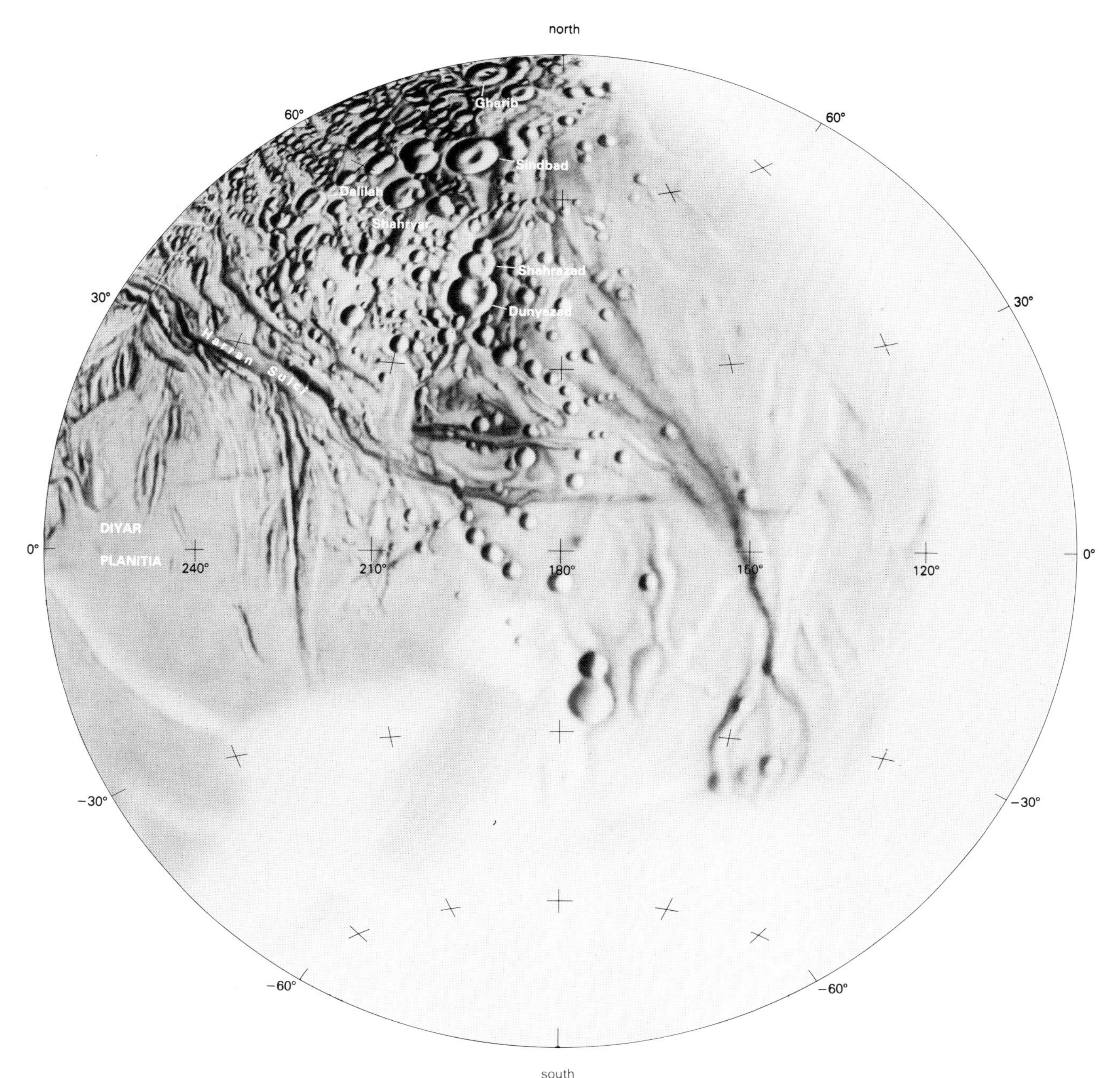

Anti-saturnian hemisphere (western)

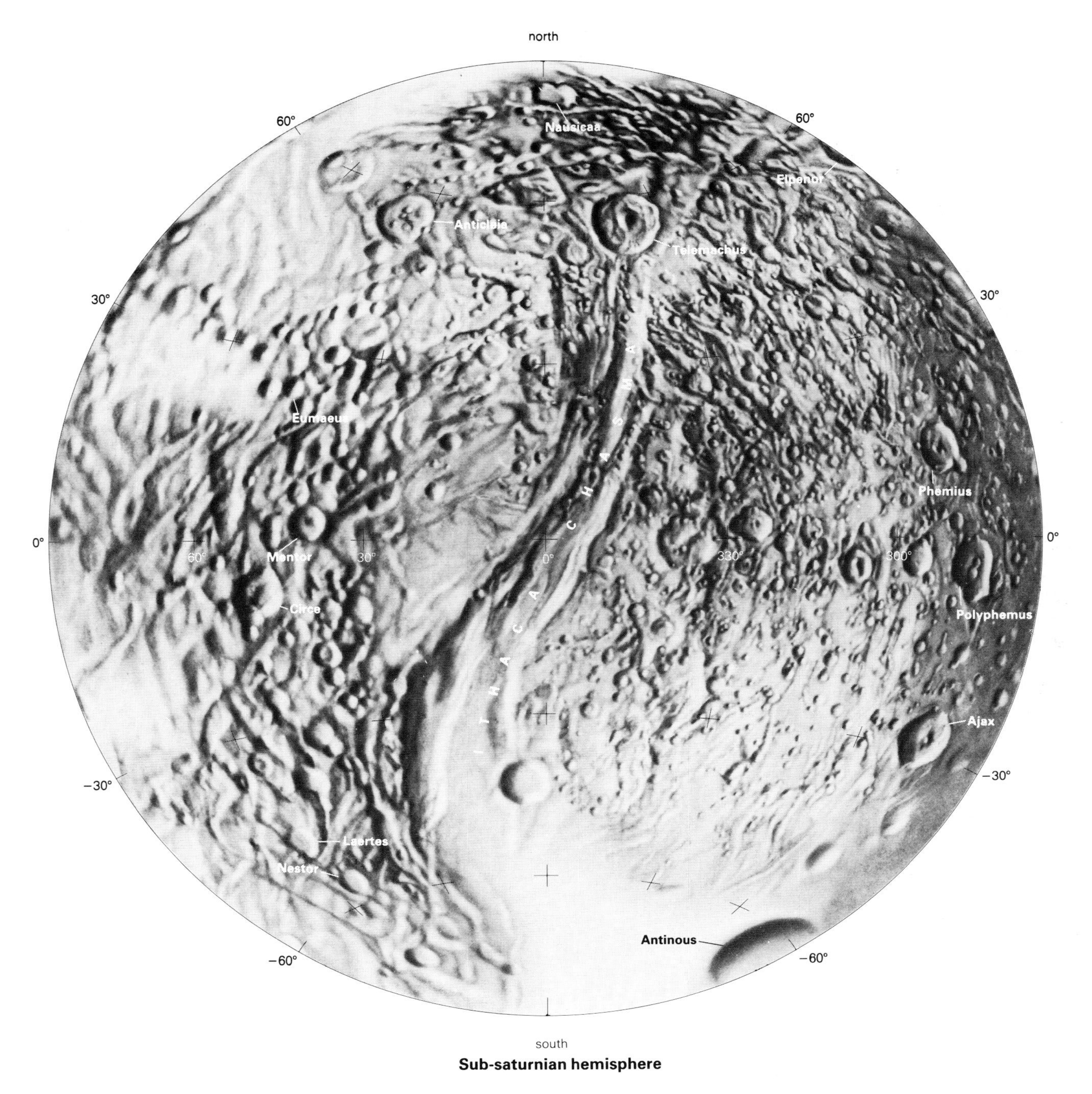

Sub-saturnian hemisphere

Figure 9.8 Physiographic map of Enceladus (from Passey 1983, copyright © 1983 Academic Press).

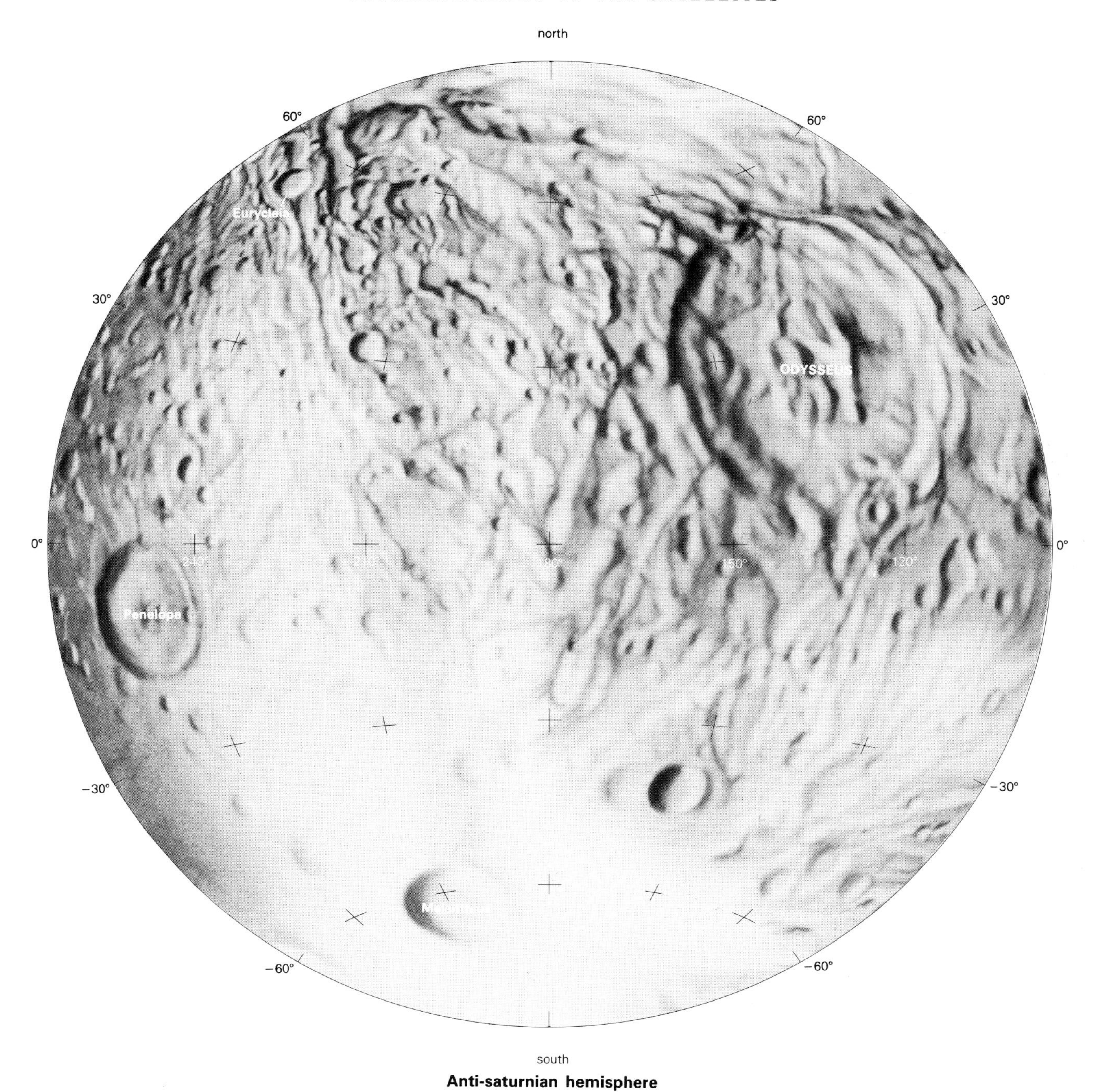

Anti-saturnian hemisphere

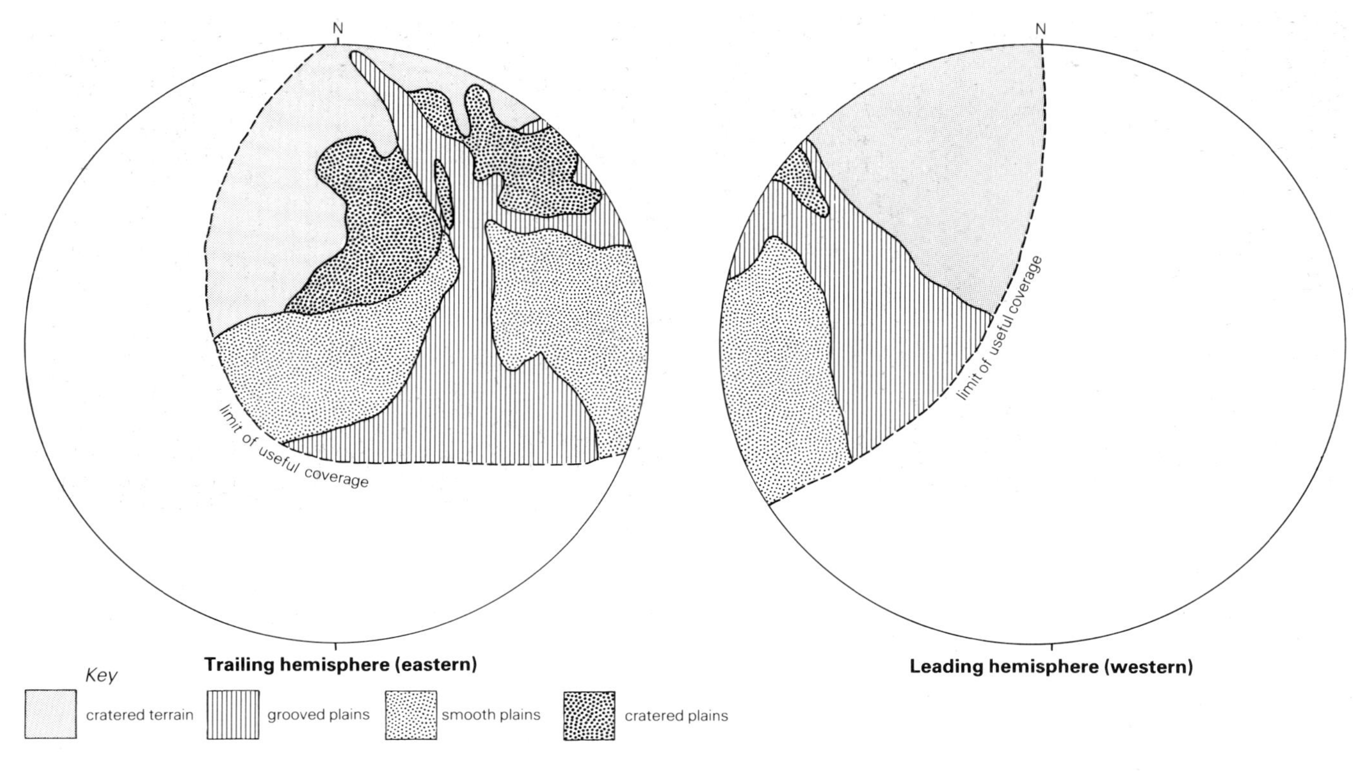

Figure 9.9 Shaded airbrush relief map of Tethys (courtesy US Geological Survey).

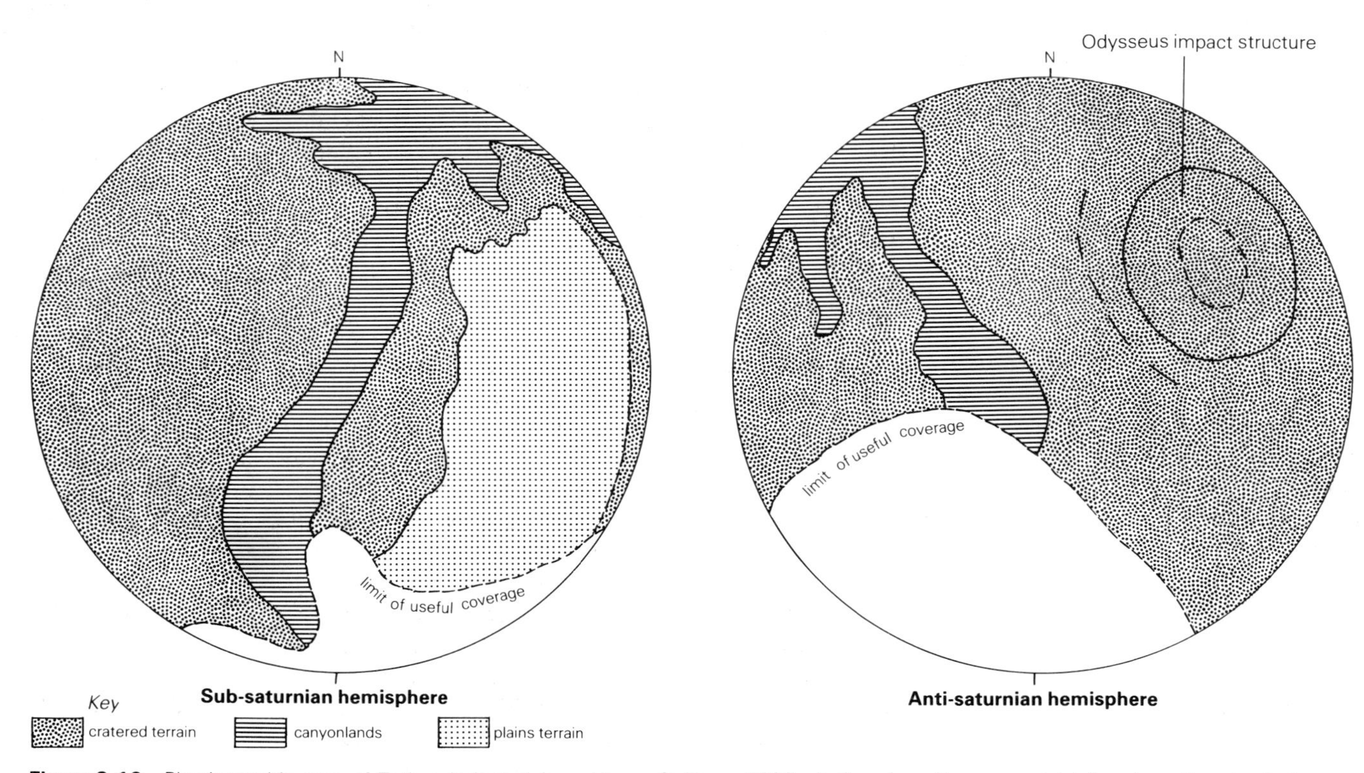

Figure 9.10 Physiographic map of Tethys (adapted from Moore & Ahern 1983, *J. Geophys. Res.*, copyright American Geophysical Union).

Figure 9.11 Views of Tethys showing impact crater Odysseus. The floor of Odysseus matches the curvature of the satellite (Voyager 2).

cent of the ejecta from Odysseus would have escaped the satellite to be placed in orbit about Saturn, and that much of the material would be swept up later by Tethys.

An impact of the size to produce Odysseus would have had a substantial effect on such a small satellite. Seismically induced degradation in the antipodal zone of the moon may have generated terrain similar to the hilly and lineated terrain antipodal to the Caloris Basin on Mercury. However, Moore and Ahern note that the smooth plains – which evidently post-date Odysseus – occur in the area antipodal to Odysseus and would have buried the seismically jostled terrain, even if it had formed.

Ithaca Chasma is a branching, terraced canyon system more than 1000 km long, 100 km wide and up to 4 km deep (Figs 9.12 & 9.13). Parts of the canyon rim are raised 500 m above the surrounding terrain. As mapped by Moore and Ahern (1983), many lineaments splay off Ithaca Chasma, forming V-shaped patterns pointing northward. The canyonland may have been created as some sort of lithospheric response to the impact that formed Odysseus. It has been observed by Smith *et al.* (1982) and Moore and Ahern (1982) that the canyonland comprising the Ithaca Chasma complex generally lies on a great circle which is perpendicular to a radius-line from the center of Odysseus (Fig. 9.14). Moore and Ahern (1983) invoked seismic disruption from the Odysseus impact as the rifting mechanism that formed the canyonland. They noted that if the Odysseus impact caused Tethys to become temporarily oblate, this could induce near-surface tensional fracturing along a narrow region that is presently occupied by the canyonland. Alternatively, models of the thermal evolution of Tethys which begin with freezing of water shows that the volume would expand ~10 percent, which equals an increase in surface area of ~7 percent. Tectonics of this magnitude could easily account for the extension to form Ithaca Chasma but this model fails to explain why the canyonland occurs only in a narrow zone.

9.2.4 Dione

Dione, discovered in 1684 by Cassini, is about the same size as Tethys but is considerably more dense. At 1.43 g cm^{-3}, it is second only to Titan in density, indicating a higher proportion of rocky material in comparison to the other icy satellites. Dione also shows a wide range of albedo patterns, including bright, wispy markings (Fig. 9.18). Geologic mapping by Plescia (1983) and Moore (1984) shows that Dione can be classified into cratered terrain and various plains units (Fig. 9.15). *Heavily cratered terrain*, consisting of rough surfaces having numerous craters >20 km in diameter, has the greatest areal extent of the region imaged by Voyager. *Cratered plains* have a lower frequency of large craters in comparison to heavily cratered terrain, whereas *smooth plains* have very few craters or other topographic features.

The trailing hemisphere of Dione is distinguished by a network of intensely bright streaks set on a dark background. These wispy features are composed of, or pass into, narrow bright lines (Fig. 9.16). In addition an elliptical ~240 km feature, named Amata, occurs in the center of the complex and may be an impact scar. The bright wispy marks, however, do not resemble bright rays from impact craters but, rather, follow irregular paths. Smith *et al.* (1981) suggest that these markings could represent deposits of frost-like material formed by the explosive release of volatiles from the interior through fractures.

The largest craters on Dione occur on the trailing hemisphere (Fig. 9.17) and are as large as 200 km across. Craters >100 km in diameter are common in the

Figure 9.12 Voyager view of Tethys showing Ithaca Chasma (arrow), a plains unit (lower right) and heavily cratered terrain (Voyager 2 JPL P-24065).

Figure 9.13 Voyager view of Tethys showing a low albedo strip running north–south across the plains unit in the center of the image. Ithaca Chasma is visible near the terminator (Voyager 2 JPL P-23948).

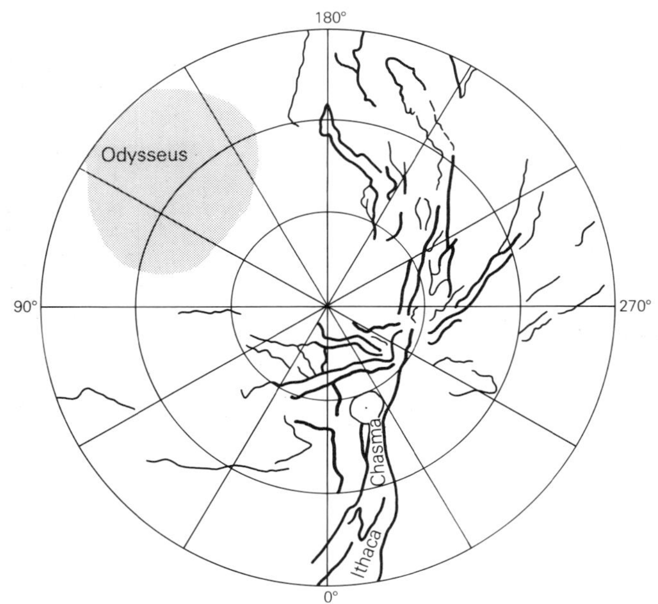

Figure 9.14 Diagram of north polar region of Tethys showing the relation of Odysseus to various lineaments (from Moore & Ahern 1983, *J. Geophys. Res.*, copyright American Geophysical Union).

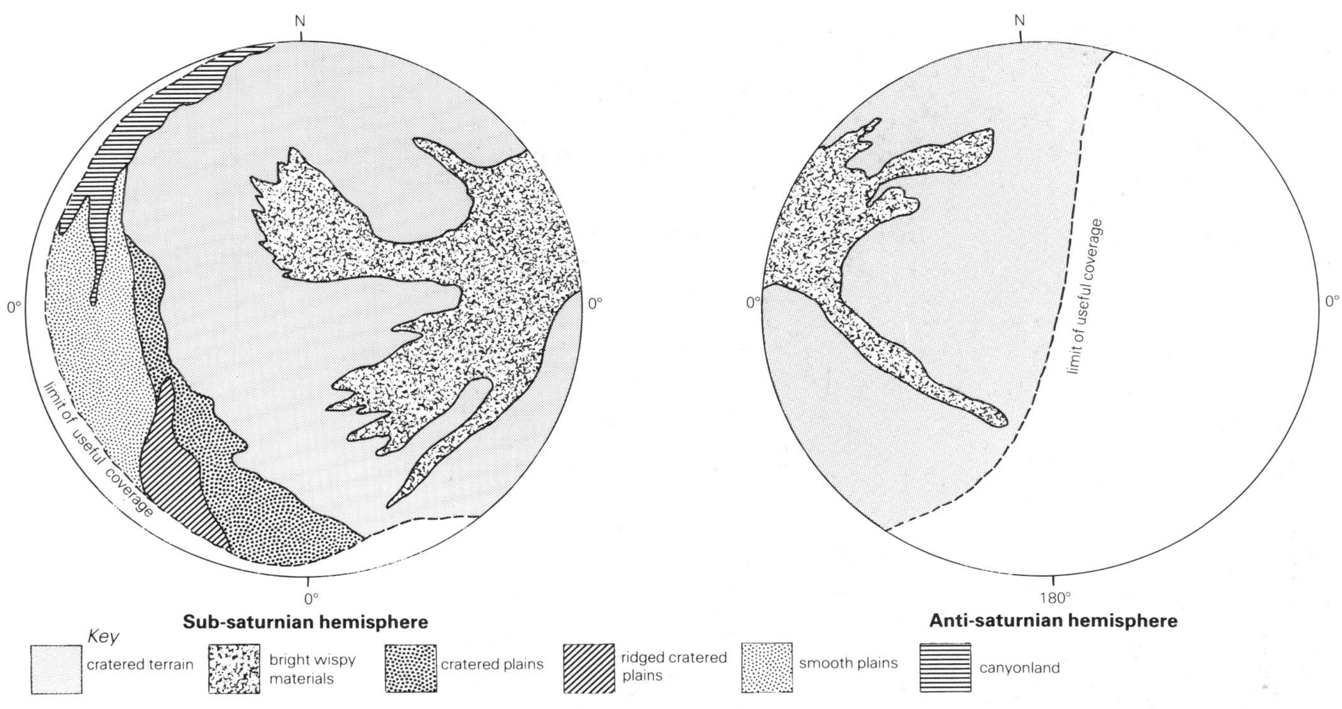

Figure 9.15 Physiographic map of Dione (modified from Plescia 1983, copyright © 1983 Academic Press).

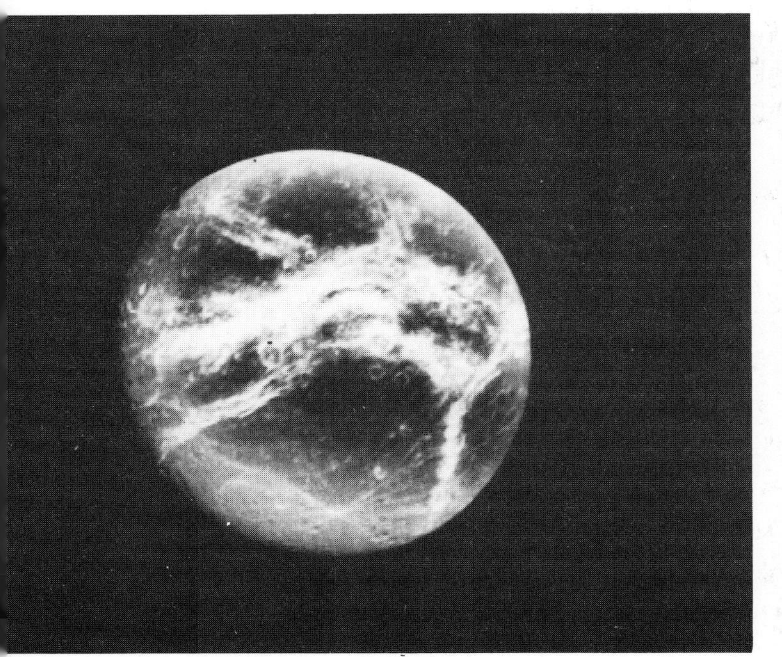

Figure 9.16 Voyager image of the trailing hemisphere of Dione showing bright, wispy streaks crossing the trailing, low-albedo hemisphere (Voyager 1 1182S1-001).

Figure 9.17 Voyager image of the trailing hemisphere of Dione showing heavily cratered terrain, the apparent association of bright wisps with troughs, and the parallel groups of some of the bright wisps related lineaments (Voyager 1 6251 + 000).

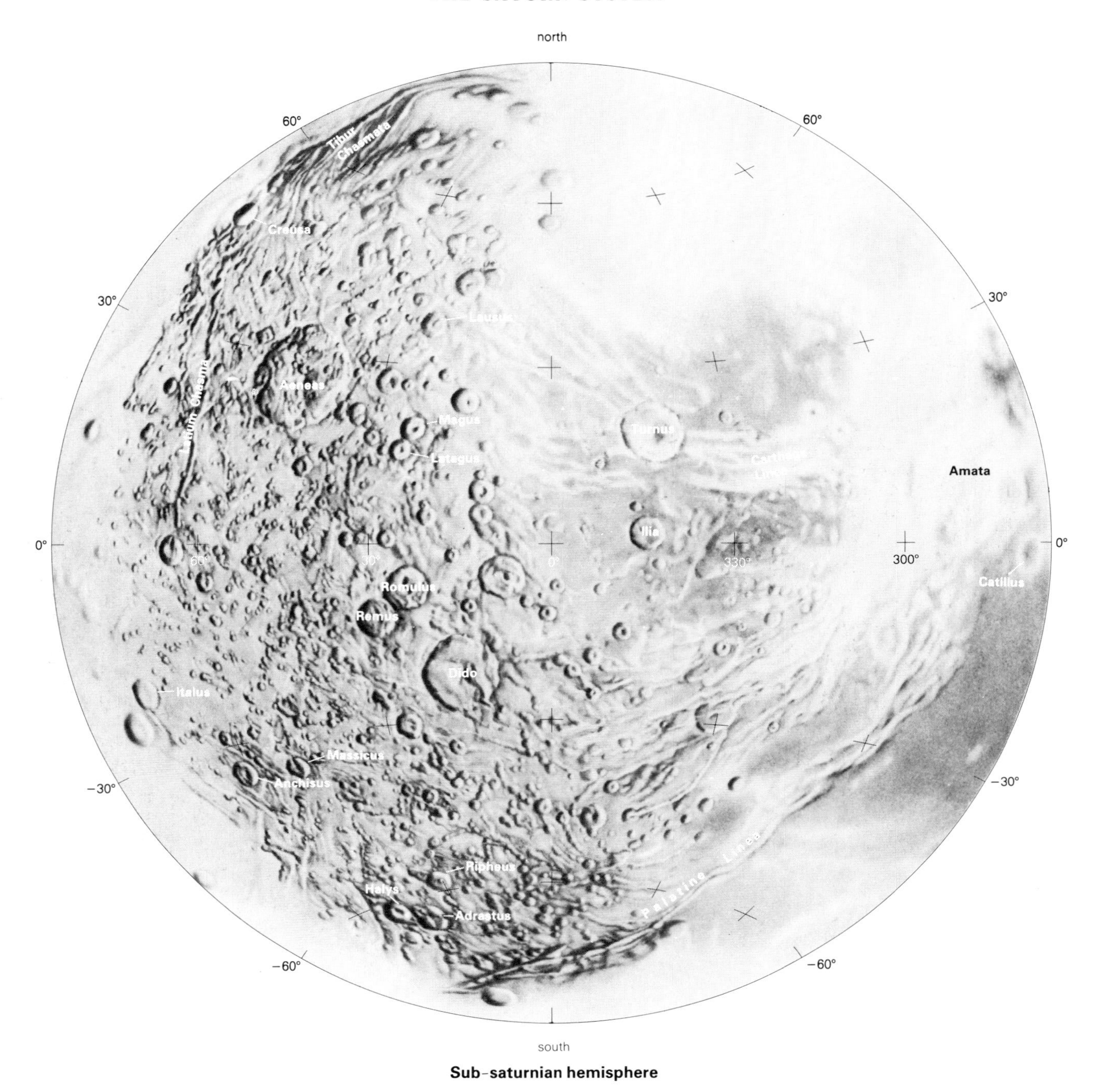

Sub-saturnian hemisphere

Figure 9.18 Shaded airbrush relief map of Dione (courtesy US Geological Survey).

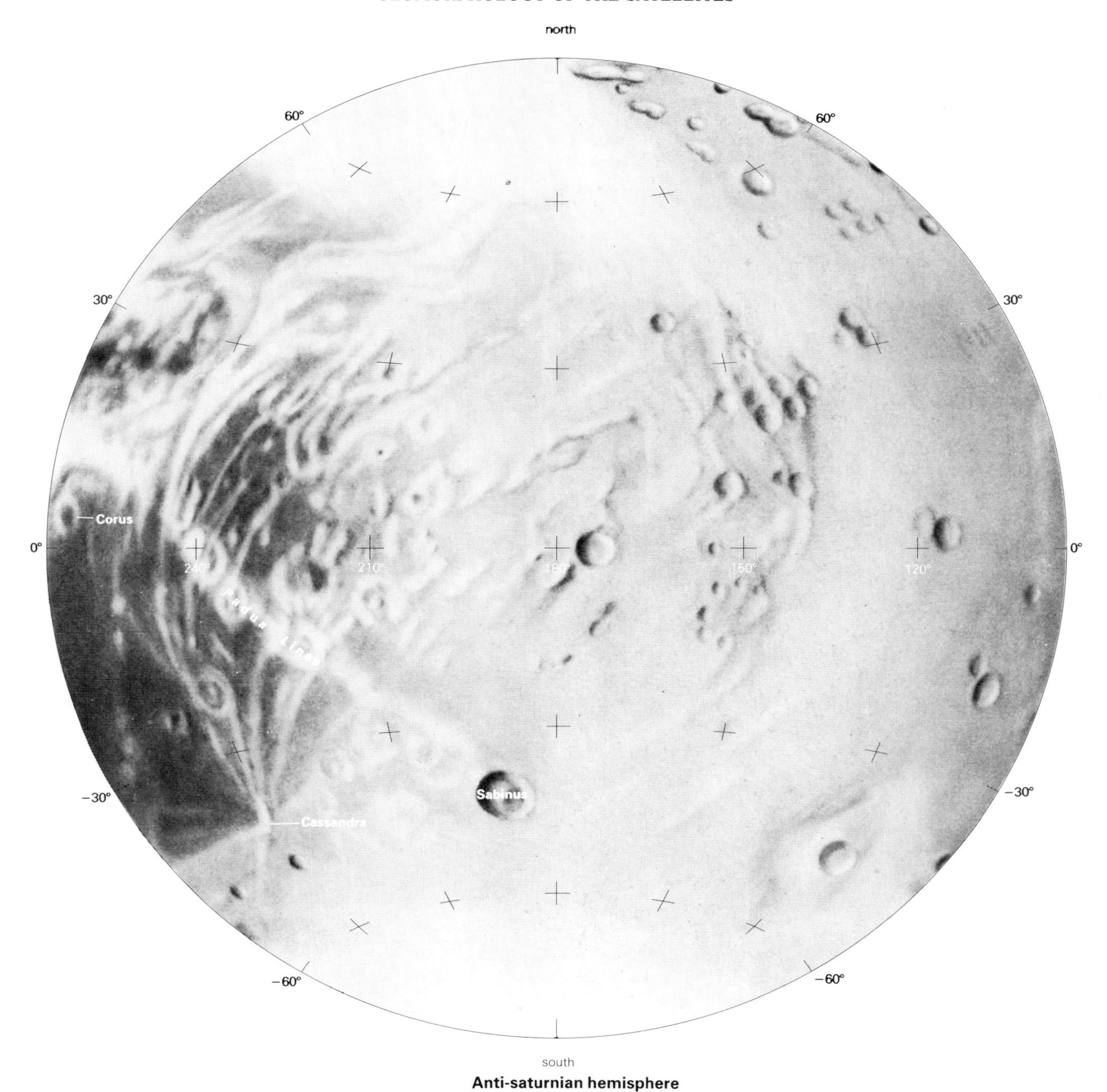

Anti-saturnian hemisphere

Figure 9.19 Voyager image of subsaturnian hemisphere of Dione showing troughs and trough networks in the northern hemisphere.

Figure 9.20 Voyager 2 image of Dione showing area beyond the terminator of Figure 9.19. Note the westward extension of the plains and the only unequivocally identifiable ray crater (Creusa) seen on a saturnian satellite (upper portion of disk) (Voyager 2 1239S2-001).

heavily cratered terrain. Most of the large craters have terraced walls and central peaks (Figs 9.19 & 9.20), while many of the smaller (< 15 km) craters are simple, bowl-shaped depressions. As is the case for all the saturnian satellites, clearly defined ejecta deposits (other than crater rays) are lacking, although this could be the result of poor image resolution. In general, craters are more shallow on Dione than on Tethys.

Tectonic features on Dione include ridges, scarps, troughs and some crater chains, as described by Jeff Moore (1984). Ridges are 50–100 km long and < 0.5 km high. These features were mapped by Plescia (1983) as lobate deposits, which he suggested to be similar to martian ejecta flow deposits, although a source impact crater was not identified on Dione. Scarps up to 100 km long occur in the heavily cratered terrain. These may be the result of faulting or they could be landforms associated with mass wasting. Troughs range in length from 30–100 km, but some exceed 500 km, and most have a pit crater at one or both ends (Fig. 9.19). Troughs may form networks, as in the smooth plains, or complex branching systems. Chains of craters, found principally on the plains, are as long as ~ 100 km and 10 km wide and may be secondary impact craters, volcanic craters or aligned pit craters of internal origin, such as collapse.

Moore (1984) analyzed the landforms on Dione and proposed the following evolutionary sequence. After the formation of a brittle lithosphere following accretion, Dione experienced global expansion resulting from heat generated by radionuclides and tidal stresses. This expansion produced a pattern of global lineaments prior to the end of heavy meteoritic bombardment. Subsequently, an ammonia–water melt was produced in the interior and was erupted onto the surface to form plains. Cooling of the interior and/or a phase change led to compression of the surface, resulting in thrust or high-angle reverse faults which formed ridges. Later events involved light impact cratering, primarily by cometary objects.

9.2.5 Rhea

Voyager 1 passed within 59 000 km of Rhea and returned the highest-resolution image available from any of the saturnian satellites, about 500 m/pixel for part of the north polar region (Fig. 9.21). Except for Titan, Rhea is the largest of the satellites, having a diameter of 1530 km. Like Dione, its trailing hemisphere is dark and has bright wispy markings (Fig. 9.22), whereas the leading hemisphere is uniformly bright, except for one large, diffuse, very bright feature thought by Smith *et al.* (1981) to be ejecta deposits. Cratered terrains are dominated by large, degraded craters and resemble the lunar highlands. None of the craters shows the "flattening" to the degree of those on Ganymede. Most of the craters ⩾25 km have central peaks and some craters show very bright patches on the walls, which Smith *et al.* (1981) attribute to fresh ice deposits or other fresh materials exposed by slumping (Fig. 9.23).

The general surface features on Rhea (Figs 9.24 & 9.25) have been described by Moore and Horner (1984) who note the presence of a ~ 450 km multi-ring basin (Fig. 9.26). The outer ring appears to consist of two concentric, inward-facing scarps, with the inner scarp being much more prominent and the outer scarp being discontinuous in places, similar to Odysseus on Tethys. The spacing of the scarps is ~ 10–15 percent of the diameter of the basin. The inner ring is slightly more than half-way between the center of the basin and the inner scarp and is composed of ridges or hills rather than scarps. The outer rim scarps may be analogous to mega-terraces, as proposed for some multi-ringed basins on the terrestrial planets.

Richard Pike and Paul Spudis (personal communication) have discovered a second, very degraded basin which may have as many as seven rings, centered at ~ 30°N, 315°W and more than twice the size of the basin discussed by Moore *et al.* (1985). The center of the basin is in the resurfaced terrain. Subsequent impact cratering of this region has muted the topography, rendering the inner rings into subtle ridges with gentle slopes.

Figure 9.21 Mosaic of Voyager images for north polar region of Rhea showing both ancient cratered terrain (on which occurs coalescing pits and linear troughs) and resurfaced terrain dominated by smaller, population II craters (and on which scarps and ridges can be seen) (Voyager 1 JPL P-23177).

Figure 9.22 View of Rhea showing differences between the leading (bright) hemisphere and trailing (dark) hemisphere (Voyager 1 1558S1-001).

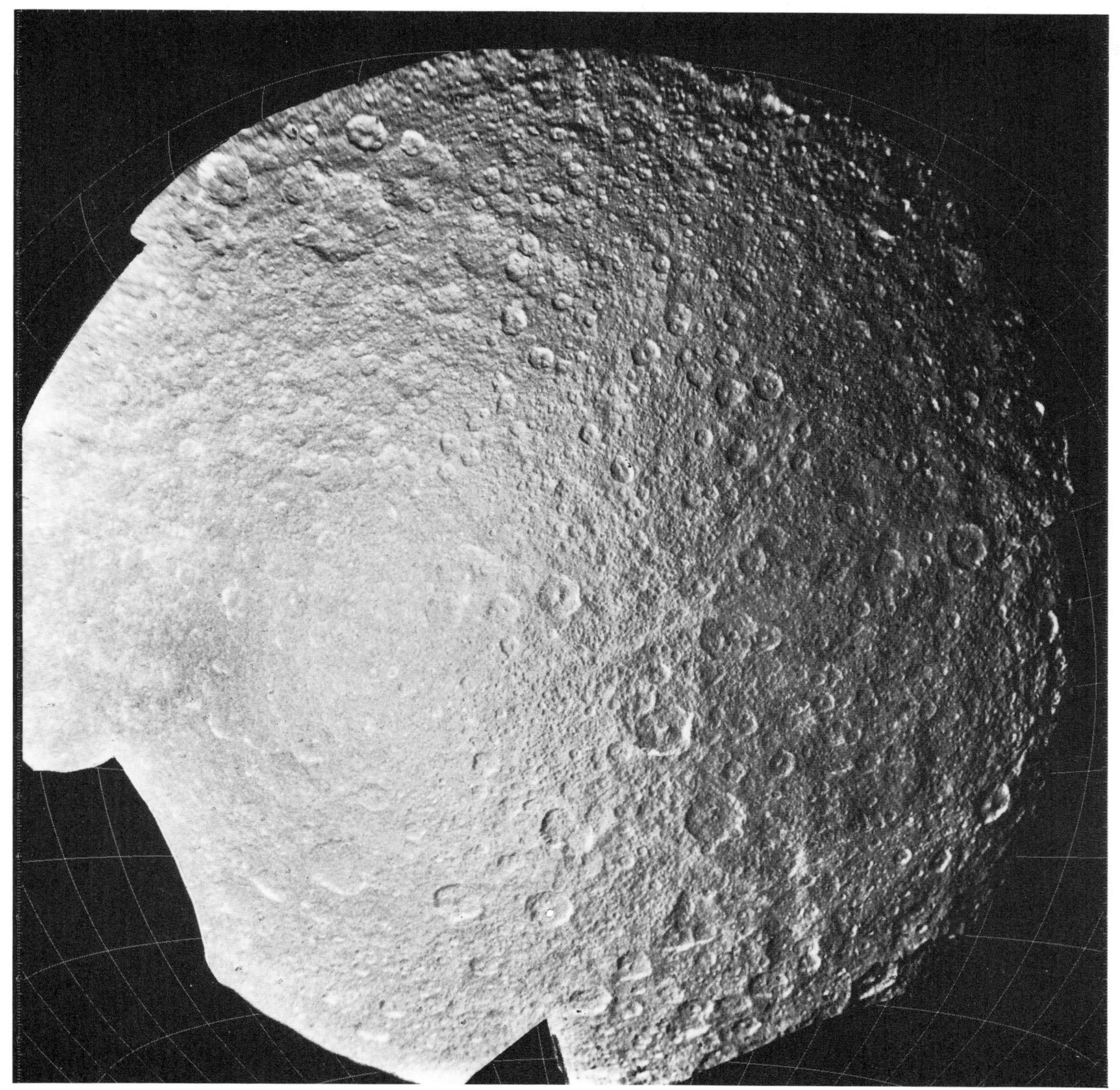

Figure 9.23 Mosaic of Rhea showing the subsaturnian hemisphere along the equatorial band and resurfaced terrain along the terminator in the northern hemisphere.

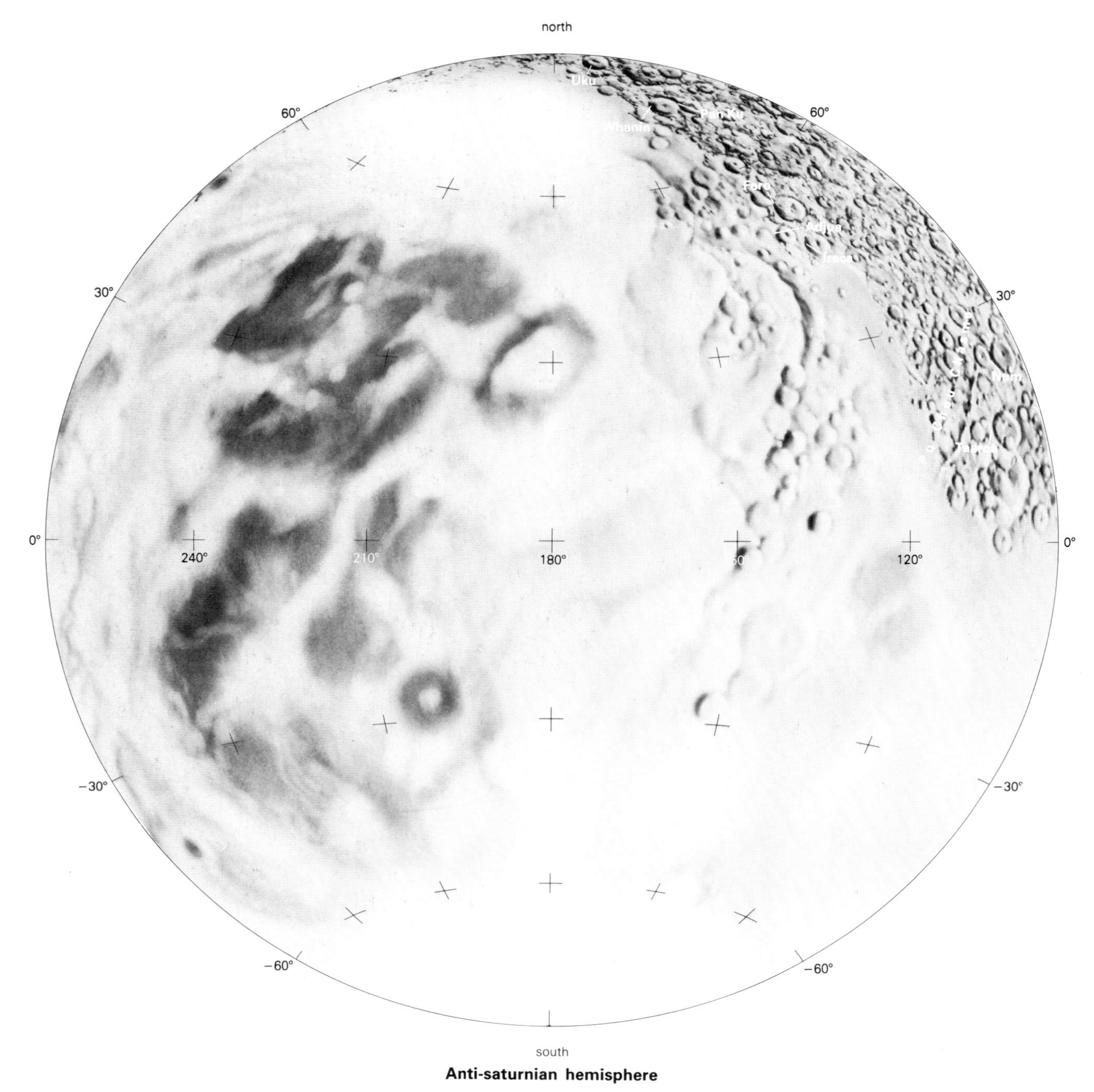

Anti-saturnian hemisphere

Figure 9.24 Shaded airbrush relief map of Rhea (courtesy US Geological Survey).

Sub-saturnian hemisphere

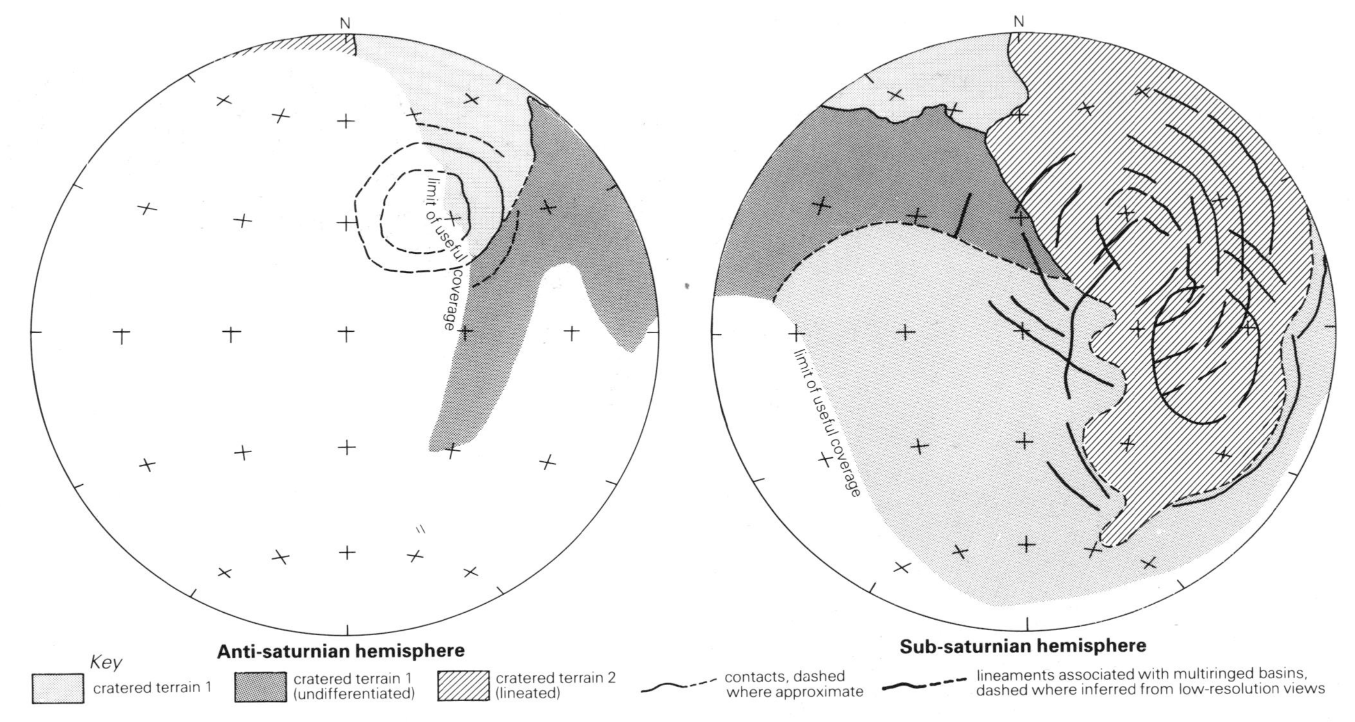

Figure 9.25 Physiographic map of Rhea (from Moore *et al.* 1984).

Some resurfaced terrains, such as those centered at 25°N, 310°W, may represent flooding by water "magmas". Parallel, N 30°E-striking, sharp-crested ridges or scarps occur in resurfaced terrain between 270°W and 360°W longitude, especially near the crater Pedn. They are ~5 km wide, up to several hundred kilometers long, and their relief averages <0.25 km. Some ~20 km craters are superimposed on the ridges or scarps, whereas other craters are transected by ridges. The rim of Pedn shares a common scarp with two lineaments on its northwest and southeast extensions, implying a common subsurface control of the scarps or ridges and the crater.

Parallel linear troughs or coalescing pits oriented at N 45°E occur in older cratered terrain between 0° to 60°N and 60° to 120°W. These features are ~5–10 km wide and up to ~1000 km long. Shadow measurements indicate a depth of <0.5 km. Nearly all troughs and many of the coalescing pits show little or no indication of raised rims. Some of the pits have raised rims, implying either an impact origin or features that were sites of pyroclastic activity. Most troughs and pits are superposed on older impact craters. As they cannot be traced into younger cratered terrain, Moore and Horner (1984) infer that trough and coalescing pit formation continued beyond the end of heavy bombardment but ended prior to the resurfacing event which mantled the older craters. The troughs and coalescing pits may, therefore, be a consequence of an episode of extensional tectonics and volcanism occurring between the end of heavy impact bombardment and the resurfacing event.

9.2.6 Titan

Titan is a remarkable satellite. Exceeding the diameter of Mercury, at 5120 km it is in the same size class as Ganymede and Callisto and, like those satellites of Jupiter, its mass suggests a composition that is about 45 percent water ice and 55 percent rocky material. What makes it remarkable is the presence of a dense atmosphere. Although the satellite was the first to be discovered in the saturnian system (found by Huygens in 1655), the presence of an atmosphere was not suspected until the mid-1900s (as reviewed by Owen 1983). In 1944, the Dutch–American astronomer, Gerard Kuiper, identified methane in the spectrum of Titan; this initiated speculations on the various chemical reactions that might occur in the atmosphere.

The presence of extensive clouds on Titan was well established by the time the Voyagers arrived. Voyager investigators hoped that the clouds would be broken and sufficiently scattered to allow glimpses of the surface. Such was not the case, and the Voyager pictures are disappointingly bland, showing a relatively uniform orange sphere with only hints of variations in the clouds. Limb views, however, show layers in the atmosphere (Fig. 9.27), which are thought to be smog-like photochemical hazes. It is likely that very little sunlight penetrates the clouds, and the surface must be a frigid, gloomy place.

The non-imaging experiments on board Voyager yielded results that show Titan to be a fascinating object. Nitrogen was found to be the principal component in the atmosphere, and Voyager also discovered ethane,

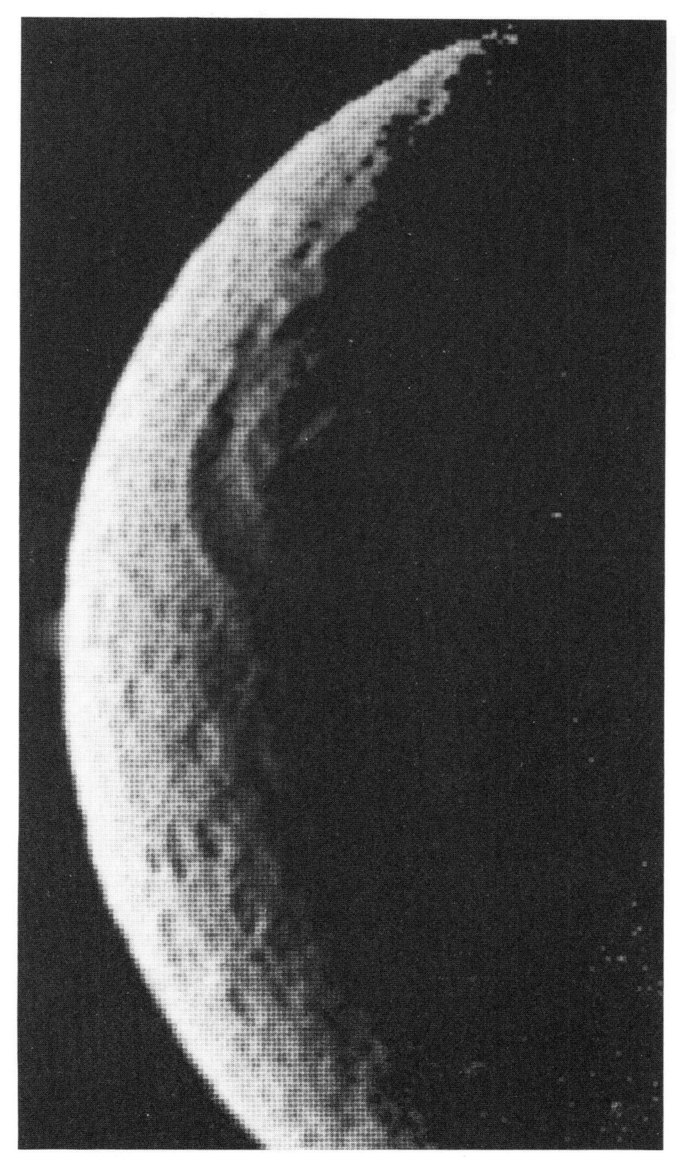

Figure 9.26 Multi-ringed basin about 450 km across found on Rhea (Voyager 1 1152S1 + 009).

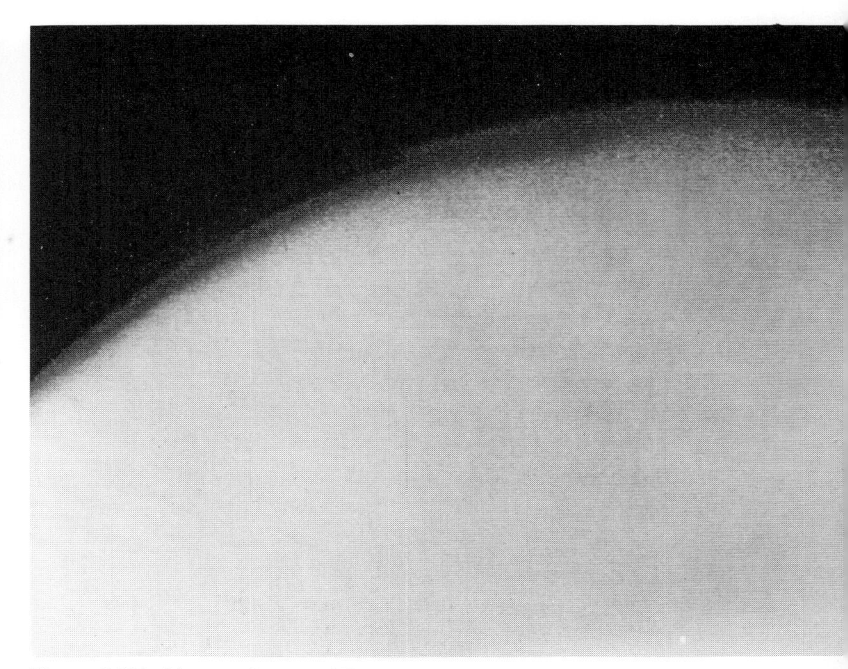

Figure 9.27 Voyager image of Titan showing structure within the atmosphere considered to be photochemical hazes (Voyager 2 JPL P-24067).

acetylene, ethylene, hydrogen cyanide and various other carbon–nitrogen components. It is thought that when Titan first formed, methane and ammonia ices were incorporated into the accreting mass. With heating and subsequent differentiation, these compounds were released from the interior and formed an early-stage atmosphere. The action of sunlight on the ammonia caused release of the nitrogen and escape of most of the hydrogen, while methane remained as a planet-wide ocean. In a 1983 conference, however, Lunine, Stevenson and Yung (personal communication) showed that methane which evaporates from the surface of this ocean would be photo-dissociated into compounds which would then recombine to form ethane and acetylene. These and other molecules would then condense and rain down onto Titan's surface.

Surface temperatures are about $-180\,^{\circ}$C, with perhaps a 3 $^{\circ}$C variation from the equator to the poles. This temperature, however, is near the triple point of ethane and methane, and it has been suggested that Titan may be covered with a liquid ethane–methane ocean to a depth of 1 km, which might have islands of solid H_2O ice rising from the ocean floor. Liquid ethane and methane may play the same role on Titan as water does on Earth, so far as surface processes are concerned. Sagan and Dermott (1982) have calculated that the present eccentricity of Titan's orbit raises large ($\sim$10 m high) tides in a global ocean that quickly erodes any surface above ocean-level, suggesting that if there is dry land, it would have to be newly formed.

We can only speculate what geologic events might take place on Titan, either at present or in the past. Fluid processes may modify solid parts of the surface and could involve rivers of liquid ethane–methane, tidal activity by an ocean, and perhaps even aeolian processes involving windblown ice particles. Whether these or other geologic processes, such as volcanism, tectonics or impact cratering, have occurred and left landforms is open to question. Similar to the situation for cloud-covered Venus, a radar-imaging system is required to view the surface of Titan; such a system has been discussed for some future spacecraft mission. In many respects, Titan resembles a primordial Earth, and the presence of organic compounds in a chemically rich atmosphere makes Titan a target of high priority, not only for atmospheric and geologic sciences, but for biology as well.

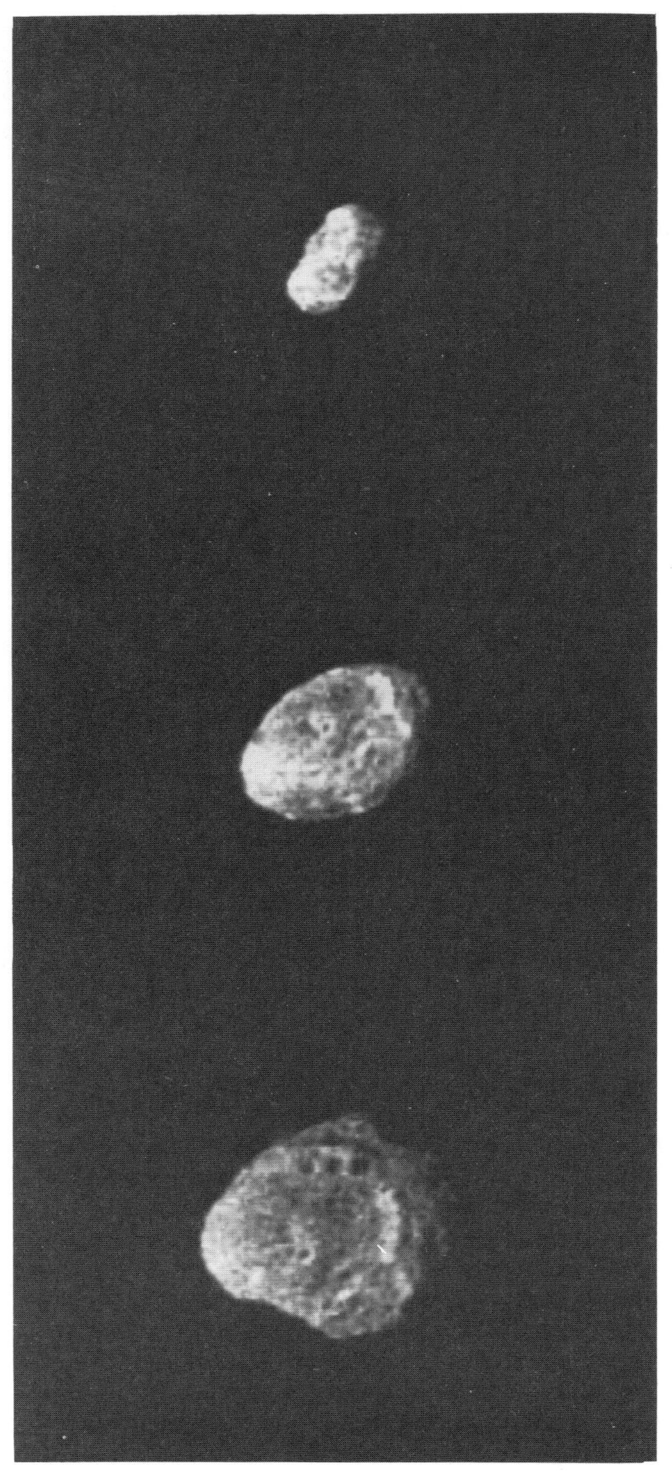

Figure 9.28 Voyager views of Hyperion showing irregular shape and the presence of large degraded craters and crater-related scarps (Voyager 2 JPL P-23932).

9.2.7 *Hyperion*

Hyperion is a small, irregularly shaped object whose elliptical orbit lies just outside that of Titan. Measuring 350 by 235 by 200 km, its shape includes both angular and rounded parts. Hyperion is the darkest object of the inner saturnian satellites and, although its mass cannot be estimated from Voyager data, it is presumed to be composed of a mixture of ice and rock.

Hyperion is heavily cratered. Large craters – some as large as 120 km in diameter with 10 km of relief – are scattered over the surface, while craters 10 km across are relatively abundant (Fig. 9.28). In addition, the surface displays segments of scarps which may be part of a connected, arcuate feature nearly 300 km long. This feature could be the remnant of a craterform that would have a diameter of >200 km.

Because of its irregular shape, heavily impacted surface and the possibility of an impact scar nearly half its size, Hyperion is considered by most investigators to be the remnant of a much larger object which was shattered by impact processes.

9.2.8 *Iapetus*

Iapetus is nearly the twin of Rhea in its diameter, but its mass suggests a much lower density, about 1.1 g cm^{-3}, and a higher proportion of ice than for Rhea. Since the satellite's discovery in the 17th century by Cassini, the albedo contrasts on Iapetus have long been recognized as distinctive. While the leading hemisphere is very dark, the trailing hemisphere is about an order of magnitude brighter and, as found on Voyager images, the boundary is fairly sharply defined (Fig. 9.29).

Surface features are difficult to see in the dark area (named *Cassini Regio*), but craters up to 120 km in diameter are clearly visible in the north polar area and in the bright terrain of the trailing hemisphere (Fig. 9.30), despite the relatively poor resolution of the images (at best, ~10 km). The large craters display central peaks and have well defined rims. Some of the craters in the bright terrain along the boundary with the dark terrain have floors covered with material that is dark and that has the same spectral properties as the dark terrain. In addition, a dark ring about 200 km across, which may be a multi-ring impact basin (Smith *et al.* 1982), was imaged by Voyager 1 (Fig. 9.31).

The dark material is about as black as tar or asphalt; its albedo is 3–5 percent. The only known materials of this darkness common in the Solar System are carbonaceous chondrites. Contact relations and the possible filling of some craters suggest that the dark material is a mantle, or surficial deposit. The origin of the dark material on Iapetus is somewhat controversial, and two general ideas, each with variations, have been proposed.

One model suggests internal sources (eruptions from the interior) and the other suggests origins by external sources. In either case, most investigators agree that the material seems to be a mantling deposit that is draped over bright terrain, as inferred from contact relations.

Because of the symmetry of the dark material in relation to the leading hemisphere, even before the Voyager mission Cook and Franklin (1970) proposed that impacts on the leading hemisphere could erode the ice components in the lithosphere and leave a lag deposit of dark, rocky materials. This idea requires a high proportion of rocky materials to ice, which does not fit the overall low density of the satellite. Steve Soter of Cornell University suggested at a conference in 1974 that Phoebe, the satellite in the next orbit outward from Saturn and a very dark object itself, was the source of dark material on Iapetus. Bombardment of Phoebe by impact cratering would eject debris which would fall inward toward Saturn. He suggested that some of this material would intersect the orbit of Iapetus and would be deposited on the leading hemisphere. The main weakness with this argument is that the spectral properties of Phoebe and the dark terrain on Iapetus do not match (Iapetus is redder), suggesting that they are not the same material.

An internal origin for the dark material was proposed by Smith *et al.* (1982). They observed that the bright terrain appears to be flooded by dark material because dark material occurs in topographically low areas, such as crater floors. In their model, materials were erupted onto the surface from the interior, perhaps as a carbonaceous-rich or organic-rich ice slurry.

Despite the additional information provided by the Voyagers, the origin of the dark material on Iapetus remains open to speculation.

9.2.9 Phoebe

Phoebe, the outermost saturnian satellite, is in a retrograde orbit inclined to the equatorial plane of Saturn. This atypical orbit suggests that Phoebe may be a captured object. Estimates of the mass of this ~220 km size satellite could not be made by Voyager because the spacecraft did not pass close enough to the satellite; however, its low albedo and general spectral properties suggest a surface consisting of carbonaceous material. Voyager images are very poor – about 20 km resolution at best – but show Phoebe to be approximately spherical and the darkest object in the saturnian system. Circular markings and indentations suggest the presence of craters.

Analysis of the Voyager color data and ground-based multicolor photometric results suggested to Voyager team members (reviewed in Smith *et al.* 1982) that Phoebe is similar to one class of asteroids that are common in the outer Solar System. Thus, Phoebe may be the first unmodified primitive object to be seen in the outer Solar System by a spacecraft.

9.2.10 Small satellites

In addition to the satellites described above, there are eight additional known satellites. All are rather small, ranging in size from 25 to 220 km (Fig. 9.32). Several have been officially named, but all of them still carry the designation that indicates the year of discovery (all were found in 1980, some of them from Voyager images).

Five fall within the orbit of Mimas (Fig. 9.1). Two (Janus, 1980S1, and Epimetheus, 1980S3) are called the co-orbital satellites because their orbits are within 50 km of each other. They are thought to be remnants of a larger object which was fragmented by an impact.

Two satellites, 1980S26 and 1980S27, are called "ring shepherds" because they bracket the F Ring and are thought to confine the orbits of the particles that make up the ring. The A Ring shepherd, Atlas (1980S28), is about 30 km across and one of the smallest known saturnian satellites.

The remaining three satellites, Calypso (1980S13), Telesto (1980S25) and the unnamed 1980S6, are known as the Lagrangian satellites. As reviewed by Morrison (1982b), it has long been known that a small object can have the same orbit and orbital speed as a larger object, so long as it maintains 60° of arc ahead or behind the larger object (these positions are called the Lagrangian points). Tethys has four such small objects in its orbit, including Calypso and Telesto; Dione has one, as shown in Figure 9.1.

9.3 Geologic processes

Collectively, the surfaces of the saturnian satellites display landforms indicative of impact cratering, tectonic deformation and processes associated with the release of fluids from the interior. Direct evidence for gradation is mostly lacking, probably owing to insufficient image resolution; however, mass wasting and impact erosion on a wide range of scales almost certainly occur now and have occurred in the past.

9.3.1 Impact cratering

As with the Galilean satellites, Ganymede and Callisto, impacts on the satellites of Saturn involve ice-rich targets. However, because the saturnian satellites are smaller and less dense (except for Titan), their internal temperatures will be much lower (Table 2.1), and the ice, because of its lower temperature, may behave more

Figure 9.29 Shaded airbrush relief map of Iapetus (courtesy US Geological Survey).

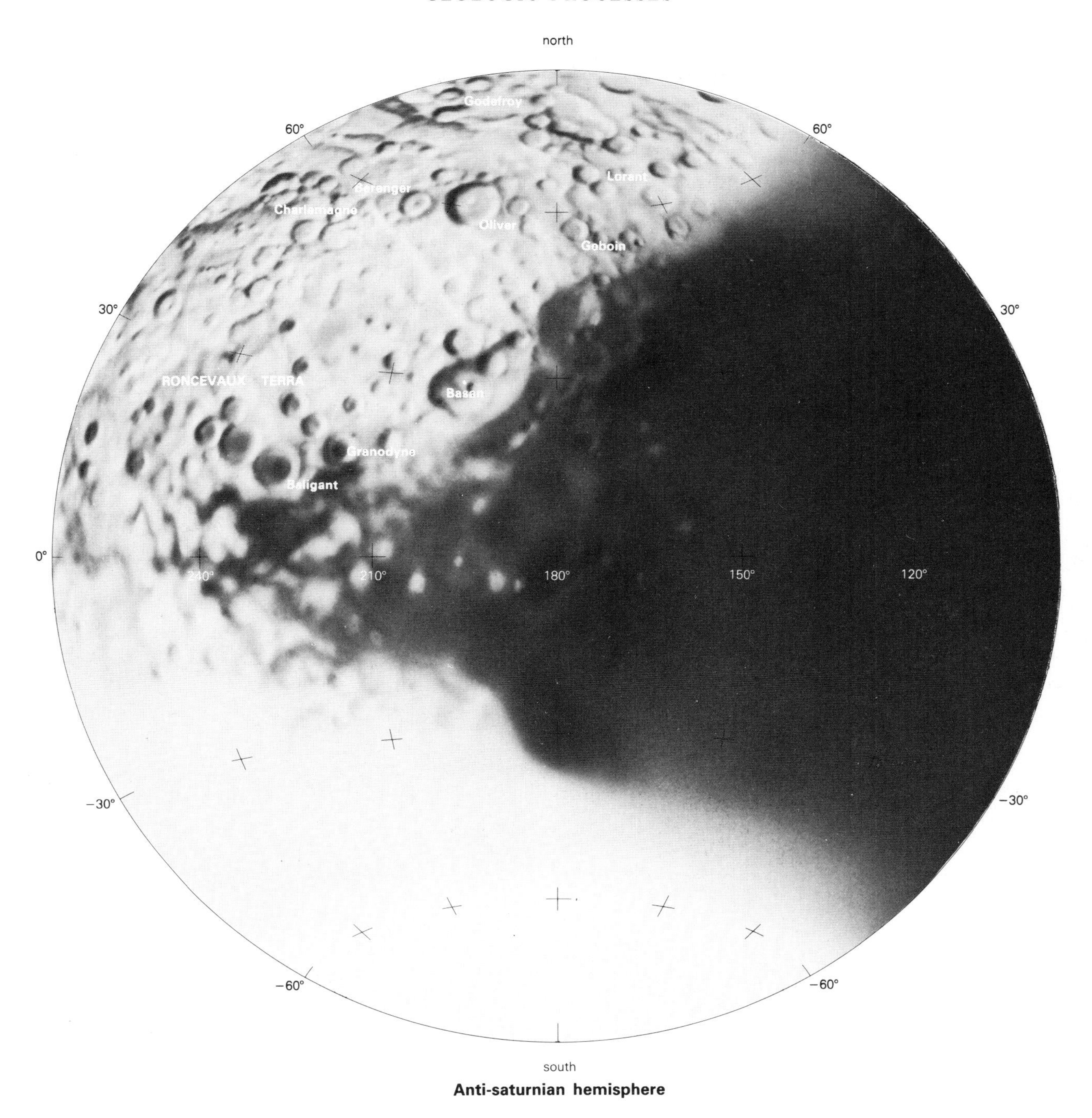

Anti-saturnian hemisphere

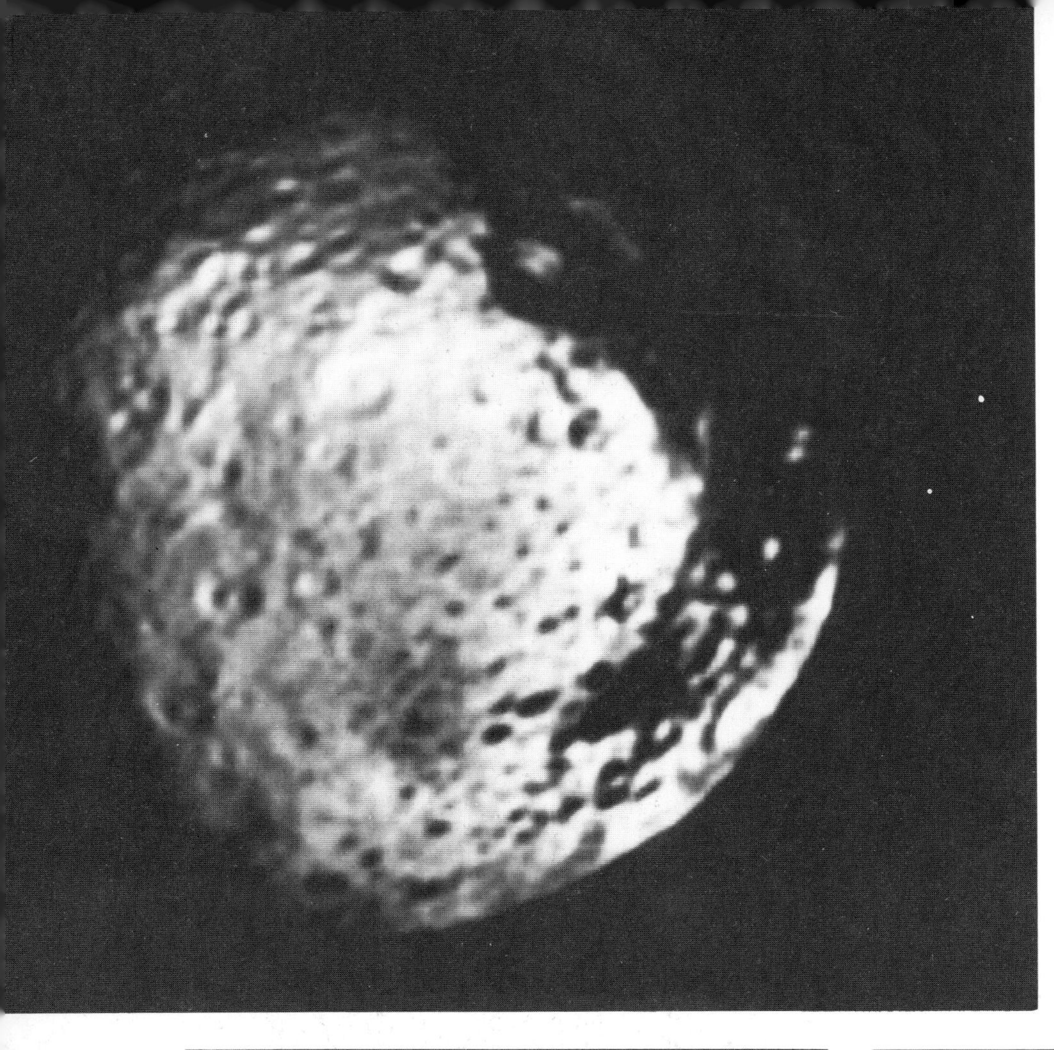

Figure 9.30 Voyager 1 view of the north polar region of Iapetus showing portions of both the bright and the dark hemispheres. The ring of dark material may indicate the presence of a large impact basin (Voyager 2 JPL P-23920).

Figure 9.31 Voyager 1 view of Iapetus showing the sharp contact between the two albedo units and the apparent "flooding" of crater floors with dark material (Voyager 1 JPL P-23105).

Figure 9.32 Mosaic of Voyager images showing some of the small saturnian satellites; upper left: 1980S26 (outer F-ring shepherd), 120 by 100 km; lower left: 1980S27 (inner F-ring shepherd), 145 by 70 km; upper right: 1980S1 (Janus), 220 by 160 km; lower right: 1980S3 (Epimetheus), 140 by 100 km (Voyager 2 JPL P-24061).

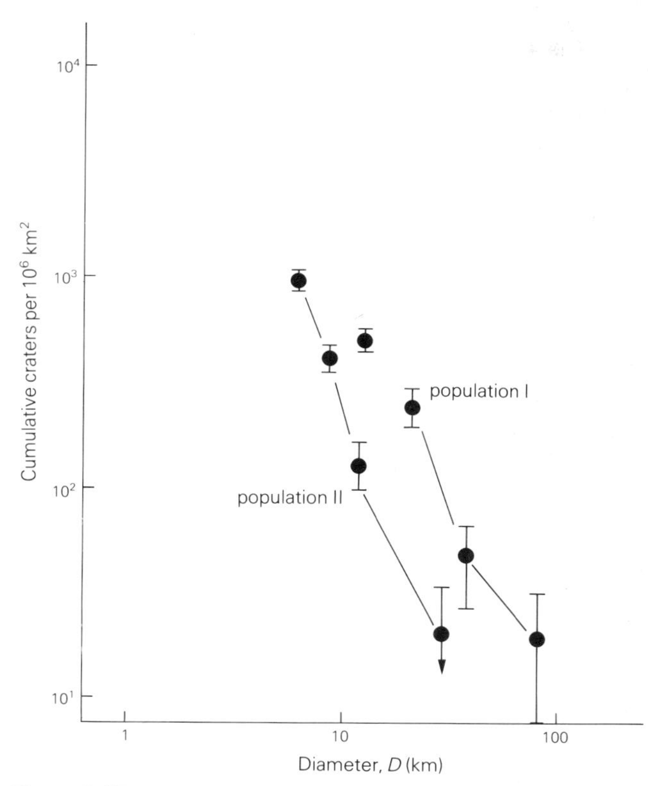

Figure 9.33 Impact crater frequency distributions for the light cratered terrain and heavily cratered terrain on Dione, difference reflecting population I (HC) and population II (LC) craters (modified from Plescia & Boyce 1982; reprinted by permission from *Nature* **295**, 3; copyright © 1982 Macmillan Journals Ltd).

like rock, both during impact and in the post-impact modification stages. Furthermore, the relatively smaller sizes of the saturnian satellites means that gravitationally driven viscous relaxation of topography would be less pronounced than on Ganymede and Callisto.

Most of the craters on the saturnian satellites are circular to polygonal in outline, are generally bowl-shaped, and have more or less lunar-like depth-to-diameter ratios. However, the 400 km Odysseus crater on Tethys, the two unnamed multi-ring basins on Rhea, and some of the craters on Enceladus appear to be topographically "relaxed". These craters are generally ancient in comparison to the other features on the respective satellites, and it is thought that they may have formed at times when the temperatures of the objects were higher, i.e. following the stage of accretion when impact-generated heat may have been retained.

Smith *et al.* (1981 & 1982) analyzed impact crater distributions on the saturnian satellites and compared them with present-day fluxes both near Saturn and throughout the rest of the Solar System. As a starting point, they considered the accepted variation in impact cratering within the Saturn system, taking into account the differences in gravitational focusing by Saturn in relation to the orbital distance for each satellite, as well as differences in mass among the satellites and the effect on impact energies (and hence differences in crater size).

Although there are uncertainties in the modeling, Smith *et al.* (1982) note that Iapetus and Phoebe would be least cratered, and that the inner satellites could be more heavily cratered by two orders of magnitude! For example, at the current rates of bombardment by comets (the primary source of bolides in the outer Solar System), Rhea would have twice the number of craters as Iapetus, while Mimas would have 20 times the craters. Moreover, there could be a ratio as large as 20 : 1 in the crater frequency from the leading hemisphere to the trailing hemisphere for the satellites.

From the considerations outlined above, Smith *et al.* (1982) recognized two distinct sets of craters on the saturnian satellites (Fig. 9.33), which they termed population I and population II craters. Population I terrain is characterized by abundant craters >200 km in diameter which are considered to represent the end of the period of heavy bombardment, approximately equivalent to the heavily cratered terrain of the inner planets.

Population II terrain has a high frequency of craters <20 km in diameter and relatively few large craters. The size-frequency of the population II craters is statistically similar to the distribution of secondary crater populations on terrestrial planets. Analyses by Smith *et al.* (1982) and Plescia and Boyce (1982 & 1983) show that both populations are present on Dione, Rhea and Tethys, while Mimas is dominated by population II craters, and Enceladus lacks population I craters entirely, evidently reflecting differences in resurfacing histories.

One of the more intriguing discoveries to come from the Voyager mission is the possibility that the satellites have been repeatedly disrupted and reaccreted. Smith *et al.* (1981 & 1982) estimate that all of the satellites from Dione inward have been struck at least once by a body with sufficient kinetic energy to form a crater equal in size to the satellite being struck. On the inner satellites, such impacts would have occurred many times. Thus, statistically, it is likely that the satellites have been broken apart repeatedly. They note that, because of the circular orbits, most of the disrupted material would accrete to form a reassembled body. This may explain the absence of population I craters on Mimas.

9.3.2 Tectonism and volcanism

Prior to the Voyager flybys of Jupiter and Saturn, models of satellite evolution by Lewis (1971) and other early studies suggested that the larger, denser satellites would experience heating by impact cratering following accretion and from radionuclides, leading to chemical differentiation in which the rocky materials would form cores, while volatiles (mostly water) would form mantles and lithospheres. Smaller or less dense objects were thought not to experience sufficient heating to lead to differentiation but would remain as cold, homogeneous mixtures of ice and rocky material. Lewis, however, noted that ammonia may well be present in the outer Solar System, and its addition could lower the melting point for the mixture.

Observations of the surface features and the inferred histories for the saturnian satellites have caused reassessment of the evolutionary models for small, volatile-rich satellites. Volcanism involving icy slushes and tectonism (Poirier 1982) that has fractured the moons, in some cases on a grand scale (Fig. 9.17), requires explanations that take into account not only the sizes, densities and relative positions of the satellites in the saturnian system, but also the very origin of the Solar System, as reviewed by Smith *et al.* (1981 & 1982).

Ellsworth and Schubert (1983) note that the four largest satellites (excluding Titan), Tethys, Dione, Rhea and Iapetus, are large enough and contain sufficient heat-producing rocky materials to have generated convection beneath a lithosphere. This process is independent of initial thermal conditions and could have generated the observed surface features. Mimas is so small that any thermal activity would have occurred early in its history due to accretionary processes. In contrast, Enceladus must experience heating from tidal stresses, although estimates by Squyres *et al.* (1983) based on present conditions indicate that these are inadequate to melt the interior.

Stevenson (1982) has assessed the role of volcanism and igneous processes on the saturnian satellites and concludes that an ammonia–water magma is required to explain the resurfacing on Enceladus. He suggests that explosive venting could occur if clathrates are involved with the heating on all the larger satellites. Contact between the magma and one or more clathrates of methane, nitrogen and argon could cause pyroclast-forming explosions to generate features such as the bright, wispy albedo patterns seen on Dione and Rhea.

9.4 The future

Besides providing many important discoveries, the Voyager mission raised a great many questions for the saturnian system. From the organic "soup" of Titan to the curious "shepherding" satellites of the rings, the Saturn system remains of high intrinsic interest to planetary science. Although not in a high priority position for current plans of Solar System exploration, a Galileo-like mission involving an orbiter and various probes would provide many answers to questions raised by Voyager.

10 Epilogue

The 1970s have been referred to as the "Golden Decade" of Solar System exploration (Murray 1983). During this decade more than 34 successful missions were flown to the Moon and planets by the United States and the Soviet Union. As political interests shifted and economic forces prevailed, the beginning of the 1980s saw a marked decline in the number of missions. However, it must also be pointed out that both the quality and quantity of scientific data have increased dramatically with time. Thus, the reduced number of missions has been offset at least partly by better and more efficient spacecraft and instrumentation.

But what does the future hold for the geologic exploration of the Solar System? Let us first examine current and planned missions for the 1980s. In the late fall of 1983 the USSR put Veneras 15 and 16 into orbit around Venus. Carrying radar-imaging systems, these two spacecraft began systematic mapping of the northern hemisphere at spatial resolutions as good as 1–2 km from an orbital altitude of 1000–2000 km. Presentation of preliminary results by V. L. Barsukov and A. T. Basilevsky at the 15th Lunar and Planetary Conference, Houston, in early 1984 indicated that the goal was to map the entire area of Venus north of 35°N latitude during the nominal lifetime of the two missions. The primary data product consists of radar images in strips covering areas ~130 km wide by 7000 km long. These strips have been mosaicked to provide regional views of Venus. As discussed in Chapter 6, the Venera radar images reveal considerable geologic detail, and their analysis will substantially increase knowledge of Venus.

Exploration of Venus will continue through the proposed joint USSR and France program designated Venera 17-18/Vega 1-2. Designed for a Halley's comet encounter in 1986, the "bus" will drop off a lander to obtain geochemical measurements of the surface and balloons which will be set adrift in the atmosphere. Plans call for the Venus encounter in the summer of 1985.

The United States' Venus program involves the Venus Radar Mapper (VRM) scheduled to begin operation in the late 1980s. At a nominal altitude of 300 km, VRM will be able to obtain radar images with 150–500 m spatial resolution for ~92 percent of the surface. The images will be obtained for strips of terrain 25 km wide by 14 800 km long.

Exploration of the outer Solar System will continue with the journey of Voyager 2. Scheduled to fly past Uranus in January 1986, this Mariner-class spacecraft will observe the uranian system in much the same manner as did earlier flybys of Jupiter and Saturn. Of particular interest will be the ring system – discovered in the 1970s – and the satellites, the four largest of which range in size from ~1100 to ~1600 km in diameter. Although mechanical malfunctions and the greatly increased distance reduce the quantity of data that can be collected, current estimates predict that about 1000 images will be obtained for the satellites and ring system. If all goes well with the spacecraft, the flight path of Voyager 2 will carry it past Neptune for an encounter in the fall of 1989. During its planned flyby, it may obtain about 500 images of the planet and satellites. Of particular interest will be Triton. Triton may have a tenuous atmosphere, seas of liquid nitrogen and land masses composed of solid methane. This ~3500 km diameter satellite ranks in size and density with the large icy satellites of Saturn and Jupiter and will provide another important data point for comparative planetology.

The Galileo mission is the last of the currently scheduled large US deep-space projects. As discussed in Section 8.6, this mission will be launched from a shuttle in 1986 and involves both an orbiter and an entry probe. During the nominal 20 month mission scheduled to begin in late 1988, more than 70 000 images will be obtained of Jupiter, its ring system and the satellites. When coupled with data from the other experiments, especially the near-infra-red mapping spectrometer, these images will take the Jupiter system out of the reconnaissance stage of exploration and enable a more complete understanding of this "miniature solar system".

Although not a deep-space project, the Edwin P. Hubble Space Telescope will be put into operation in 1986. Primarily designed for astronomical observation, this shuttle-operated project will also be useful for planetary research through the Planetary Camera. Using essentially the same imaging system as the Galileo mission (CCD detectors), Space Telescope will be able to monitor such geologic processes as active dust storms on Mars and volcanic eruptions on Io.

In addition to the joint Soviet-French mission to Venus and Halley's comet, both the European Space Agency (ESA) and the Japanese plan missions to

Halley's comet. Named *Giotto*, the ESA mission involves a spin-stabilized spacecraft carrying an imaging system and will encounter the comet in March, 1986. *Planet A*, Japan's mission, will carry an ultraviolet imaging system and will encounter the comet at about the same time. Although the US is not planning a Halley mission, the sun probe *ISEE-2*, which was launched in 1978 to monitor the solar wind, will be diverted to intercept the comet Giacobini-Zinner in 1985.

In 1983, NASA released a carefully considered plan for Solar System exploration through the year 2000 (SSEC 1983). A key element of this plan is a series of missions, called the Core Program, which has focused objectives. Unlike many previous missions which addressed a wide range of scientific questions, these missions would be smaller in scope and cost and would be aimed at one or two key problems. The Venus Radar Mapper, for example, will be the first in this series. The principal instrument is the radar-imaging system and the objective is to characterize the morphology of the surface of Venus.

Next in line in the Core Program is the Mars Geoscience/Climatology Orbiter (MGCO). Scheduled for launch in 1990, this mission is the first of a new class of spacecraft, the Planetary Observers, which are low-cost spacecraft developed for other purposes, such as communications or weather monitoring. MGCO would focus on determining the global surface composition of Mars and the role of water in the martian climate. Two other missions in the initial Core Program are the Comet Rendezvous/Asteroid Flyby and the Titan Probe/Radar Mapper, both planned for launches in the 1988 to 1992 period.

Parallel with the Core Program and the comet missions discussed above, the European Space Agency is considering various planetary missions including: (a) the *Kepler* mission, which involves an orbiter at Mars to explore the upper atmosphere and interactions with the solar wind; (b) the *Polo* mission, which is a polar lunar orbiter designed to map the chemical elemental abundances on the Moon; and (c) the *Agora* mission to fly to an asteriod. The Soviets apparently have several missions under study for the period through the early 1990s. These include a Mars Orbiter/Phobos Rendezvous, a Lunar Polar Orbiter and various advanced Venera missions for Venus.

The US, the Soviets and ESA have all considered more ambitious plans for future missions, either through separate efforts or joint programs. These include sample returns from Mars, asteroids and other solid-surface objects, and planetary outposts on the Moon and Mars staffed by scientists, engineers and technicians.

While the "shopping list" of planned, potential and postulated missions for planetary exploration is impressive, we must recognize that selection and final approval are in many ways as hazardous as space travel and are usually driven by political and financial considerations. Nonetheless, whichever missions are ultimately flown, geoscience figures prominently in the analyses of the data for the solid-surface objects and we look forward to an exciting era of continued exploration of the Solar System.

References

Anderson, D. L. 1981. Plate tectonics on Venus. *Geophys. Res. Lett.* **8**, 309–11.

Arvidson, R. E., K. A. Goettel and C. M. Hohenberg 1980. A post-Viking view of Martian geologic evolution. *Rev. Geophys. Space Phys.* **18**, 565–603.

Bagnold, R. A. 1941. *The physics of blown sand and desert dunes*. London: Methuen.

Baker, V. R. 1982. *The channels of Mars*. Austin: University of Texas Press.

Baker, V. R. 1984. Planetary geomorphology. *J. Geol Educ.* (in press).

Baker, V. R. and R. C. Kochel 1979. Martian channel morphology: Maja and Kasei Valles. *J. Geophys. Res.* **84**, 7961–83.

Baker, V. R. and D. Nummedal (eds) 1978. *The channeled scabland: a guide to the geomorphology of the Columbia Basin, Washington*. Washington, DC: NASA, Planet. Geol. Prog.

Baldwin, R. B. 1949. *The face of the Moon*. Chicago: University of Chicago Press.

Baldwin, R. B. 1963. *The measure of the Moon*. Chicago: University of Chicago Press.

Barsukov, V. L., V. P. Volkov and I. L. Khodakovsky 1982. The crust of Venus: theoretical models of chemical and mineral composition. *J. Geophys. Res.* **87**, suppl. A3–A9.

Basilevsky, A. T., R. O. Kuzmin, O. V. Nikolaeva, A. A. Pronin, L. B. Ronca, V. S. Avduevsky, G. R. Uspensky, Z. P. Cheremukhina, V. V. Semenchenko, Yu. S. Tyuflin and S. A. Kadnichansky 1984. The surface of Venus as revealed by the Venera landings – Part II (in press).

Batson, R. M. 1981. Status and future of extraterrestrial mapping programs. *NASA CR 3390*.

Beatty, J. K., B. O'Leary and A. Chaikin (eds) 1981. *The new Solar System*. Cambridge: Cambridge University Press and Cambridge, Mass: Sky.

Belton, M. J. S., D. M. Hunten and R. M. Goody 1968. Quantitative spectroscopy of Venus in the region 8,000–11,000 Å. In *The atmospheres of Venus and Mars*, J. C. Brandt and M. B. McElroy (eds), 69–98. New York: Gordon and Breach.

Bloom, A. L. 1978. *Geomorphology*. Englewood Cliffs: Prentice-Hall.

Blunck, J. 1977. *Mars and its satellites: a detailed commentary on the nomenclature*. Hicksville: Exposition Press.

Boyce, J. M., R. J. Pike and P. D. Spudis 1984. Basin ring spacing on the planets: new data from Venus. *Lunar Planet. Sci.* **15** (in press).

Burns, J. A. (ed.) 1977. *Planetary satellites*. Tucson: University of Arizona Press.

BVSP (Basaltic Volcanism Study Project) 1981. *Balsatic volcanism on the terrestrial planets*. New York: Pergamon Press.

Campbell, D. B. and B. A. Burns 1980. Earth-based radar imagery of Venus. *J. Geophys. Res.* **85**, 8271–81.

Campbell, D. B., J. W. Head, J. K. Harmon and A. A. Hine 1983. Venus: identification of banded terrain in the mountains of Ishtar Terra. *Science* **221**, 644–6.

Carr, M. H. 1973. Volcanism on Mars. *J. Geophys. Res.* **78**, 4049–62.

Carr, M. H. 1979. Formation of martian flood features by release of water from confined aquifers. *J. Geophys. Res.* **84**, 2995–3007.

Carr, M. H. 1980. The morphology of the martian surface. *Space Sci. Rev.* **25**, 231–84.

Carr, M. H. 1981. *The surface of Mars*. New Haven and London: Yale University Press.

Carr, M. H. 1983. The geology of the terrestrial planets. *U.S. National Report, 1979–1982*, Am. Geophys. Union, 160–72.

Carr, M. H. and G. G. Schaber 1977. Martian permafrost features. *J. Geophys. Res.* **82**, 4039–54.

Carr, M. H., H. Masursky, R. G. Strom and R. J. Terrile 1979. Volcanic features of Io. *Nature* **280**, 729–33.

Carr, M. H., L. S. Crumpler, J. A. Cutts, R. Greeley, J. E. Guest and H. Masursky 1977. Martian impact craters and emplacement of ejecta by surface flow. *J. Geophys. Res.* **82**, 4055–65.

Cassen, P., S. J. Peale and R. T. Reynolds 1980. On the comparative evolution of Ganymede and Callisto. *Icarus* **41**, 232–9.

Chapman, C. R. and K. L. Jones 1977. Cratering and obliteration history of Mars. *Ann. Rev. Earth Planet. Sci.* **5**, 515–40.

Cintala, M. J., C. A. Wood and J. W. Head 1977. The effects of target characteristics on fresh crater morphology: preliminary results for the Moon and Mercury. *Proc. Lunar Sci. Conf. 8th*, 3409–25.

Classen, J. 1977. Catalogue of 230 certain, probable, possible, and doubtful impact structures. *Meteoritics* **12**, 61–78.

CLHC 1979. *Proceedings of the conference on the lunar highlands*. Houston: Lunar Planetary Institute.

Clifford, S. M. and D. Hillel 1983. The stability of ground ice in the equatorial region of Mars. *J. Geophys. Res.* **88**, 2456–74.

Clow, G. D. and M. H. Carr 1980. Stability of sulfur slopes on Io. *Icarus* **44**, 268–79.

CMRB 1980. *Proceedings of the conference on multi-ring basins*. Houston: Lunar Planetary Institute.

Colin, L. 1980. The Pioneer Venus program. *J. Geophys. Res.* **85**, 7575–98.

Colwell, R. N. (ed.) 1983. *Manual of remote sensing*, vols. 1 and 2, 2nd edn. Falls Church: American Society of Photogrammetry.

Condit, C. D. and P. S. Chavez 1979. Basic concepts of computerized digital image processing for geologists. *US Geol Surv. Bull.*, no. 1462.

Cook, A. F. and F. A. Franklin 1970. An explanation of the light curve of Iapetus. *Icarus* **13**, 282–91.

Cotton, C. A. 1952. *Volcanoes as landscape forms*. Whitcomb and Tombs (reprinted 1969, New York: Hafner).

Croft, S. K. 1983. A proposed origin for palimpsests and anomalous pit craters on Ganymede and Callisto. *J. Geophys. Res.* **88**, suppl., 71–89.

Cruikshank, D. P., J. Degewij and B. H. Zellner 1982. The outer satellites of Jupiter. In *Satellites of Jupiter*, D. Morrison (ed.), 129–46. Tucson: University of Arizona Press.

Crumpler, L. S. and J. C. Aubele 1978. Structural evolution of Arsia Mons, Pavonis Mons, and Ascraeus Mons: Tharsis region of Mars. *Icarus* **34**, 496–511.

Cutts, J. A. and K. R. Blasius 1981. Origin of martian outflow channels: the eolian hypothesis. *J. Geophys. Res.* **86**, 5075–102.

Cutts, J. A. and B. H. Lewis 1982. Models of climate cycles recorded in martian polar layered deposits. *Icarus* **50**, 216–44.

Cutts, J. A., T. W. Thompson and B. H. Lewis 1981. Origin of bright ring-shaped craters in radar images of Venus. *Icarus* **48**, 428–52.

Dalrymple, G. B., E. A. Silver and E. D. Jackson 1973. Origin of the Hawaiian Islands. *Am. Scientist* **61**, 294–308.

Davies, M. E. and R. M. Batson 1975. Surface coordinates and cartography of Mercury. *J. Geophys. Res.* **80**, 2417–30.

Davies, M. E. and B. C. Murray 1971. *The view from space: photographic exploration of the planets*. New York: Columbia University Press.

Davies, M. E., S. E. Dwornik, D. E. Gault and R. G. Strom 1978. *Atlas of Mercury*. NASA SP-423.

Davis, W. M. 1926. Biographical memoir of Grove Karl Gilbert, 1843–1918. *Memoir Nat. Acad. Sci.* **21**, 5th Mem., 303.

DeHon, R. A. 1978. In search of ancient astroblemes: Mercury. In *Rep. Planet. Geol. Prog. – 1977–1978*, NASA TM-79729, 150–2.

DeHon, R. A. 1979. Thickness of the western Mare Basalts. *Proc. Lunar Planet. Sci. Conf. 10th*, 2935–55.

Dence, M. R. 1972. The nature and significance of terrestrial impact structures. *Proc. 24th Int. Geol. Congr.*, Montreal, Section 15, 77–89.

Dence, M. R. 1976. Notes toward an impact model for the Imbrium basin. *Interdisc. Stud. Imbrium Consort.* **1**, 147–55.

Donahue, T. M., J. H. Hoffman, R. R. Hodges, Jr and A. J. Watson 1982. Venus was wet: a measurement of the ratio of deuterium to hydrogen. *Science* **216**, 630–3.

Dzurisin, D. 1978. The tectonic and volcanic history of Mercury as inferred from studies of scarps, ridges, troughs, and other lineaments. *J. Geophys. Res.* **83**, 4883–906.

El-Baz, F. 1979. Scientific exploration of the Moon. *Interdisc. Sci. Rev.* **4**, 239–61.

Ellsworth, K. and G. Schubert 1983. Saturn's icy satellites: thermal and structural models. *Icarus* **54**, 490–510.

Eppler, D. and G. Heiken 1975. Lunar crater chains of non-impact origin. *Proc. Lunar Planet. Sci. Conf. 6th*, 2571–83.

Fanale, F. P. 1976. Martian volatiles: their degassing history and geochemical fate. *Icarus* **28**, 179–202.

Fanale, F. P., R. H. Brown, D. P. Cruikshank and R. N. Clark 1979. Significance of absorption features in Io's IR reflectance spectrum. *Nature* **280**, 761–3.

Farmer, C. B., D. W. Davies, A. L. Holland, D. D. La Porte and P. E. Doms 1977. Mars: water vapor observations from the Viking orbiters. *J. Geophys. Res.* **82**, 4225–48.

Fielder, G. 1961. *Structure of the Moon's surface*. London: Pergamon.

Fielder, G. 1965. *Lunar geology*. Chester Springs: Dufour.

Fimmel, R. O., W. Swindell and E. Burgess 1977. *Pioneer odyssey*. NASA SP-396.

Fink, J. H., S. O. Park and R. Greeley 1983. Cooling and deformation of sulfur flows. *Icarus* **56**, 38–50.

Finnerty, A. A., G. A. Ransford, D. C. Pieri and K. D. Collerson 1980. Is Europa surface cracking due to thermal evolution?. *Nature* **289**, 24–7.

Florensky, C. P., L. B. Ronca, A. T. Basilevsky, G. A. Burba, O. V. Nikolaeva, A. A. Pronin, A. M. Trakhtman, V. P. Volkov, and V. V. Zazetsky 1977. The surface of Venus as revealed by Soviet Venera 9 and 10. *Geol Soc. Am. Bull.* **88**, 1537–45.

Florensky, C. P., A. T. Basilevsky, V. P. Kryuchkov, R. O. Kusmin, O. V. Nikolaeva, A. A. Pronin, I. M. Chernaya, Y. S. Tyuflin, A. S. Selivanov, M. K. Naraeva and L. B. Ronca 1983. Venera 13 and Venera 14: Sedimentary rocks on Venus? *Science* **221**, 57–9.

Francis, P. W. and C. A. Wood 1982. Absence of silicic volcanism on Mars: implications for crustal composition and volatile abundance. *J. Geophys. Res.* **87**, 9881–9.

French, B. M. and N. M. Short (eds) 1968. *Shock metamorphism of natural materials*. Baltimore: Mono Book.

Frey, H. and B. L. Lowry 1979. Large impact basins on Mercury and relative crater production rates. *Proc. Lunar Planet. Sci. Conf. 10th*, 2669–87.

Frey, H., B. L. Lowry and S. A. Chase 1979. Pseudocraters on Mars. *J. Geophys. Res.* **84**, 8075–86.

Garvin, J. B., J. W. Head and L. Wilson 1982. Magma vesiculation and pyroclastic volcanism on Venus. *Icarus* **52**, 365–72.

Garvin, J. B., J. W. Head, M. T. Zuber and P. Helfenstein 1984. Venus: the nature of the surface from Venera panoramas. *J. Geophys. Res.* (in press).

Gault, D. E. 1970. Saturation and equilibrium conditions for impact cratering on the lunar surface: criteria and implications. *Radio Sci.* **5**, 273–91.

Gault, D. E. 1974. Impact craters. In *Primer in lunar geology*, R. Greeley and P. E. Schultz (eds), 137–75. NASA TM X 62359.

Gault, D. E. and R. Greeley 1978. Exploratory experiments of impact craters formed in viscous-liquid targets: analogs for martian rampart craters? *Icarus* **34**, 486–95.

Gault, D. E. and J. A. Wedekind 1978. Experimental studies of oblique impact. *Proc. Lunar Planet. Sci. Conf. 9th*, 3843–75.

Gault, D. E., W. L. Quaide and V. R. Oberbeck 1968. Impact

cratering mechanics and structures. In *Shock metamorphism of natural materials*, B. M. French and N. M. Short (eds), 87–99. Baltimore: Mono Book.
Gault, D. E., J. A. Burns, P. Cassen and R. G. Strom 1977. Mercury. *Ann. Rev. Astron. Astrophys.* **15**, 97–126.
Gault, D. E., J. E. Guest, J. B. Murray, D. Dzurisin and M. C. Malin 1975. Some comparisons of impact craters on Mercury and the Moon. *J. Geophys. Res.* **80**, 2444–60.
Gilbert, G. K. 1893. The Moon's face – a study of the origin and its features. *Phil Soc. Washington Bull.* **12**, 241–92.
Glass, B. P. 1982. *Introduction to planetary geology*. Cambridge: Cambridge University Press.
Glen, J. W. 1974. *The physics of ice*. Cold Regions Research and Engineering Laboratory, US Army, Monograph II-C2a.
Golombek, M. P. 1979. Structural analysis of lunar grabens and the shallow crustal structure of the Moon. *J. Geophys. Res.* **84**, 4657–66.
Golombek, M. P. 1982. Constraints on the expansion of Ganymede and the thickness of the lithosphere. *J. Geophys. Res.* **87**, suppl., 77–83.
Greeley, R. 1971. Lunar Hadley Rille: considerations of its origin. *Science* **172**, 722–5.
Greeley, R. 1976. Modes of emplacement of basalt terrains and an analysis of mare volcanism in the Orientale Basin. *Proc. Lunar Sci. Conf. 7th*, 2747–59.
Greeley, R. and M. H. Carr (eds) 1976. *A geological basis for the exploration of the planets*, NASA SP-417.
Greeley, R. and D. E. Gault 1979. Endogenic craters on basaltic lava flows: size frequency distributions. *Proc. Lunar Planet. Sci. Conf. 10th*, 2919–33.
Greeley, R. and J. D. Iversen 1985. *Wind as a geological process*. Cambridge: Cambridge University Press, Planetary Science Series.
Greeley, R. and P. Spudis 1978a. Mare volcanism in the Herigonius region of the Moon. *Proc. Lunar Planet. Sci. Conf. 9th*, 3333–49.
Greeley, R. and P. D. Spudis 1978b. Volcanism in the cratered terrain hemisphere of Mars. *Geophys. Res. Lett.* **5**, 453–5.
Greeley, R. and P. D. Spudis 1981. Volcanism on Mars. *Rev. Geophys. Space Phys.* **19**, 13–41.
Greeley, R. and E. Theilig 1978. Small volcanic constructs in the Chryse Planitia region of Mars. In *Rep. Planet. Geol. Prog. – 1977–1978*, NASA TM-79729, 202.
Greeley, R., J. H. Fink, D. E. Gault and J. E. Guest 1982. Experimental simulation of impact cratering on icy satellites. In *Satellites of Jupiter*, D. Morrison (ed.), 340–78. Tucson: University of Arizona Press.
Greeley, R., J. Fink, D. E. Gault, D. B. Snyder, J. E. Guest and P. H. Schultz 1980. Impact cratering in viscous targets: laboratory experiments. *Proc. Lunar Planet. Sci. Conf. 11th*, 2075–97.
Greeley, R., J. Iversen, R. Leach, J. Marshall, B. White and S. Williams 1984. Windblown sand on Venus: preliminary results of laboratory simulations. *Icarus* (in press).
Greeley, R., E. Theilig, J. E. Guest, M. H. Carr, H. Masursky and J. A. Cutts 1977. Geology of Chryse Planitia. *J. Geophys. Res.* **82**, 4093–109.
Green, J. 1965. Hookes and spurrs in selenology. In *Geological problems in lunar research*, *Ann. New York Acad. Sci.* **123**, 373–402.
Green, J. and N. M. Short 1971. *Volcanic landforms and surface features: a photographic atlas and glossary*. New York: Springer.
Grieve, R. A. F. and J. W. Head 1982. The impact cratering process on Venus. *Lunar Planet. Sci.* **13**, 285–6.
Guest, J. E. 1971. Centers of igneous activity in maria. In *Geology and physics of the Moon*, G. Fielder (ed.), 41–53. Amsterdam: Elsevier.
Guest, J. E. and D. E. Gault 1976. Crater populations in the early history of Mercury. *Geophys. Res. Lett.* **3**, 121–3.
Guest, J. E. and R. Greeley 1977. *Geology on the Moon*. London and Crane: Wykeham, and New York: Russak.
Guest, J. E. and J. B. Murray 1976. Volcanic features of the nearside equatorial lunar maria. *J. Geol Soc. Lond.* **132**, 251–8.
Guest, J. E. and W. P. O'Donnell 1977. Surface history of Mercury: a review. *Vistas Astron.* **20**, 273–300.
Guest, J. E., P. S. Butterworth and R. Greeley 1977. Geological observations in the Cydonia region of Mars from Viking. *J. Geophys. Res.* **82**, 4111–20.
Guest. J., with P. Butterworth, J. Murray and W. O'Donnell 1979. *Planetary geology*. New York: Halsted.

Hale, W. S. and R. A. F. Grieve 1982. Volumetric analysis of complex lunar craters: implications for basin ring formation. *J. Geophys. Res.* **87**, A65–A76.
Hale, W., J. W. Head and E. M. Parmentier 1980. Origin of the Valhalla ring structure: alternative models. In *Conference on multi-ring basins*, 30–2. Houston: Lunar Planetary Institute.
Hall, J. L., S. C. Solomon and J. W. Head 1981. Lunar floor-fractured craters: evidence for viscous relaxation of crater topography. *J. Geophys. Res.* **86**, 9537–52.
Hanel, R., B. Conrath, F. M. Flasar, V. Kunde, W. Maguire, J. Pearl, J. Pirraglia, R. Samuelson, L. Herath, M. Allison, D. Cruikshank, D. Gautier, P. Gierasch, L. Horn, R. Koppany and C. Ponnamperuma 1981. Infrared observations of the saturnian system from Voyager 1. *Science* **212**, 192–200.
Harris, S. A. 1977. The aureole of Olympus Mons, Mars. *J. Geophys. Res.* **82**, 3099–107.
Hartmann, W. K. 1983. *Moon and planets*. Belmont: Wadsworth.
Hartmann, W. K., R. G. Strom, S. J. Weidenschilling, K. R. Blasius, A. Woronow, M. R. Dence, R. A. F. Grieve, J. Diaz, C. R. Chapman, E. M. Shoemaker and K. L. Jones 1981. Chronology of planetary volcanism by comparative studies of planetary cratering. In *Basaltic volcanism on the terrestrial planets*, Basaltic Volcanism Study Project, 1048–127. New York: Pergamon Press.
Hawke, B. R. and J. W. Head 1978. Lunar KREEP volcanism: geological evidence for history and mode of emplacement. *Proc. Lunar Planet. Sci. Conf. 9th*, 3285–309.
Head, J. W. 1976. Lunar volcanism in space and time. *Rev. Geophys. Space Phys.* **14**, 265–300.
Head, J. W. and A. Gifford 1980. Lunar mare domes: classification and modes of origin. *Moon Planets* **22**, 235–58.
Head, J. W. and S. C. Solomon 1981. Tectonic evolution of

the terrestrial planets. *Science* **213**, 62–76.

Head, J. W., S. E. Yuter and S. C. Solomon 1981a. Topography of Venus and Earth: a test for the presence of plate tectonics. *Am. Scientist* **69**, 614–23.

Head, J. W. III, W. B. Bryan, R. Greeley, J. E. Guest, P. H. Schultz, R. S. J. Sparks, G. P. L. Walker, J. L. Whitford-Stark, C. A. Wood and M. H. Carr 1981b. Distribution and morphology of basaltic deposits on planets. In *Basaltic volcanism on the terrestrial planets*, Basaltic Volcanism Study Project, 701–800. New York: Pergamon Press.

Helfenstein, P. and E. M. Parmentier 1980. Fractures on Europa: possible response of an ice crust to tidal deformation. *Proc. Lunar Planet. Sci. Conf. 11th*, 1987–98.

Hodges, C. A. 1973. Mare ridges and lava lakes. *Apollo 17 Prelim. Sci. Rep.*, NASA SP-330, 31.12–31.21.

Hodges, C. A. and H. J. Moore 1979. The subglacial birth of Olympus Mons and its aureoles. *J. Geophys. Res.* **84**, 8061–74.

Hodges, C. A. and D. E. Wilhelms 1978. Formation of lunar basin rings. *Icarus* **34**, 294–323.

Horner, V. M. and R. Greeley 1982. Pedestal craters on Ganymede. *Icarus* **51**, 549–62.

Hörz, F. 1978. How thick are lunar mare basalts? *Proc. Lunar Planet. Sci. Conf. 9th*, 3311–31.

Howard, A. D. 1967. Drainage analysis in geological interpretation: a summation. *Am. Assoc. Petrol. Geol. Bull.* **51**, 2246–59.

Howard, K. A. 1973. Avalanche mode of motion: implications from lunar examples. *Science* **180**, 1052–5.

Howard, K. A., J. W. Head and G. A. Swann 1972. Geology of Hadley Rille. *Proc. Lunar Sci. Conf. 3rd*, 1–14.

Howard, K. A., D. E. Wilhelms and D. H. Scott 1974. Lunar basin formation and highland stratigraphy. *Rev. Geophys. Space Phys.* **12**, 309–27.

Hunten, D. M., L. Colin, T. M. Donahue and V. I. Moroz (eds) 1983. *Venus*. Tucson: University of Arizona Press.

Inge, J. L. and P. M. Bridges 1976. Applied photo interpretation for airbursh cartography. *Photogram. Engng Remote Sens.* **42**, 749–60.

Johnson, T. V. 1978. The Galilean satellites of Jupiter: four worlds. *Ann. Rev. Earth Planet. Sci.* **6**, 93–125.

Johnson, T. V. and T. R. McGetchin 1973. Topography on satellite surfaces and the shape of asteroids. *Icarus* **18**, 612–20.

Johnson, T. V. and L. A. Soderblom 1982. Volcanic eruptions on Io: implications for surface evolution and mass loss. In *Satellites of Jupiter*, D. Morrison (ed.), 634–46. Tucson: University of Arizona Press.

Johnson, T. V., A. F. Cook II, C. Sagan and L. A. Soderblom 1979. Volcanic resurfacing rates and implications for volatiles on Io. *Nature* **280**, 746–50.

Kieffer, H. H. and F. D. Palluconi 1979. *The climate of the martian polar cap*. NASA CP-2072, 45–6.

Kieffer, S. W. 1982. Dynamics and thermodynamics of volcanic eruptions: implications for the plumes on Io. In *Satellites of Jupiter*, D. Morrison (ed.), 647–723. Tucson: University of Arizona Press.

King, E. A. 1976. *Space geology*. New York: John Wiley.

King, J. S. and J. R. Riehle 1974. A proposed origin of the Olympus Mons escarpment. *Icarus* **23**, 300–17.

King, L. C. 1967. *The morphology of the Earth*. Edinburgh: Oliver and Boyd.

Klein, H. P. 1979. The search for life on Mars. *Rev. Geophys. Space Phys.* **17**, 1655–62.

Kozai, Y. 1976. Masses of satellites and oblateness parameters of Saturn. *Publn Astron. Soc. Jpn* **28**, 675–91.

Kuiper, G. P. 1962. Infrared spectra of stars and planets. I. Photometry of the infrared spectrum of Venus, 1–2.5 microns. *Comm. Lunar Planet. Lab.* **1**, 83–117.

Lanzerotti, L. J., W. L. Brown, J. M. Poate and W. M. Augustyniak 1978. On the contribution of water products from Galilean satellites to the jovian magnetosphere. *Geophys. Res. Lett.* **5**, 155–8.

Leake, M. 1981. The intercrater plains of Mercury and the Moon: their nature, origin, and role in terrestrial planet evolution. Ph.D. dissertation, Univ. of Arizona, Tucson, AZ, also in *Adv. Planet. Geol.*, NASA TM 84894, Sept. 1982, 1–535.

Leopold, L. B., M. G. Wolman and J. P. Miller 1964. *Fluvial processes in geomorphology*. San Francisco: W. H. Freeman.

Lewis, J. S. 1971. Satellites of the outer planets: their physical and chemical nature. *Icarus* **15**, 174–85.

Lopes, R. M. C., J. E. Guest and C. J. Wilson 1980. Origin of the Olympus Mons aureole and perimeter scarp. *Moon Planets* **22**, 221–34.

Lopes, R., J. E. Guest, K. Hiller and G. Neukum 1982. Further evidence for a mass movement origin of the Olympus Mons aureole. *J. Geophys. Res.* **87**, 9917–28.

Lowman, P. D. Jr 1981. *A global tectonic activity map with orbital photographic supplement*. NASA TM-82073.

Lucchitta, B. K. 1976. Mare ridges and related highland scarps – result of vertical tectonism? *Proc. Lunar Planet. Sci. Conf. 7th*, 2761–82.

Lucchitta, B. K. 1979. Landslides in Valles Marineris, Mars. *J. Geophys. Res.* **84**, 8097–113.

Lucchitta, B. K. 1981. Mars and Earth: comparison of cold-climate features. *Icarus* **45**, 264–303.

Lucchitta, B. K. 1982. Ice sculpture in the martian outflow channels. *J. Geophys. Res.* **87**, 9951–73.

Lucchitta, B. K. and H. M. Ferguson 1983. Chryse Basin channels: low-gradients and ponded flows. *J. Geophys. Res.* **88**, suppl., A553–68.

Lucchitta, B. K. and J. L. Klockenbrink 1981. Ridges and scarps in the equatorial belt of Mars. *Moon Planets* **24**, 415–29.

Lucchitta, B. K. and L. A. Soderblom 1982. The geology of Europa. In *Satellites of Jupiter*, D. Morrison (ed.), 521–55. Tucson: University of Arizona Press.

Lucchitta, B. K. and J. A. Watkins 1978. Age of graben systems on the Moon. *Proc. Lunar Planet. Sci. Conf. 9th*, 3459–72.

Lucchitta, B. K., L. A. Soderblom and H. M. Ferguson 1981. Structures on Europa. *Proc. Lunar Planet. Sci.* **12B**, 1555–67.

Lunine, J. I., G. Neugebauer and B. M. Jakosky 1982.

Infrared observations of Phobos and Deimos from Viking. *J. Geophys. Res.* **87**, 10297–305.

McCauley, J. F. 1977. Orientale and Caloris. *Phys. Earth Planet. Inter.* **15**, 220–50.

McCauley, J. F. 1978. *Geologic map of the Coprates quadrangle of Mars*. US Geol Survey, Misc. Inv. Map I-897.

McCauley, J. F., B. A. Smith and L. A. Soderblom 1979. Erosional scarps on Io. *Nature* **280**, 736–8.

McCauley, J. F., J. E. Guest, G. G. Schaber, N. J. Trask and R. Greeley 1981. Stratigraphy of the Caloris Basin, Mercury. *Icarus* **47**, 184–202.

Macdonald, G. A. 1967. Forms and structures of extrusive basaltic rocks. In *Basalts*, H. H. Hess and A. Poldervaart (eds), 1–61. New York: John Wiley Interscience.

Macdonald, G. A. 1972. *Volcanoes*. Englewood Cliffs, NJ: Prentice-Hall.

McEwen, A. S. and L. A. Soderblom 1983. Two classes of volcanic plumes on Io. *Icarus* **55**, 191–217.

McGetchin, T. R. and J. W. Head 1973. Lunar cinder cones. *Science* **180**, 68–71.

McGill, G. E. 1983. The geology of Venus. *Episodes* **4**, 10–17.

McGill, G. E., S. J. Steenstrup, C. Barton and P. G. Ford 1981. Continental rifting and the origin of Beta Regio, Venus. *Geophys. Res. Lett.* **8**, 737–40.

McGill, G. E., J. L. Warner, M. C. Malin, R. E. Arvidson, E. Eliason, S. Nozette and R. D. Reasenberg 1983. Topography, surface properties, and tectonic evolution. In *Venus*, D. M. Hunten, L. Colin, T. M. Donahue and V. I. Moroz (eds), 69–130. Tucson: University of Arizona Press.

McKee, E. D. 1979. Introduction to a study of global sand seas. In *A study of global sand seas*, E. D. McKee (ed.). US Geol Surv. Prof. Paper 1052, 1–19.

McKinnon, W. B. and H. J. Melosh 1980. Evolution of planetary lithospheres: evidence from multiringed structures on Ganymede and Callisto. *Icarus* **44**, 454–71.

McLaughlin, D. 1954. Volcanism and aeolian deposition on Mars. *Geol Soc. Am. Bull.* **65**, 715–17.

Malin, M. C. 1976. Observations of intercrater plains on Mercury. *Geophys. Res. Lett.* **3**, 581–4.

Malin, M. C. 1978. Surfaces of Mercury and the Moon: effects of resolution and lighting conditions on the discrimination of volcanic features. *Proc. Lunar Planet. Sci. Conf. 9th*, 3395–409.

Malin, M. C. 1980. Domes on Ganymede. In *Rep. Planet. Geol. Prog. – 1980*, NASA TM-82385, 67.

Malin, M. C. and D. Dzurisin 1977. Landform degradation on Mercury, the Moon, and Mars: evidence from crater depth/diameter relationships. *J. Geophys. Res.* **82**, 376–88.

Malin, M. C. and D. Dzurisin 1978. Modification of fresh crater landforms: evidence from the Moon and Mercury. *J. Geophys. Res.* **83**, 233–43.

Malin, M. C. and R. S. Saunders 1977. Surface of Venus. Evidence of diverse landforms from radar observations. *Science* **196**, 987–90.

Masursky, H. 1973. An overview of geological results from Mariner 9. *J. Geophys. Res.* **78**, 4009–30.

Masursky, H., G. G. Schaber, L. A. Soderblom and R. G. Strom 1979. Preliminary geological mapping of Io. *Nature* **280**, 725–9.

Masursky, H., J. M. Boyce, A. L. Dial, G. G. Schaber and M. E. Strobell 1977. Classification and time of formation of martian channels based on Viking data. *J. Geophys. Res.* **82**, 4016–38.

Masursky, H., E. Eliason, P. G. Ford, G. E. McGill, G. H. Pettengill, G. G. Schaber and G. Schubert 1980. Pioneer Venus radar results: geology from images and altimetry. *J. Geophys. Res.* **85**, 8232–60.

Matson, D. L., G. A. Ransford and T.V. Johnson 1981. Heat flow from Io. *J. Geophys. Res.* **86**, 1664–72.

MCWG 1983. Channels and valleys on Mars. *Geol Soc. Am. Bull.* **94**, 1035–54.

Melosh, H. J. 1977. Global tectonics of a despun planet. *Icarus* **31**, 221–43.

Melosh, H. J. 1982. A simple mechanical model of Valhalla Basin, Callisto. *J. Geophys. Res.* **87**, 1880–90.

Moore, H. J., J. M. Boyce, G. G. Schaber and D. H. Scott 1980. *Lunar remote sensing and measurements*. US Geol Surv. Prof. Pap. 1046-B.

Moore, J. M. 1984. The tectonic and volcanic history of Dione. *Icarus* **59**, 205–20.

Moore, J. M. and J. L. Ahern 1982. Tectonic and geological history of Tethys. *Lunar Planet. Sci.* **13**, 538–9.

Moore, J. M. and J. L. Ahern 1983. The geology of Tethys. *J. Geophys. Res.* **88**, A577–84.

Moore, J. M. and V. M. Horner 1984. The geomorphologic features on Rhea. *Lunar and Planet. Sci.* **15**, 560–1.

Moore, J. M., V. M. Horner and R. Greeley 1985. Geomorphology of Rhea: implications for geologic history and surface processes. *J. Geophys. Res.* **90**, 785–96.

Morabito, L. A., S. P. Synnott, P. N. Kupferman and S. A. Collins 1979. Discovery of currently active extraterrestrial volcanism. *Science* **204**, 972.

Morgan, P. and R. J. Phillips 1983. Hot spot heat transfer: its application to Venus and implications to Venus and Earth. *J. Geophys. Res.* **88**, 8305–17.

Moroz, V. I. 1983. Summary of preliminary results of the Venera 13 and Venera 14 missions. In *Venus*, D. M. Hunten, L. Colin, T. M. Donahue and V. I. Moroz (eds), 45–68. Tucson: University of Arizona Press.

Morris, E. C. 1982. Aureole deposits of the martian volcano Olympus Mons. *J. Geophys. Res.* **87**, 1164–78.

Morris, E. C. and J. R. Underwood 1978. Polygonal fractures of the martian plains. In *Rep. Planet. Geol. Prog. – 1977–1978*, NASA TM-79729, 97–9.

Morrison, D. 1982a. Introduction to the satellites of Jupiter. In *Satellites of Jupiter*, D. Morrison (ed.), 3–43. Tucson: University of Arizona Press.

Morrison, D. 1982b. *Voyages to Saturn*. NASA SP-451.

Morrison, D. 1983. Outer planets satellites. *US Nat. Rep. 1979–1982*, 151–9.

Morrison, D. and J. Samz 1980. *Voyage to Jupiter*. NASA SP-439.

Mouginis-Mark, P. 1979. Martian fluidized crater morphology: variations with crater size, latitude, altitude, and target material. *J. Geophys. Res.* **84**, 8011–22.

Mouginis-Mark, P. 1981. Ejecta emplacement and modes of formation of martian fluidized ejecta craters. *Icarus* **45**,

60–76.

Murray, B. C. 1975. The Mariner 10 pictures of Mercury: an overview. *J. Geophys. Res.* **80**, 2342–4.

Murray, B. C. 1983. *The planets – readings from Scientific American*. San Francisco: W. H. Freeman.

Murray, B. C. and E. Burgess 1977. *Flight to Mercury*. New York: Columbia University Press.

Murray, B. C., M. C. Malin and R. Greeley 1981. *Earthlike planets*. San Francisco: W. H. Freeman.

Murray, B. C., R. G. Strom, N. J. Trask and D. E. Gault 1975. Surface history of Mercury: implications for terrestrial planets. *J. Geophys. Res.* **80**, 2508–14.

Mutch, T. A., R. E. Arvidson, J. W. Head, K. L. Jones and R. S. Saunders 1976. *The geology of Mars*. Princeton: Princeton University Press.

Nelson, R. M., D. C. Pieri, D. Nash and S. M. Baloga 1982. Reflection spectrum of liquid sulfur and its implication for Io. In *Rep. Planet. Geol. Prog. – 1982*, NASA TM-85127, 12-15.

Neukum, G. and K. Hiller 1981. Martian ages. *J. Geophys. Res.* **86**, 3097–121.

Neukum, G. and D. U. Wise 1976. Mars: a standard crater curve and possible new time scale. *Science* **194**, 1381–7.

Newburn, R. 1978. Planetary data. *Proc. Lunar Planet. Sci. Conf. 9th*, end papers.

Newburn, R. and D. Matson 1978. Satellite data. *Proc. Lunar Planet. Sci. Conf. 9th*, end papers.

Nozette, S. and J. S. Lewis 1982. Venus: chemical weathering of igneous rocks and buffering of atmospheric composition. *Science* **216**, 181–3.

Nummedal, D. 1978. The role of liquefaction in channel development on Mars. In *Rep. Planet. Geol. Prog. – 1977–1978*, NASA TM-79729, 257–9.

Oberbeck, V. R. and W. L. Quaide 1968. Genetic implications of lunar regolith thickness variations. *Icarus* **9**, 446–65.

Oberbeck, V. R., W. L. Quaide, R. E. Arvidson and H. R. Aggarwal 1977. Comparative studies of lunar, martian, and mercurian craters and plains. *J. Geophys. Res.* **82**, 1681–98.

Orton, G. 1978. Planetary atmospheres. *Proc. Lunar Planet. Sci. Conf. 9th*, end papers.

Owen, T. 1983. Titan. In *The planets*, B. Murray (ed.), 84–93. San Francisco: W. H. Freeman.

Parmentier, E. M. and J. W. Head 1981. Viscous relaxation of impact craters on icy planetary surfaces: determination of viscosity variation with depth. *Icarus* **47**, 100–11.

Parmentier, E. M., S. W. Squyres, J. W. Head and M. L. Allison 1982. The tectonics of Ganymede. *Nature* **295**, 290–3.

Passey, Q. R. 1983. Viscosity of the lithosphere of Enceladus. *Icarus* **53**, 105–20.

Passey, Q. R. and E. M. Shoemaker 1982. Craters and basins on Ganymede and Callisto: morphological indicators of crustal evolution. In *Satellites of Jupiter*, D. Morrison (ed.), 379–434. Tucson: University of Arizona Press.

Peale, S. J., P. Cassen and R. T. Reynolds 1979. Melting of Io by tidal dissipation. *Science* **203**, 892–4.

Pechmann, J. C. 1980. The origin of polygonal troughs on the northern plains of Mars. *Icarus* **42**, 185–210.

Peterson, J. E. 1978. Antipodal effects of major basin-forming impacts on Mars. *Lunar Planet. Sci.* **9**, 885–6.

Pettengill, G. H., E. Eliason, P. G. Ford, G. B. Loriot, H. Masursky and G. E. McGill 1980. Pioneer Venus radar results: altimetry and surface properties. *J. Geophys. Res.* **85**, 8261–70.

Phillips, R. J. and E. R. Ivins 1979. Geophysical observations pertaining to solid-state convection in the terrestrial planets. *Phys. Earth Planet. Interiors* **19**, 107–48.

Phillips, R. J. and K. Lambeck 1980. Gravity fields of the terrestrial planets: long-wavelength anomalies and tectonics. *Rev. Geophys. Space Phys.* **18**, 27–76.

Phillips, R. J. and M. C. Malin 1983. The interior of Venus and tectonic implications. In *Venus*, D. M. Hunten, L. Colin, T. M. Donahue and V. I. Moroz (eds), 159–214. Tucson: University of Arizona Press.

Phillips, R. J., W. Kaula, G. McGill and M. C. Malin 1981. Tectonics and evolution of Venus. *Science* **212**, 879–87.

Pieri, D. C. 1979. *Geomorphology of martian valleys*. Ph.D. dissertation, Cornell University.

Pieri, D. C. 1980. Lineament and polygon patterns on Europa. *Nature* **289**, 17–21.

Pieri, D. 1983. The ancient rivers of Mars. *The Planet. Rep.* **3**, 4–7.

Pieri, D. C., S. M. Baloga and R. M. Nelson 1982. Colors of lava flows at Ra Patera, Io. In *Rep. Planet. Geol. Prog. – 1982*, NASA TM-85127, 16–19.

Pieri, D. C., S. M. Baloga, R. M. Nelson and C. Sagan 1984. Sulfur flows of Ra Patera, Io. Submitted to *Icarus*.

Pike, R. J. 1974. Depth/diameter relations of fresh lunar craters: revision from spacecraft data. *Geophys. Res. Lett.* **1** (7), 291–4.

Pike, R. J. 1976. *Geologic map of the Rima Hyginus region of the Moon*. US Geol Surv. Misc. Inv. Map I-945.

Pike, R. J. 1980. *Apollo 15–17 orbital investigations – geometric interpretation of lunar craters*. US Geol Surv. Prof. Paper 1046C.

Pilcher, C. B., S. T. Ridgway and T. B. McCord 1972. Galilean satellites: identification of water frost. *Science* **178**, 1087–9.

Plescia, J. B. 1983. The geology of Dione. *Icarus* **56**, 255–77.

Plescia, J. B. and J. M. Boyce 1982. Crater densities and geological histories of Rhea, Dione, Mimas and Tethys. *Nature* **295**, 285–90.

Plescia, J. B. and J. M. Boyce 1983. Crater numbers and geological histories of Iapetus, Enceladus, Tethys and Hyperion. *Nature* **301**, 666–70.

Plescia, J. B. and R. S. Saunders 1979. The chronology of martian volcanoes. *Proc. Lunar Planet. Sci. Conf. 10th*, 2841–59.

Poirier, J. P. 1982. Rheology of ices: a key to the tectonics of the ice moons of Jupiter and Saturn. *Nature* **299**, 683–8.

Pollack, J. B. and C. Sagan 1965. The microwave phase effect of Venus. *Icarus* **4**, 62–103.

Pollack, J. B., J. A. Burns and M. E. Tauber 1979. Gas drag in primordial circumplanetary envelopes: a mechanism for satellite capture. *Icarus* **37**, 587–611.

Ransford, G. A., A. A. Finnerty and K. D. Collerson 1980. Europa's petrological thermal history. *Nature* **289**, 21–4.

Rittmann, A. 1962. *Volcanoes and their activity* (translated by E. A. Vincent from the German 2nd edn). New York: Interscience.

Roddy, D. J. 1977. Large-scale impact and explosion craters: comparisons of morphological and structural analogs. In *Impact and explosion cratering: planetary and terrestrial implications*, D. J. Roddy, R. O. Pipen and R. B. Merrill (eds), 185–246. New York: Pergamon Press.

Roddy, D. J., R. O. Pepin and R. B. Merrill (eds) 1977. *Impact and explosion cratering: planetary and terrestrial implications*. New York: Pergamon Press.

Rossbacher, L. A. and S. Judson 1981. Ground ice on Mars: inventory, distribution, and resulting landforms. *Icarus* **45**, 39–59.

Ryder, G. and P. D. Spudis 1980. Volcanic rocks in the lunar highlands. *Proc. Conf. Lunar Highlands Crust*, J. J. Papike and R. B. Merrill (eds), 353–75. New York: Pergamon Press.

Ryder, G. and G. J. Taylor 1976. Did mare-type volcanism commence early in lunar history? *Proc. Lunar Planet. Sci. Conf. 7th*, 1741–55.

Sagan, C. 1961. The planet Venus. *Science* **133**, 849–58.

Sagan, C. 1979. Sulfur flows on Io. *Nature* **280**, 750–3.

Sagan, C. and S. F. Dermott 1982. The tide in the seas of Titan. *Nature* **300**, 731–3.

Sagan, C., J. Veverka, P. Fox, R. Dubisch, J. Lederberg, E. Levinthal, L. Quam, R. Tucker, J. B. Pollack and B. A. Smith 1972. Variable features on Mars: preliminary Mariner 9 television results. *Icarus* **17**, 346–72.

Sagan, C., J. Veverka, P. Fox, R. Dubisch, R. French, P. Gierasch, L. Quam, J. Lederberg, E. Levinthal, R. Tucker, B. Eross and J. B. Pollack 1973. Variable features on Mars, 2, Mariner 9 global results. *J. Geophys. Res.* **78**, 4163–96.

Saunders, R. S. and M. C. Malin 1977. Geologic interpretation of new observations of the surface of Venus. *Geophys. Res. Lett.* **4**, 547–50.

Schaber, G. G. 1973. Lava flows in Mare Imbrium: geologic evaluation from Apollo orbital photography. *Proc. Lunar Sci. Conf. 4th*, 73–92.

Schaber, G. G. 1980. The surface of Io: geologic units, morphology and tectonics. *Icarus* **43**, 302–33.

Schaber, G. G. 1982a. Syrtis Major: a low-relief volcanic shield. *J. Geophys. Res.* **87**, 9852–66.

Schaber, G. G. 1982b. The geology of Io. In *Satellites of Jupiter*. D. Morrison (ed.), 556–97. Tucson: University of Arizona Press.

Schaber, G. G. 1982c. Venus: limited extension and volcanism along zones of lithospheric weakness. *Geophys. Res. Lett.* **9**, 499–502.

Schaber, G. G. and J. M. Boyce 1977. Probable distribution of large impact basins on Venus: comparison with Mercury and the Moon. In *Impact and explosion cratering*, D. J. Roddy, R. O. Pepin and R. B. Merrill (eds), 603–12. New York: Pergamon Press.

Schaber, G. G., J. M. Boyce and N. J. Trask 1977. Moon–Mercury: large impact structures, isostasy and average crustal viscosity. *Phys. Earth Planet. Int.* **15**, 189–201.

Schonfeld, E. 1977. Martian volcanism. *Lunar Sci.* **8**, 843–5.

Schultz, P. H. 1976a. Floor-fractured craters. *Moon* **15**, 241–73.

Schultz, P. H. 1976b. Introduction to planetary surfaces. In *NASA short course in planetary geology*, P. H. Schultz and R. Greeley (eds), 1–14.

Schultz, P. H. 1976c. *Moon morphology*. Austin: University of Texas Press.

Schultz, P. H. 1977. Endogenic modification of impact craters on Mercury. *Phys. Earth Planet. Interiors* **15**, 202–19.

Schultz, P. H. 1981. Impact cratering on Venus. *Int. Conf. Venus Environment*, 6.

Schultz, P. H. and D. E. Gault 1975. Seismic effects from major basin formation on the Moon and Mercury. *Moon* **12**, 159–77.

Schultz, P. H. and D. E. Gault 1979. Atmospheric effects on martian ejecta emplacement. *J. Geophys. Res.* **84**, 7669–87.

Schultz, P. H. and A. B. Lutz-Garihan 1982. Grazing impacts on Mars: a record of lost satellites. *J. Geophys. Res.* **87**, A84–A96.

Schultz, P. H. and P. D. Spudis 1979. Evidence for ancient mare volcanism. *Proc. Lunar Planet. Sci. Conf. 10th*, 2899–918.

Schultz, P. H. and P. D. Spudis 1983. Beginning and end of lunar mare volcanism. *Nature* **302**, 233–6.

Schultz, P. H., R. A. Schultz and J. Rogers 1982. The structure and evolution of ancient impact basins on Mars. *J. Geophys. Res.* **87**, 9803–20.

Scott, D. H. 1977. Moon–Mercury: relative preservation states of secondary craters. *Phys. Earth Planet. Int.* **15**, 173–8.

Scott, D. H. 1982. Volcanoes and volcanic provinces: Martian western hemisphere. *J. Geophys. Res.* **87**, 9839–51.

Scott, D. H. and M. H. Carr 1978. *Geologic map of Mars*. US Geol Survey, Misc. Inv. Map I-1083.

Scott, D. H. and K. L. Tanaka 1981. Mars: paleostratigraphic restoration of buried surfaces in Tharsis Montes. *Icarus* **45**, 304–19.

Scott, D. H., J. F. McCauley and M. N. West 1978. *Geologic map of the west side of the Moon*. US Geol Surv. Misc. Geol. Inv. Map I–1034.

Sharp, R. P. 1973. Mars: south polar pits and etched terrains. *J. Geophys. Res.* **78**, 4222–30.

Sharp, R. P. 1980. Geomorphological processes on terrestrial planetary surfaces. *Ann. Rev. Earth Planet. Sci.* **8**, 231–61.

Sharp, R. P. and M. C. Malin 1975. Channels on Mars. *Geol Soc. Am. Bull.* **86**, 593–609.

Sharpe, C. F. S. 1968. *Landslides and related phenomena*. New York: Cooper Square.

Shoemaker, E. M. 1963. Impact mechanics at Meteor Crater, Arizona. In *The Solar System, the moon, meteorites, and comets*, vol. 4, B. M. Middlehurst and G. P. Kuiper (eds), 301–36. Chicago: University of Chicago Press.

Shoemaker, E. M. and R. J. Hackman 1962. Stratigraphic basis for a lunar time scale. In *The Moon – International Astronomical Union Symposium 14*, A. Kopal and Z. K. Mikhailov (eds), 289–300. New York: Academic Press.

Shoemaker, E. M. and R. F. Wolfe 1982. Cratering time scales for the Galilean satellites. In *Satellites of Jupiter*, D. Morrison (ed.), 277–339. Tucson: University of Arizona

Press.
Shoemaker, E. M., B. K. Lucchitta, J. B. Plescia, S. W. Squyres and D. E. Wilhelms 1982. The geology of Ganymede. In *Satellites of Jupiter*, D. Morrison (ed.), 435–520. Tucson: University of Arizona Press.
Sill, G. T. and R. N. Clark 1982. Composition of the surfaces of the Galilean satellites. In *Satellites of Jupiter*, D. Morrison (ed.), 174–212. Tucson: University of Arizona Press.
Silver, L. T. and P. H. Schultz (eds) 1982. *Geological implications of impacts of large asteroids and comets on the Earth*. Geol Soc. Am. Spec. Paper 190.
Sjogren, W. L. 1979. Mars gravity: high resolution results from Viking orbiter 2. *Science*, **203**, 1006–9.
Sjogren, W. L., J. Lorell, L. Wong and W. Downs 1975. Mars gravity field based on a short-arc technique. *J. Geophys. Res.* **80**, 2899–908.
Sleep, N. H. and R. J. Phillips 1979. An isostatic model for the Tharsis province, Mars. *Geophys. Res. Lett.* **6**, 803–6.
Smith, B. A., L. Soderblom, R. Beebe, J. Boyce, G. Briggs, A. Bunker, S. A. Collins, C. J. Hansen, T. V. Johnson, J. L. Mitchell, R. J. Terrile, M. Carr, A. F. Cook II, J. Cuzzi, J. B. Pollack, G. E. Danielson, A. Ingersoll, M. E. Davies, G. E. Hunt, H. Masursky, E. Shoemaker and D. Morrison 1981. Encounter with Saturn: Voyager 1 imaging science results. *Science* **212**, 163–91.
Smith, B. A., L. A. Soderblom, R. Beebe, J. Boyce, G. Briggs, M. Carr, S. A. Collins, A. F. Cook II, G. E. Danielson, M. E. Davies, G. E. Hunt, A. Ingersoll, T. V. Johnson, H. Masursky, J. McCauley, D. Morrison, T. Owen, C. Sagan, E. M. Shoemaker, R. Strom, V. E. Suomi and J. Veverka 1979a. The Galilean satellites and Jupiter: Voyager 2 imaging science results. *Science* **206**, 927–50.
Smith, B. A., L. Soderblom, T. V. Johnson, A. P. Ingersoll, S. A. Collins, E. M., Shoemaker, G. E. Hunt, H. Masursky, M. H. Carr, M. E. Davies, A. F. Cook II, J. Boyce, G. E. Danielson, T. Owen, C. Sagan, R. F. Beebe, J. Veverka, R. G. Strom, J. F. McCauley, D. Morrison, G. A. Briggs and V. E. Suomi 1979b. The Jupiter system through the eyes of Voyager 1. *Science* **204**, 951–72.
Smith, B. A., L. Soderblom, R. Batson, P. Bridges, J. Inge, H. Masursky, E. Shoemaker, R. Beebe, J. Boyce, G. Briggs, A. Bunker, S. A. Collins, C. J. Hansen, T. V. Johnson, J. L. Mitchell, R. J. Terrile, A. F. Cook II, J. Cuzzi, J. B. Pollack, G. E. Danielson, A. P. Ingersoll, M. E. Davies and G. E. Hunt 1982. A new look at the Saturn system: the Voyager 2 images. *Science* **215**, 504–37.
Smith, E. I. 1976. Comparison of the crater morphology-size relationship for Mars, Moon, and Mercury. *Icarus* **28**, 543–50.
Soderblom, L. A. 1980. The Galilean moons of Jupiter. *Scientific Am.* **242**, 88–100.
Soderblom, L. and T. V. Johnson 1983. The moons of Saturn. In *The planets*, B. Murray (ed.), 95–107. San Francisco: W. H. Freeman.
Soderblom, L. A. and D. B. Wenner 1978. Possible fossil H_2O liquid–ice interfaces in the martian crust. *Icarus* **34**, 622–37.
Soderblom, L. A., R. A. West, B. M. Herman, T. J. Kreidler and C. D. Condit 1974. Martian planetwide crater distributions: implications for geologic history and surface processes. *Icarus* **22**, 239–63.
Solomon, S. C. 1976. Some aspects of core formation in Mercury. *Icarus* **28**, 509–21.
Solomon, S. C. 1981. The geophysics of Mars: whence the Tharsis plateau? *Nature* **294**, 304–5.
Solomon, S. C. and J. W. Head 1982a. Evolution of the Tharsis province of Mars: the importance of heterogeneous lithospheric thickness and volcanic construction. *J. Geophys. Res.* **87**, 9755–74.
Solomon, S. C. and J. W. Head 1982b. Mechanisms for lithospheric heat transport on Venus: implications for tectonic style and volcanism. *J. Geophys. Res.* **87**, 9236–46.
Solomon, S. C., S. K. Stephens and J. W. Head 1982. On Venus impact basins: viscous relaxation of topographic relief. *J. Geophys. Res.* **87**, 7763–71.
Spudis, P. 1978. Composition and origin of the Apennine Bench Formation. *Proc. Lunar Planet. Sci. Conf. 9th*, 3379–94.
Spudis, P. D. 1982. *The geology of lunar multi-ring basins*. Unpublished Ph.D. dissertation, Arizona State University.
Spudis, P. and R. Greeley 1976. *Surficial geology of Mars: a study in support of a penetrator mission to Mars.* NASA TM X-73184.
Spudis, P. D. and M. E. Strobell 1984. New identification of ancient multi-ring basins on Mercury and implications for geologic evolution. *Lunar Planet. Sci.* **15**, 814–15.
Spurr, J. E. 1944. *Geology applied to selenology: I. the Imbrium plain region of the Moon.* Lancaster: Science Press.
Spurr, J. E. 1945. *Geology applied to selenology: II. the features of the Moon*. Lancaster: Science Press.
Spurr, J. E. 1948. *Geology applied to selenology: III. lunar catastrophic history*. Concord: Rumford Press.
Spurr, J. E. 1949. *Geology applied to selenology: IV. the shrunken Moon.* Concord: Rumford Press.
Squyres, S. W. 1979. The distribution of lobate debris aprons and similar flows on Mars. *J. Geophys. Res.* **84**, 8087–96.
Squyres, S. W. 1980. Volume changes in Ganymede and Callisto and the origin of grooved terrain. *Geophys. Res. Lett.* **7**, 593–6.
Squyres, S. W., R. T. Reynolds, P. M. Cassen and S. J. Peale 1983. The evolution of Enceladus. *Icarus* **53**, 319–31.
SSEC (Solar System Exploration Committee) 1983. *Planetary exploration through the year 2000: a core program.* Washington, DC: NASA.
Stephens, S. K., S. C. Solomon and J. W. Head 1983. On the age of Venus highland topography: constraints from the viscous relaxation of relief. *Lunar Planet. Sci.* **14**, 747–8.
Stevenson, D. J. 1982. Volcanism and igneous processes in small icy satellites. *Nature* **298**, 142–4.
Strain, P. L. and F. El-Baz 1980. The geology and morphology of Ina. *Proc. Lunar Planet. Sci. Conf. 11th*, 2437–46.
Strom, R. G. 1979. Mercury: a post-Mariner 10 assessment. *Space Sci. Rev.* **24**, 3–70.
Strom, R. G. and N. M. Schneider 1982. Volcanic eruption plumes on Io. In *Satellites of Jupiter*, D. Morrison (ed.), 598–633. Tucson: University of Arizona Press.

Strom, R. G., N. J. Trask and J. E. Guest 1975b. Tectonism and volcanism on Mercury. *J. Geophys. Res.* **80**, 2478–507.

Strom, R. G., R. J. Terrile, H. Masursky and C. Hansen 1979. Volcanic eruption plumes on Io. *Nature* **280**, 733–6.

Strom, R. G., B. C. Murray, M. J. S. Belton, G. E. Danielson, M. E. Davies, D. E. Gault, B. Hapke, B. O'Leary, N. Trask, J. E. Guest, J. Anderson and K. Klaasen 1975a. Preliminary imaging results from the second Mercury encounter. *J. Geophys. Res.* **80**, 2345–56.

Stuart-Alexander, D. E. and K. A. Howard 1970. Lunar maria and circular basins – a review. *Icarus* **12**, 440–56.

Surkov, Y. A. 1983. Studies of Venus rocks by Veneras 8, 9 and 10. In *Venus*, D. M. Hunten, L. Colin, T. M. Donahue and V. I. Moroz (eds), 154–8. Tucson: University of Arizona Press.

Surkov, Y. A., L. P. Moskalyeva, O. P. Shcheglov, V. P. Kharyukova, O. S. Manvelyan, V. S. Kirichenko and A. D. Dudin 1983. Determination of the elemental composition of rocks on Venus by Venera 13 and Venera 14 (preliminary results). *J. Geophys. Res.* **88**, suppl., A481–93.

Tauber, M. E. and D. B. Kirk 1976. Impact craters on Venus. *Icarus* **28**, 351–7.

Tauber, M. E., D. B. Kirk and D. E. Gault 1978. An analytic study of impact ejecta trajectories in the atmospheres of Venus, Mars, and Earth. *Icarus* **33**, 529–36.

Taylor, S. R. 1982. *Planetary science: a lunar perspective.* Houston: Lunar Planetary Institute.

Terrile, R. J., T. V. Johnson, L. A. Soderblom and R. G. Strom 1981. Variable features on Io. In *Rep. Planet. Geol. Prog. – 1981*, NASA TM-84211, 29–31.

Theilig, E. and R. Greeley 1979. Plains and channels in the Lunae Planum–Chryse Planitia region of Mars. *J. Geophys. Res.* **84**, 7994–8010.

Thomas, P. 1980. Plain formation on Mercury: tectonic implications. *Moon Planets* **22**, 261–8.

Thomas, P. and J. Veverka 1979. Seasonal and secular variation of wind streaks on Mars: an analysis of Mariner 9 and Viking data. *J. Geophys. Res.* **84**, 8131–46.

Thomas, P. and J. Veverka 1982. Amalthea. In *Satellites of Jupiter*, D. Morrison (ed.), 147–73. Tucson: University of Arizona Press.

Thomas, P. G., P. Masson and L. Fleitout 1982. Global volcanism and tectonism on Mercury: comparison with the Moon. *Earth Planet. Sci. Lett.* **58**, 95–103.

Thomas, P., J. Veverka and T. Duxbury 1978. Origin of the grooves on Phobos. *Nature* **273**, 282–4.

Trask, N. J. and J. E. Guest 1975. Preliminary geologic terrain map of Mercury. *J. Geophys. Res.* **80**, 2461–77.

Tsoar, H., R. Greeley and A. R. Peterfreund 1979. Mars: the north polar sand sea and related wind patterns. *J. Geophys. Res.* **84**, 8167–82.

Twidale, C. R. 1976. *Analysis of landforms.* Brisbane: John Wiley.

Tyler, G. L., V. R. Eshleman, J. D. Anderson, G. S. Levy, G. F. Lindal, G. E. Wood and T. A. Croft 1981. Radio science investigations of the Saturn system with Voyager 1: preliminary results. *Science* **212**, 201–6.

Urey, H. C. 1952. *The planets: their origin and development.* New Haven: Yale University Press.

Varnes, D. J. 1978. Slope movement types and processes. In *Landslides analysis and control*, R. L. Schuster and R. J. Krisek (eds), 11–33. Washington DC: National Academy of Sciences.

Veverka, J. and P. Thomas 1979. Phobos and Deimos: a preview of what asteroids are like? In *Asteroids*, T. Gehrels (ed.), 628–51. Tucson: University of Arizona Press.

Walker, G. P. L. 1973. Lengths of lava flows. *Phil Trans R. Soc. Lond. A* **274**, 107–18.

Ward, A. W. 1979. Yardangs on Mars: evidence of recent wind erosion. *J. Geophys. Res.* **84**, 8147–66.

Warner, J. L. 1983. Sedimentary processes and crustal cycling on Venus. *J. Geophys. Res.* **88**, suppl. A495–A500.

Weertman, J. 1979. Height of mountains on Venus and the creep properties of rocks. *Phys. Earth Planet. Int.* **19**, 197–207.

Wentworth, C. K. and G. A. Macdonald 1953. Structures and forms of basaltic rocks in Hawaii. *US Geol Surv. Bull.* **994**.

Whitaker, E. A. 1972. An unusual mare feature. In *Apollo 15 Prelim. Sci. Rep.* NASA SP-289, 25-84 through 25-85.

Whitford-Stark, J. L. 1982. Factors influencing the morphology of volcanic landforms: an Earth–Moon comparison. *Earth Sci. Rev.* **18**, 109–68.

Wilhelms, D. E. 1976. Mercurian volcanism questioned. *Icarus* **28**, 551–8.

Wilhelms, D. E. 1980. *Stratigraphy of part of the lunar nearside.* US Geol Surv. Prof. Paper 1046-A.

Wilhelms, D. E. and D. E. Davis 1971. Two former faces of the Moon. *Icarus* **15**, 368–72.

Wilhelms, D. E. and J. F. McCauley 1971. *Geologic map of the nearside of the Moon.* US Geol Surv. Misc. Geol. Inv. Map I-703.

Williams, S. H. and J. M. Moore 1984. Sediment gravity flows on Venus. *Lunar Planet. Sci.* **15**, 918–19.

Wilshire, H. G., T. W. Offield, K. A. Howard and D. Cummings 1972. *Geology of the Sierra Madera cryptoexplosion structure, Pecos County, Texas.* US Geol Surv. Prof. Paper 599-H.

Wilson, L. and J. W. Head 1981. Ascent and eruption of basaltic magma on the Earth and Moon. *J. Geophys. Res.* **86**, 2971–3001.

Wilson, L. and J. W. Head III 1983. A comparison of volcanic eruption processes on Earth, Moon, Mars, Io and Venus. *Nature* **302**, 663–9.

Wise, D. U., M. P. Golombek and G. E. McGill 1979. Tectonic evolution of Mars. *J. Geophys. Res.* **84**, 7934–9.

Wood, C. 1979. *Venusian volcanism: environmental effects of style and landforms.* NASA TM-80339, 244–6.

Wood, C. A. and J. W. Head 1976. Comparison of impact basins on Mercury, Mars, and the Moon. *Proc. Lunar Sci. Conf. 7th*, 3629–51.

Wood, C. A., J. W. Head and M. J. Cintala 1978. Interior morphology of fresh martian craters: the effects of target characteristics. *Proc. Lunar Planet. Sci. Conf. 9th*, 3691–709.

Woronow, A., R. G. Strom and M. Gurnis 1982. Interpreting the cratering record: Mercury to Ganymede and Callisto. In *Satellites of Jupiter*, D. Morrison (ed.), 237–76. Tucson: University of Arizona Press.

Young, R. A., W. J. Brennan, R. W. Wolfe and D. J. Nichols 1973. Volcanism in the lunar maria. In *Apollo 17 Prelim. Sci. Rep.* NASA SP-330, 31-1 through 31-11.

Index

Numbers in *italics* refer to text figures.